A U R O R A

DOVER MODERN MATH ORIGINALS

Dover Publications is pleased to announce the publication of the first volumes in our new Aurora Series of original books in mathematics. In this series we plan to make available exciting new and original works in the same kind of well-produced and affordable editions for which Dover has always been known.

Aurora titles currently in the process of publication are:

Optimization in Function Spaces by Amol Sasane. (978-0-486-78945-3)

The Theory and Practice of Conformal Geometry by Steven G. Krantz. (978-0-486-79344-3)

Numbers: Histories, Mysteries, Theories by Albrecht Beutelspacher. (978-0-486-80348-7)

Elementary Point-Set Topology: A Transition to Advanced Mathematics by André L. Yandl and Adam Bowers. (978-0-486-80349-4)

Additional volumes will be announced periodically.

Elementary Point-Set Topology

A Transition to Advanced Mathematics

André L. Yandl
Seattle University

Adam Bowers
University of California San Diego

DOVER PUBLICATIONS, INC.
Mineola, New York

Bibliographical Note

Elementary Point-Set Topology: A Transition to Advanced Mathematics is a new work, first published by Dover Publications, Inc., in 2016.

Library of Congress Cataloging-in-Publication Data

Names: Yandl, André L. | Bowers, Adam (Adam Roman)
Title: Elementary point-set topology : a transition to advanced mathematics
  / André L. Yandl, Seattle University, Adam Bowers, University of California
  San Diego.
Description: Mineola, New York : Dover Publications, Inc., [2016] | Series:
  Aurora Dover modern math originals | Includes index.
Identifiers: LCCN 2015050597 | ISBN 9780486803494 | ISBN 048680349X
Subjects: LCSH: Point set theory. | Propositional calculus. | Topology. |
  Calculus.

Classification: LCC QA603 .Y36 2016 | DDC 514/.322—dc23 LC record avail-
able at http://lccn.loc.gov/2015050597

Manufactured in the United States by RR Donnelley
80349X01   2016
www.doverpublications.com

Preface

About this book

As the title indicates, this book is about topology. In particular, this book is an introduction to the basics of what is often called *point-set* topology (also known as *general* topology). However, as the subtitle suggests, this book is intended to serve another purpose as well. A primary goal of this text, in addition to introducing students to an interesting subject, is to bridge the gap between the elementary calculus sequence and more advanced mathematics courses. For this reason, the focus of the text is on learning to read and write proofs rather than providing an advanced treatment of the subject itself.

During its infancy, this book consisted of a set of notes covering the basic topics of point-set topology and proof writing. These early topics make up the content of Chapters 1, 2, 3, and selected sections from Chapters 4, 5, and 6. For many years, the notes were used in the topology classes at Seattle University. The students had completed their calculus sequence and were using the topology course to help them make the transition into courses on abstract mathematics— in particular, the analysis sequence.

The desire to make this introduction to topology intuitive and accessible to our students has led to several innovations that we feel make our approach to the subject unique. Over the years, we collected feedback from many students and found they had difficulty seeing the connection between a product of n sets as a set of ordered n-tuples and as a set of functions. So we revised Chapter 4 several times until we had it in a form accessible to most students. We stressed the fact that it is desirable to have the projection functions be continuous, which helped the students appreciate the definition of the product topology. We introduced in Chapter 5 the idea of a function splitting a set which resulted in giving simpler proofs of theorems on connectedness.

Another aspect of this text which distinguishes it from most introductory topology textbooks is the content of Chapter 7, which demonstrates applications of topological concepts to other areas of mathematics. We decided to choose applications based on a single concept: the fixed point property. The applications are quite far-reaching, and include solving differential equations and proving the Fundamental Theorem of Algebra.

Layout and style

This text contains the standard content for an introductory point-set topology class, as well as an introduction to techniques of proof writing. This means the text can be used either for a "transitions to advanced mathematics" course or a standard topology course. The additional topics, which are not generally included in introductory topology books, make this text suitable also as a supplement for a more advanced topology course.

Chapter 1 introduces some elementary concepts in logic and basic techniques of proof, some elementary set theory, and an introduction to cardinal arithmetic. In Chapter 2, topological spaces and metric spaces are defined and a brief treatment of Euclidean space is given. The motivation for the definition of a topology is based on the notion of open sets in basic calculus. The need for the definition of Hausdorff space is shown by stressing that the concept is essential to prove that the limit of a convergent sequence is unique. Continuity and homeomorphism are presented in Chapter 3. Again the definitions are based on the familiar concept of continuity in calculus. Product spaces are discussed in Chapter 4. In Chapter 5, we treat connectedness and consider the special case of connectedness on the real line, leading to the proof of the Intermediate Value Theorem. Different forms of compactness are treated in Chapter 6, where we try to make the student appreciate that compactness is the vehicle that takes us from the infinite to the finite. Using compactness in the space of real numbers, we prove some important theorems from calculus. Our aim in Chapter 7 is to give the student an appreciation of the fact that topology can serve as a powerful tool in other branches of mathematics. We present a proof of the Fundamental Theorem of Algebra using Brouwer's Fixed-Point Theorem. We also prove Picard's Existence Theorem for Differential Equations using the Banach Fixed-Point Theorem for Contractions.

Although in principle we would avoid introducing a term, or a topic, in a textbook unless that term, or topic, is used later in the book, we have gone against this general rule. We feel that it is extremely useful and important for the student to be given ample opportunity to provide simple proofs as often as possible. It is our experience that one of the greatest difficulties encountered by students is to write a proof, even if it does not require much more that checking that some definitions are satisfied. Therefore, we introduce a number of terms in the main text, or in the exercises, even though those terms will not be used later, simply to give the students an opportunity to write proofs that are straightforward.

Whenever possible, we stress that the concepts covered in topology are abstractions of what the student learned in calculus. We strongly feel that the student will have a better understanding of abstract notions if concrete, familiar examples are tied to them. Throughout the book, we have used both a narrative and a formal symbolic style. We believe it is important for the student to be exposed, as often as possible, to both styles. It is our hope that the student will then be at ease when reading or writing mathematical proofs, using either sentences or mathematical symbols.

How to use this book

This book could be used in a number of different courses.

- *Elementary Topology, a Transition to Abstract Mathematics*
  (One quarter/semester.) Intended for students who have not been exposed to writing proofs. This course would cover Chapters 1, 2, 3, 4 and Sections 5.1, 5.3, 6.1, and 6.2. (These are the topics covered in the original set of notes at Seattle University.)

- *Introduction to Topology*
  (One quarter/semester.) Intended for students who are familiar with proof writing. This course would cover Chapters 2, 3, 4, 5, Sections 6.1, 6.2, 6.4, 6.5, and selected sections from Chapter 7.

- *Introduction to Abstract Mathematics via Topology*
  (Two quarters/semesters.) Intended for students whose only mathematical exposure has been the Calculus sequence, Linear Algebra, Differential Equations. This sequence would cover the whole book.

- A reference for a number of Topology courses. Since the treatment of product spaces (Chapter 4), connectedness (Chapter 5), and applications (Chapter 7) is very likely different from the standard approach of those topics in other topology texts, this book could serve as an excellent reference for a number of Topology courses.

Acknowledgments

The first author wishes to express his appreciation to the many students who have made positive suggestions for improvement, in particular Colleen O'Meara Brajcich whose excellent set of classroom notes served as the starting point of the manuscript.

The second author would like to express his gratitude to the first author for giving him the opportunity to be involved in the creation of this text. He would also like to show his appreciation to Minerva Catral, Rachel Schwell, and Elliott Neyme for useful discussions and support.

André L. Yandl
Seattle, WA

Adam Bowers
La Jolla, CA

Contents

Preface 3

List of Figures 8

List of Symbols 10

1   Mathematical Proofs and Sets   13

   1.1   Introduction to Elementary Logic 13

   1.2   More Elementary Logic 17

   1.3   Quantifiers 23

   1.4   Methods of Mathematical Proof 29

   1.5   Introduction to Elementary Set Theory 40

   1.6   Cardinality 53

   1.7   Cardinal Arithmetic 61

2   Topological Spaces   67

   2.1   Introduction 67

   2.2   Topologies 69

   2.3   Bases 74

   2.4   Subspaces 79

   2.5   Interior, Closure, and Boundary 82

   2.6   Hausdorff spaces 92

   2.7   Metric Spaces 96

   2.8   Euclidean Spaces 105

3   Continuous Functions   111

   3.1   Review of the Function Concept 111

   3.2   More on Image and Inverse Image 118

   3.3   Continuous Functions 124

   3.4   More on Continuous Functions 133

   3.5   More on Homeomorphism 142

4   Product Spaces   149

   4.1   Products of Sets 149

   4.2   Product Spaces 156

   4.3   More on Product Spaces 162

5 Connectedness   **167**
   5.1 Introduction to Connectedness 167
   5.2 Products of Connected Spaces 173
   5.3 Connected Subsets of the Real Line 176

6 Compactness   **185**
   6.1 Introduction to Compactness 185
   6.2 Compactness in the Space of Real Numbers 191
   6.3 The Product of Compact Spaces 194
   6.4 Compactness in Metric Spaces 198
   6.5 More on Compactness in Metric Spaces 205
   6.6 The Cantor Set 210

7 Fixed Point Theorems and Applications   **217**
   7.1 Sperner's Lemma 217
   7.2 Brouwer's Fixed Point Theorem 221
   7.3 The Fundamental Theorem of Algebra 226
   7.4 Function Spaces 233
   7.5 Contractions 239

Index   **251**

List of Figures

1.4.a	An increasing function	31
1.4.b	The Pigeonhole Principle	35
1.5.a	Venn diagram: subset	41
1.5.b	Venn diagram: complement of a set	44
1.5.c	Venn diagram: difference of sets	44
1.5.d	Venn diagrams: unions and intersections	46
1.5.e	Venn diagrams: De Morgan's Laws	48
1.5.f	Venn diagrams: Distributive Law	49
1.5.g	Venn diagram: symmetric difference	53
1.6.a	Relations on subsets of the unit interval	55
2.3.a	Equivalent topologies	78
2.5.a	Interior of a set	82
2.5.b	Neighborhood of a point	85
2.5.c	Boundary point	85
2.5.d	Interior point	86
2.6.a	Hausdorff space	93
2.7.a	Open ball in a metric space	98
2.7.b	Bounded set	101
2.7.c	Open ball in the space of continuous functions	103
2.8.a	Norm of a vector in $\mathbb{R}^2$	106
2.8.b	Illustration for the Cauchy-Schwarz Inequality	107
3.1.a	The function concept	111
3.1.b	Restriction of a function	115
3.3.a	Illustration to accompany Exercise 17 in Section 3.3	133
3.4.a	The Pasting Lemma (Part 1)	142
3.4.b	The Pasting Lemma (Part 2)	142
3.5.a	The Fixed Point Property	144
3.5.b	Regular topological space	146
3.5.c	Normal topological space	146
3.5.d	An example of a retract	147
4.1.a	Projections in $\mathbb{R}^2$	150
4.1.b	Illustrations for Example 4.1.4	154

4.2.a   The product topology 157

5.1.a   A function that splits a topological space 170
5.1.b   Connected subsets and splitting functions 171
5.3.a   Connectedness for intervals 178
5.3.b   The Intermediate Value Theorem 179
5.3.c   Illustration to accompany Exercise 18 in Section 5.3 183
5.3.d   Illustration to accompany Exercise 19 in Section 5.3 183

6.1.a   A cover which admits a subcover 186
6.3.a   Illustration to accompany Exercise 8 in Section 6.3. 198
6.6.a   The Cantor Set 211

7.1.a   A graph with five vertices, four edges, and one loop. 218
7.1.b   A graph with two vertices and two edges. 219
7.1.c   Partitioning a triangle 220
7.1.d   Sperner's Lemma 220
7.2.a   Linear dependence 222
7.2.b   Linear independence 223
7.2.c   A point in a triangle 225
7.2.d   Centroid of a triangle 226
7.3.a   Complex numbers 227

List of Symbols

$\mathscr{A}, \mathscr{B}, \mathscr{C}, \ldots$ Collections of sets are denoted by upper case letters in script

$\mathscr{R}, \mathscr{S}, \mathscr{T}, \ldots$ Common letters used for topologies

$\boldsymbol{x}, \boldsymbol{y}, \boldsymbol{z}, \ldots$ Vectors are denoted by lower case letters in bold

$\mathbb{N}$	The set of natural numbers (or positive integers)
$\mathbb{E}$	The set of even positive integers
$\mathbb{Z}$	The set of integers
$\mathbb{Q}$	The set of rational numbers
$\mathbb{R}$	The set of real numbers
$\mathbb{R}^+$	The set of positive real numbers
$\mathbb{R}^n$	The set of n-dimensional vectors (ordered n-tuples) with entries in $\mathbb{R}$
$\mathbb{C}$	The set of complex numbers
$\mathscr{H}$	Hilbert space
$\mathscr{K}$	Cantor set
$\emptyset$	The empty set

$\in$	Membership of an element in a set
$\subseteq$	Inclusion of one set in another
$\subset$	Proper inclusion of one set in another
$=$	Equality
$\cup$	Union of sets
$\cap$	Intersection of sets

$\forall$	Universal quantifier ("for all")
$\exists$	Existential quantifier ("there exists at least one")
$\exists!$	Unique existence quantifier ("there exists a unique")
$\wedge$	Conjunction ("and")
$\vee$	Disjunction ("inclusive or")
$\neg$	Negation
$\Rightarrow$	Implication
$\Leftrightarrow$	Double implication, equivalence ("if and only if")
$\oplus$	Exclusive or
$\sim$	Equivalence relation
$[x]$	Equivalence class represented by x
A'	Complement of the set A
$A \setminus B$	Complement of B with respect to A ("A without B")
$A \triangle B$	Symmetric difference of the sets A and B
$A \times B$	Cartesian product of the sets A and B
$\mathscr{P}(A)$	Power set of A (the collection of all subsets of A)
2^A	Power set of A (the collection of all subsets of A)
A^B	The set of functions from B to A
$|A|$	Cardinality of the set A
$\aleph_0$	Aleph-naught, the cardinality of the set $\mathbb{N}$
$|A|^{|B|}$	Cardinality of the set A^B
A°	Interior of the set A
$\overline{A}$	Closure of the set A
$d(A)$	Diameter of the set A
sup	Supremum of a set
inf	Infimum of a set
$\mathscr{T}_d$	Metric topology generated by the metric d

$n!$ n-factorial (the product of the first n positive integers)

$\binom{n}{k}$ Binomial coefficient ("n choose k")

$\boldsymbol{x} \cdot \boldsymbol{y}$ Dot product (or scalar product) of the vectors $\boldsymbol{x}$ and $\boldsymbol{y}$

$\|\boldsymbol{x}\|$ Norm (or length) of the vector $\boldsymbol{x}$

Q.E.D. *Quod erat demonstrandum* (denotes the end of a proof)

Chapter 1

Mathematical Proofs and Sets

1.1 Introduction to Elementary Logic

In everyday language, terms are defined using other terms, which inevitably leads to circular definitions. In mathematics, we begin with some terms that are *undefined terms*. Using these, we then introduce *defined terms*. Using undefined and defined terms, we make statements that are universally accepted as either true or false. Such statements are called *propositions*. We must stress the fact that a statement may be a proposition even though it may not be possible to verify whether it is true or false. For example, the statement "President John F. Kennedy drank exactly three glasses of wine on his twenty first birthday" is either true or false, hence a proposition; however, we may never be able to determine if it is true or if it is false.

In a mathematical theory, propositions that are accepted as true without proof are called *axioms* or *postulates*. The truth of all other propositions must be proved using the axioms as a start and rules of logic that have been carefully stated. Propositions that have to be proved true are called *theorems*. Theorems that are stated and proved as preliminaries for the main theorem are called *lemmas*. Theorems that follow from the main theorem are called *corollaries*. In mathematics, once we accept the truth of the original assumptions (the axioms), we must accept the truth of all theorems that are derived from them. Research mathematicians state propositions they suspect to be true. These are called *conjectures*. When the truth of a conjecture has been demonstrated, the conjecture becomes a theorem. However, if someone succeeds in demonstrating one situation in which the conjecture is not true, then this situation is called a *counterexample* to the conjecture, and the conjecture is regarded as false.

In this section, and the next two, we introduce the rules of logic and methods of proof that will be followed in proving theorems. Some propositions are simple in the sense that they do not contain other propositions as components.

However, new propositions can be formed using simple propositions and logical connectives. It is convenient to define compound propositions by displaying their truth values for all possible truth values of their components in tables called *truth tables*. We will see later that if a compound proposition has n components, then there are 2^n possible combinations of truth values of the components. For example, if a proposition has 3 components, then there are 8 possible combinations of truth values for these components. (See Example 1.1.5.)

Definitions 1.1.1. Given the propositions p and q, the *conjunction* of p and q (denoted $p \wedge q$), the *disjunction* of p and q (denoted $p \vee q$), and the *negation* of p (denoted $\neg p$) are defined by the truth tables below.

p	q	$p \wedge q$
T	T	T
T	F	F
F	T	F
F	F	F

p	q	$p \vee q$
T	T	T
T	F	T
F	T	T
F	F	F

p	$\neg p$
T	F
F	T

A conjunction is intended to express "and," whereas a disjunction is used to express "or." (When we say "or," we use it to mean "*inclusive* or" as opposed to "*exclusive* or," because we allow both propositions to be true at the same time.) A negation is used to mean "not."

It is important to note that in the definitions above, p and q are variable propositions that have no truth values until p and q are replaced by propositions whose truth values are known.

Definition 1.1.2. A *propositional form* is an expression involving letters that represent simple propositions and a finite number of connectives such as $\wedge$, $\vee$, $\neg$, and others to be defined later.

Just as we do when we perform arithmetical operations with signed numbers, we avoid writing many parentheses by adhering to the following rule: First, $\neg$ is applied to the simplest proposition following it. Second, $\wedge$ connects the simplest propositions on each side of it. Third, $\vee$ connects the simplest propositions on each side of it. For example, the negation of "$p \wedge q$" must be written "$\neg(p \wedge q)$" and not "$\neg p \wedge q$." However, "$[(\neg p) \wedge q] \vee r$" may be written "$\neg p \wedge q \vee r$."

Example 1.1.3. Let p be the proposition "Normal cats have four legs" and let q be the proposition "Normal eagles have three wings." For each of the following, find the truth values:

(a) "Normal cats have four legs and normal eagles have three wings,"

(b) "Normal cats have four legs or normal eagles have three wings,"

(c) "It is not the case that normal cats have four legs."

Solution. Observe first that p is true and q is false.

(a) This proposition is the conjunction $p \wedge q$. It is false since one of the components is false.

(b) This proposition is the disjunction $p \vee q$ and is true since one of its components is true.

(c) This proposition is the negation $\neg p$. This proposition is false since it is the negation of a true statement.

Example 1.1.4. Is the proposition "The authors of this text play tennis poorly" the negation of the proposition "The authors of this text play tennis well?"

Solution. The answer is "No," since both propositions are false. The authors of this text do not play tennis at all.

We often encounter propositional forms with more than two simple components. In the next example, we will make up a truth table for a proposition with three simple components.

Example 1.1.5. Make a truth table for the proposition $\neg p \wedge (q \vee r)$.

Solution. This proposition has three simple components and two main components: $\neg p$ and $q \vee r$. In the truth table, we first determine the truth values of the main components, and then using the definition of conjunction, we fill in the column for $\neg p \wedge (q \vee r)$.

p	q	r	$\neg p$	$q \vee r$	$\neg p \wedge (q \vee r)$
T	T	T	F	T	F
T	T	F	F	T	F
T	F	T	F	T	F
T	F	F	F	F	F
F	T	T	T	T	T
F	T	F	T	T	T
F	F	T	T	T	T
F	F	F	T	F	F

Definition 1.1.6. Two propositional forms are *equivalent* if and only if they have the same truth values for all possible combinations of truth values of their components.

Example 1.1.7. Show that the propositional forms $\neg(p \wedge q)$ and $\neg p \vee \neg q$ are equivalent.

Solution. Since each propositional form has the two simple components p and q, there are four possible combinations of truth values for p and q. In the truth table below, we show the truth values of $\neg(p \wedge q)$ and $\neg p \vee \neg q$ in Columns 4 and 7, respectively.

p	q	$p \wedge q$	$\neg(p \wedge q)$	$\neg p$	$\neg q$	$\neg p \vee \neg q$
T	T	T	F	F	F	F
T	F	F	T	F	T	T
F	T	F	T	T	F	T
F	F	F	T	T	T	T

Observe that the same truth values appear on the same line of Columns 4 and 7. Thus the propositional forms $\neg(p \wedge q)$ and $\neg p \vee \neg q$ are equivalent.

Exercises

1. Which of the following are propositions?

    (a) Paris is the capital of France.

    (b) The first author of this text ate pancakes on January 5, 1952.

    (c) Where are my glasses?

    (d) For all real numbers x, $x^2 < 0$.

    (e) This statement is false.

2. Make a truth table for each of the following.

    (a) $p \vee \neg p$

    (b) $\neg(p \vee q)$

    (c) $p \vee (q \wedge r)$

    (d) $p \wedge (q \vee r)$

    (e) $\neg(p \wedge \neg q)$

3. Determine which of the following pairs of propositional forms are equivalent.

    (a) $\neg(p \vee q)$ and $\neg p \wedge \neg q$

    (b) $\neg p \wedge q$ and $\neg(p \wedge q)$

(c) $p \vee \neg q$ and $\neg p \vee q$

(d) $\neg p \vee \neg q$ and $\neg(p \wedge q)$

(e) $(p \vee q) \vee r$ and $p \vee (q \vee r)$

(f) $(p \wedge q) \wedge r$ and $p \wedge (q \wedge r)$

(g) $p \wedge (q \vee r)$ and $(p \wedge q) \vee (p \wedge r)$

(h) $p \vee (q \wedge r)$ and $(p \vee q) \wedge (p \vee r)$

(i) $p \vee (\neg q \wedge r)$ and $(\neg p \vee q) \wedge (p \vee \neg r)$

4. Suppose p and q denote true propositions and r and s denote false propositions. What is the truth value of each of the following?

   (a) $\neg p \vee r \wedge q$

   (b) $p \wedge q \vee r \wedge \neg s$

   (c) $(p \vee \neg q) \wedge (\neg r \vee s)$

   (d) $(\neg r \vee \neg s) \wedge (\neg p \wedge q)$

1.2 More Elementary Logic

The student likely will remember that most theorems stated and proved in elementary mathematics are of the form "If p, then q." In this propositional form, p is called the *hypothesis* (or *antecedent*) and q is called the *conclusion* (or *consequent*). This propositional form is called a *conditional proposition* or an *implication* and is abbreviated $p \Rightarrow q$ (read "p implies q"). The symbol $\Rightarrow$ is called an *implication symbol*. We define the conditional proposition by means of a truth table.

Definition 1.2.1. Let p and q be propositions. The truth values for the conditional proposition $p \Rightarrow q$ are given in the table below.

p	q	$p \Rightarrow q$
T	T	T
T	F	F
F	T	T
F	F	T

Observe that $p \Rightarrow q$ is false only in the case where p is true and q is false.

Theorem 1.2.2. *If p and q are propositions, then $p \Rightarrow q$, $\neg p \vee q$, and $\neg(p \wedge \neg q)$ are equivalent.*

Proof. The truth values of the three propositions are displayed in Columns 3, 5, and 8 of the truth table below.

p	q	$p \Rightarrow q$	$\neg p$	$\neg p \vee q$	$\neg q$	$p \wedge \neg q$	$\neg(p \wedge \neg q)$
T	T	T	F	T	F	F	T
T	F	F	F	F	T	T	F
F	T	T	T	T	F	F	T
F	F	T	T	T	T	F	T

Observe how the entries of Columns 3, 5, and 8 are identical. Thus, the three propositions $p \Rightarrow q$, $\neg p \vee q$, and $\neg(p \wedge \neg q)$ are equivalent.   Q.E.D.

The symbol Q.E.D. that appears at the end of the proof abbreviates the Latin phrase *quod erat demonstrandum*, which means "what was to be demonstrated." (On occasion, students will jokingly claim that it abbreviates "quite easily done.")

There are three propositions that are closely related to $p \Rightarrow q$. We now introduce these propositions.

Definitions 1.2.3. Let p and q denote two propositions. The proposition $q \Rightarrow p$ is the *converse* of $p \Rightarrow q$, the proposition $\neg p \Rightarrow \neg q$ is its *inverse*, and $\neg q \Rightarrow \neg p$ is its *contrapositive*.

Theorem 1.2.4. *Let p and q be propositions. The implication $p \Rightarrow q$ and its contrapositive $\neg q \Rightarrow \neg p$ are equivalent.*

Proof. The following truth table indicates the truth values for the implication $p \Rightarrow q$ as well as its contrapositive, converse, and inverse.

p	q	$p \Rightarrow q$	$\neg p$	$\neg q$	$\neg q \Rightarrow \neg p$	$q \Rightarrow p$	$\neg p \Rightarrow \neg q$
T	T	T	F	F	T	T	T
T	F	F	F	T	F	T	T
F	T	T	T	F	T	F	F
F	F	T	T	T	T	T	T

Columns 3 and 6 are identical, and so $p \Rightarrow q$ and $\neg q \Rightarrow \neg p$ are equivalent. Q.E.D.

In the above truth table, Columns 7 and 8 both differ from Column 3. This shows that an implication is not equivalent to its converse or inverse. However, Columns 7 and 8 are identical, which means that the converse and inverse are equivalent to each other.

We have stated earlier that two propositional forms are equivalent if they have the same truth values for all possible combinations of truth values of their components. It is not surprising then, that when the propositions p and q are equivalent, we say p is true *if and only if q* is true. Formally, we have the following definition.

Definition 1.2.5. Let p and q be propositions. The propositional form $p \Leftrightarrow q$ (read "p if and only if q") is true precisely when p and q have the same truth values, or when p and q are equivalent. We call this propositional form an *equivalence* or a *double implication*. The truth table for $p \Leftrightarrow q$ is given below.

p	q	$p \Leftrightarrow q$
T	T	T
T	F	F
F	T	F
F	F	T

Definition 1.2.6. A *tautology* is a propositional form that is true for all possible truth values of its components.

A simple example of a tautology is $p \Leftrightarrow p$, where p is any proposition. No matter the truth value of p, both sides of $p \Leftrightarrow p$ will necessarily have the same truth value. Therefore, $p \Leftrightarrow p$ is always true.

Example 1.2.7. Let p be a proposition. Show that $p \Leftrightarrow \neg(\neg p)$ is a tautology.

Solution. We construct a truth table:

p	$\neg p$	$\neg(\neg p)$	$p \Leftrightarrow \neg(\neg p)$
T	F	T	T
F	T	F	T

Since the entries of Column 4 are all Ts, the propositional form is a tautology.

Because of Example 1.2.7, whenever we encounter $\neg(\neg p)$ in an argument, we can replace it by the equivalent proposition p. As we shall soon see, when we are faced with the task of solving a problem, we must have the ability to replace propositions by equivalent propositions until we get a proposition that yields a solution to the problem.

Example 1.2.8. Solve the equation $x^2 + 12x = 13$, where x is a real number.

Solution. When we write $x^2 + 12x = 13$, it represents the proposition "x satisfies the equation $x^2 + 12x = 13$." This proposition may be true or false depending on x. Then

$$[x^2 + 12x = 13] \Leftrightarrow [x^2 + 12x - 13 = 0] \Leftrightarrow [(x-1)(x+13) = 0].$$

The equation $(x-1)(x+13) = 0$ is satisfied when either factor equals zero, hence

$$[(x-1)(x+13) = 0] \Leftrightarrow [(x-1 = 0) \vee (x+13 = 0)] \Leftrightarrow [(x = 1) \vee (x = -13)].$$

Thus, the equation is satisfied if $x = 1$ or $x = -13$.

The following theorem contains a number of pairs of equivalent propositional forms.

Theorem 1.2.9. *Let p, q, and r denote propositions. Each of the following is a tautology:*

(a) $[p \Leftrightarrow q] \Leftrightarrow [(p \Rightarrow q) \wedge (q \Rightarrow p)]$

(b) $[\neg(p \Rightarrow q)] \Leftrightarrow [p \wedge \neg q]$

(c) $[p \Rightarrow \neg q] \Leftrightarrow [\neg(p \wedge q)]$

(d) $[\neg(p \vee q)] \Leftrightarrow [\neg p \wedge \neg q]$ .... *(De Morgan's Law)*

(e) $[\neg(p \wedge q)] \Leftrightarrow [\neg p \vee \neg q]$ .... *(De Morgan's Law)*

(f) $[p \vee (q \wedge r)] \Leftrightarrow [(p \vee q) \wedge (p \vee r)]$ .... *(Distributive Law)*

(g) $[p \wedge (q \vee r)] \Leftrightarrow [(p \wedge q) \vee (p \wedge r)]$ .... *(Distributive Law)*

Proof. The proof of (e) was given in Example 1.1.7. We prove (c) as an illustration and leave the proof of the other parts for the exercises. Consider the truth table below.

p	q	$\neg q$	$p \Rightarrow \neg q$	$p \wedge q$	$\neg(p \wedge q)$	$[p \Rightarrow \neg q] \Leftrightarrow [\neg(p \wedge q)]$
T	T	F	F	T	F	T
T	F	T	T	F	T	T
F	T	F	T	F	T	T
F	F	T	T	F	T	T

All entries in Columns 4 and 6 are identical. This means that the propositional forms $p \Rightarrow \neg q$ and $\neg(p \wedge q)$ have the same truth values for all combinations of truth values for p and q. Thus, all entries in Column 7 are Ts, and so the propositional form $[p \Rightarrow \neg q] \Leftrightarrow [\neg(p \wedge q)]$ is a tautology.   Q.E.D.

Theorem 1.2.10 (The Law of Contradiction). *If p denotes a proposition, then $\neg(p \wedge \neg p)$ is a tautology.*

Proof. Consider the truth table below.

p	$\neg p$	$p \wedge \neg p$	$\neg(p \wedge \neg p)$
T	F	F	T
F	T	F	T

Both entries in Column 4 are Ts. Thus, $\neg(p \wedge \neg p)$ is a tautology.   Q.E.D.

Theorem 1.2.11 (The Law of Excluded Middle). *If p denotes a proposition, then $p \vee \neg p$ is a tautology.*

Proof. Consider the truth table below.

p	$\neg p$	$p \vee \neg p$
T	F	T
F	T	T

Both entries in Column 3 are Ts. Thus, $p \vee \neg p$ is a tautology.   Q.E.D.

Theorem 1.2.12 (The Law of Syllogism). *If p, q, and r denote propositions, then* $[(p \Rightarrow q) \wedge (q \Rightarrow r)] \Rightarrow (p \Rightarrow r)$ *is a tautology.*

Proof. This propositional form involves three simple propositions as components. Therefore, the following truth table contains $2^3 = 8$ rows.

p	q	r	$p \Rightarrow q$	$q \Rightarrow r$	$(p \Rightarrow q) \wedge (q \Rightarrow r)$	$p \Rightarrow r$	(6) $\Rightarrow$ (7)
T	T	T	T	T	T	T	T
T	T	F	T	F	F	F	T
T	F	T	F	T	F	T	T
T	F	F	F	T	F	F	T
F	T	T	T	T	T	T	T
F	T	F	T	F	F	T	T
F	F	T	T	T	T	T	T
F	F	F	T	T	T	T	T

The implication (6) $\Rightarrow$ (7) in the top of Column 8 represents the propositional form $[(p \Rightarrow q) \wedge (q \Rightarrow r)] \Rightarrow (p \Rightarrow r)$, where the antecedent is in Column 6 and the consequent is in Column 7. Since all entries in Column 8 are Ts, the propositional form $[(p \Rightarrow q) \wedge (q \Rightarrow r)] \Rightarrow (p \Rightarrow r)$ is a tautology.   Q.E.D.

As we shall see in the next section, when writing a proof, we use known facts from earlier steps in the proof, or known past results. The truth of a proposition q may be deduced if we know that both propositions p and $p \Rightarrow q$ are true. The justification is in the next theorem.

Theorem 1.2.13 (*Modus Ponens*). *If p and q denote propositions, then the propositional form* $[p \wedge (p \Rightarrow q)] \Rightarrow q$ *is a tautology.*

Proof. Once again, we prove the theorem by constructing a truth table.

p	q	$p \Rightarrow q$	$p \wedge (p \Rightarrow q)$	$[p \wedge (p \Rightarrow q)] \Rightarrow q$
T	T	T	T	T
T	F	F	F	T
F	T	T	F	T
F	F	T	F	T

Since all entries in Column 5 are Ts, the propositional form is a tautology. Q.E.D.

The tautology in Thereom 1.2.13 represents a logical argument known as *Modus Ponens*, or "Affirming the Antecedent." In this form of argument, we suppose that p implies q (that is $p \Rightarrow q$). Therefore, if we assert that p is true, then we have no choice but to accept that q is also true. In other words, if both $p \Rightarrow q$ and p are true, then q is true, as well.

This form of argument is the most frequently used in mathematical literature. Theorems are usually stated in the form $p \Rightarrow q$. In order to apply the theorem, we observe that we know that the hypothesis p is true, and then conclude (from the theorem) that q is also true.

There are many different ways of reading some of the symbols we have introduced. The conditional proposition $p \Rightarrow q$ may be read as any of the following: "p implies q," or "if p, then q," or "p is sufficient for q," or "q is necessary for p," or "p only if q." The double implication $p \Leftrightarrow q$ may be read as: "p is equivalent to q," or "p if and only if q," or "p is necessary and sufficient for q." The abbreviation "p iff q" instead of $p \Leftrightarrow q$ is also common in the literature.

The symbol $\wedge$ is used to conjoin two simple propositions. Any conjunction, such as "and," "also," "but," or "however," is represented by $\wedge$. We use $\vee$ to mean "inclusive or." The statement $p \vee q$ means either p or q, or both.

Exercises

1. For each of the following, write the inverse, the converse, and the contrapositive and determine how the truth value of each relates to the truth value of the original statement.

    (a) If George is the King of England, then George is a man.
    (b) If Tom is a normal cat, then Tom has four legs.
    (c) If Michael plays basketball, then Michael is over six feet tall.
    (d) If Olga is Swedish, then she is Scandinavian.
    (e) If $a = b$ and $b = c$, then $a = c$.

2. Prove the remaining parts of Theorem 1.2.9: (a), (b), (d), (f), and (g).

3. Prove that $[(p \Rightarrow q) \wedge \neg q] \Rightarrow \neg p$ is a tautology. (This tautology represents an argument form called *Modus Tollens*, or "Denying the Consequent.")

4. Prove that $[p \Rightarrow (q \Rightarrow r)] \Leftrightarrow [(p \wedge q) \Rightarrow r]$ is a tautology.

5. Prove that $[(p \wedge \neg q) \Rightarrow (s \wedge \neg s)] \Leftrightarrow (p \Rightarrow q)$ is a tautology.

6. Write each of the following using the symbol $\Rightarrow$ or $\Leftrightarrow$.

    (a) Continuity of a function f at a is necessary for its differentiability at a.
    (b) For a continuous function to attain a maximum value on an interval, it is sufficient for that interval to be both closed and bounded.
    (c) A necessary condition for the square of an integer to be odd is for the integer itself to be odd.

(d) In order for a triangle to be isosceles, it is necessary and sufficient that two of its angles be congruent.

(e) Having two wings is a necessary condition for a pigeon to be a normal bird.

7. "I will marry you," John told Betty, "but only if I get a job." John did get a job, but did not marry Betty. Betty sued John for breach of promise. Did she have a case? Justify your answer.

8. The symbol $\oplus$ is sometimes used to mean "*exclusive or.*" An "exclusive or" is used to mean "one or the other, but not both;" that is, $p \oplus q$ means "either p or q, but not both p and q," where p and q are simple propositions.

(a) Create a truth table for $p \oplus q$.

(b) Use only the symbols $\wedge$, $\vee$, and $\neg$ (and p and q) to write down a propositional form that means $p \oplus q$.

1.3 Quantifiers

Since we wish to keep our introduction to Elementary Logic as brief as possible, we will not discuss "quantifiers" in detail. However, the student must be very careful when dealing with the negation of quantifiers. Thus, we will present several examples as illustrations. The student is undoubtedly familiar with the process of solving equations. An equation is an example of an open sentence, which we now discuss. The sentence $x^2 < 25$ is not a proposition because it is neither true nor false. It becomes a proposition only when x is replaced by a specific number. Such a sentence is called an *open sentence*. Whenever we have an open sentence, it involves at least one variable. We must have a *replacement set* for the variable. That is, we must have a set of objects that may be used as replacements for the variable. Often the replacement set is not specified but is understood from the context. However, often it must be specified because the solution set depends on the choice of the replacement set. For example, consider the open sentence $(x^2 - 2)(2x - 1)(x + 2)(x - 3) = 0$. If the replacement set is changed from the set of positive integers, to the set of integers, to the set of rational numbers, to the set of real numbers, then the solution sets are $\{3\}$, $\{-2, 3\}$, $\{-2, \frac{1}{2}, 3\}$, and $\{-2, -\sqrt{2}, \frac{1}{2}, 3, \sqrt{2}\}$, respectively.

Definition 1.3.1. Two open sentences are said to be *equivalent with respect to a replacement set* provided that their solution sets are the same when that replacement set is used.

Example 1.3.2. Let $p(x)$ be the open sentence "$(2x+1)(x-1)(x-3) = 0$" and let $q(x)$ be the open sentence "$(2x + 5)(x - 1)(x - 3) = 0$." Show that

$p(x)$ is equivalent to $q(x)$ if the replacement set is the set of integers, but not if the replacement set is the set of rational numbers.

Solution. If the set of integers is used as the replacement set, then the solution set for both $p(x)$ and $q(x)$ is the set $\{1,3\}$. Thus, $p(x)$ and $q(x)$ are equivalent with respect to the set of integers. If the set of rational numbers is used as the replacement set, however, then the solution set for $p(x)$ is $\{-\frac{1}{2}, 1, 3\}$ and the solution set for $q(x)$ is $\{-\frac{5}{2}, 1, 3\}$. Since the solution sets are different, $p(x)$ and $q(x)$ are not equivalent with respect to the set of rational numbers.

Definitions 1.3.3. If $p(x)$ is an open sentence with replacement set S, the sentence "for all x, $p(x)$" is true precisely when the solution set of $p(x)$ is the replacement set S. This sentence is abbreviated as $(\forall x)p(x)$. The sentence "for some x, $p(x)$" is true when the solution set for $p(x)$ has at least one member. This sentence is abbreviated as $(\exists x)p(x)$. The symbol $\forall$ is read "for all" and is called the *universal quantifier*. The symbol $\exists$ is read "there exists" and is called the *existential quantifier*.

Example 1.3.4. Let the replacement set be the set of all real numbers. Which of the following are true?

(a) $(\exists x)(x^2 < 0)$

(b) $(\forall x)(x \leq 4)$

(c) $(\exists x)(\sin x = \frac{1}{2})$

(d) $(\forall x)(|x| \geq 0)$

Solution. (a) The sentence $(\exists x)(x^2 < 0)$ is false because the solution set of $x^2 < 0$ is empty.

(b) The sentence $(\forall x)(x \leq 4)$ is false because, for example, the real number 5 is not a solution of $x \leq 4$, and so the solution set is not the set of all real numbers.

(c) The sentence $(\exists x)(\sin x = \frac{1}{2})$ is true because $\pi/6$ is in the solution set.

(d) The sentence $(\forall x)(|x| \geq 0)$ is true because the solution set of $|x| \geq 0$ is the set of all real numbers. We should note, however, that $(\forall x)(|x| > 0)$ is false since 0 is not in the solution set of $|x| > 0$.

One of the most common mistakes made in everyday language, as well as in elementary mathematics classes, is to incorrectly write the negation of a quantified expression. Consider the following example: Recently a phone company was running an advertisement on a Seattle radio station which described a new package and then added: "This package is not available in all areas." This sentence is equivalent to the statement "There are no areas where this package is available." We assume they must have meant "There are some areas where this package is not available."

The following theorem should serve as a guide to expressing the negation of propositions involving quantifiers.

Theorem 1.3.5. *If $p(x)$ is an open sentence with variable x and replacement set S, then*

(a) $\neg[(\exists x)p(x)]$ is equivalent to $(\forall x)(\neg p(x))$, and

(b) $\neg[(\forall x)p(x)]$ is equivalent to $(\exists x)(\neg p(x))$.

Proof. We will argue by replacing statements with equivalent statements until we reach the desired conclusion.

(a) $\neg[(\exists x)p(x)]$ is true $\Leftrightarrow$ $(\exists x)p(x)$ is false $\Leftrightarrow$ it is false that the solution set of $p(x)$ is nonempty $\Leftrightarrow$ the solution set of $p(x)$ is empty $\Leftrightarrow$ $p(x)$ is false for all x in S $\Leftrightarrow$ $\neg p(x)$ is true for all x in S $\Leftrightarrow$ the solution set of $\neg p(x)$ is S $\Leftrightarrow$ $(\forall x)(\neg p(x))$.

(b) $\neg[(\forall x)p(x)]$ is true $\Leftrightarrow$ $(\forall x)p(x)$ is false $\Leftrightarrow$ it is false that the solution set of $p(x)$ is S $\Leftrightarrow$ $p(x)$ is false for at least one x in S $\Leftrightarrow$ $\neg p(x)$ is true for at least one x in S $\Leftrightarrow$ the solution set of $\neg p(x)$ is nonempty $\Leftrightarrow$ $(\exists x)(\neg p(x))$.   Q.E.D.

Example 1.3.6. Let the replacement set be the set of real numbers. State the negation of "Every rational number has a square root."

Solution. It would be technically correct to say that the negation of the given statement is "It is not the case that every rational number has a square root." This is not exactly the most natural way of saying it, however, so we will use the preceding results to rephrase it in an equivalent, but more natural way.

Symbolically, we express the statement as

$$(\forall x)\big[x \text{ is rational} \Rightarrow (\exists y)(y^2 = x)\big].$$

The negation of this statement is

$$\neg(\forall x)\big[x \text{ is rational} \Rightarrow (\exists y)(y^2 = x)\big].$$

(This is essentially the negation we stated at the beginning of the solution.) By Theorem 1.3.5(b), this negation is equivalent to

$$(\exists x)\Big(\neg\big[x \text{ is rational} \Rightarrow (\exists y)(y^2 = x)\big]\Big).$$

By Theorem 1.2.9(b), we can write this as

$$(\exists x)\big[x \text{ is rational} \wedge \neg(\exists y)(y^2 = x)\big].$$

By Theorem 1.3.5(a), $\neg(\exists y)(y^2 = x)$ is equivalent to $(\forall y)\big[\neg(y^2 = x)\big]$, or alternately $(\forall y)(y^2 \neq x)$. Therefore, the negation of the original statement is (equivalent to) the statement

$$(\exists x)\big[x \text{ is rational} \wedge (\forall y)(y^2 \neq x)\big].$$

In everyday language, this would read "There is a rational number that has no real square root."

Note that this last statement is true because, for example, -1 is a rational number that has no real square root. Of course, the original statement is false, so naturally the negation is true.

There are instances when the solution set of an open sentence has a single element. We introduce a special symbol to indicate this situation.

Definition 1.3.7. Let $p(x)$ be an open sentence with variable x and replacement set S. The proposition $(\exists! x)p(x)$ is true precisely when the solution set of $p(x)$ with respect to S is a single element. The proposition $(\exists! x)p(x)$ is read "There is one and only one x such that $p(x)$," or "There is a unique x such that $p(x)$." The symbol $\exists!$ is called the *unique existence quantifier*.

Example 1.3.8. Determine the truth of the proposition $(\exists! x)(x^2 = 9)$ in the following two cases:

(a) The replacement set is the set of real numbers.

(b) The replacement set is the set of positive real numbers.

Solution. (a) The solution set is $\{-3, 3\}$, and so the statement is false.

(b) The solution set is $\{3\}$, and so the statement is true.

Frequently, the replacement set will not be explicitly stated, but can be determined from the context.

Example 1.3.9. Suppose f is a real valued function defined over all real numbers, and let x be a real number. State the negation of "For any $\epsilon > 0$, there exists a $\delta > 0$ such that $|f(x) - f(y)| < \epsilon$ whenever $|x - y| < \delta$."

Solution. Much like Example 1.3.6, we could state the negation simply as "It is not the case that ..." followed by the original statement. That is of course what it means to negate a statement, but we wish to find a more natural way of phrasing it, and so we will express the statement symbolically and then apply the results of this section.

The "variables" in this open sentence are ϵ and δ. Therefore, the replacement set is the set of all positive real numbers. Written symbolically, the statement we wish to negate is

$$(\forall \epsilon)\Big[(\exists \delta)\Big(|x - y| < \delta \Rightarrow |f(x) - f(y)| < \epsilon\Big)\Big].$$

Therefore, the negation is

$$\neg(\forall \epsilon)\Big[(\exists \delta)\Big(|x - y| < \delta \Rightarrow |f(x) - f(y)| < \epsilon\Big)\Big].$$

We now seek to rephrase the negation in an equivalent, but more natural way. By Theorem 1.3.5(b), the negation becomes

$$(\exists \epsilon)\Big[\neg(\exists \delta)\Big(|x - y| < \delta \Rightarrow |f(x) - f(y)| < \epsilon\Big)\Big].$$

And by Theorem 1.3.5(a),

$$(\exists \epsilon)\Big[(\forall \delta)\neg\Big(|x - y| < \delta \Rightarrow |f(x) - f(y)| < \epsilon\Big)\Big].$$

By Theorem 1.2.9(b), this is equivalent to

$$(\exists \epsilon)\Big[(\forall \delta)\Big(|x - y| < \delta \wedge \neg[|f(x) - f(y)| < \epsilon]\Big)\Big],$$

which can also be written as

$$(\exists \epsilon)\Big[(\forall \delta)\Big(|x - y| < \delta \wedge |f(x) - f(y)| \geq \epsilon\Big)\Big].$$

In words, this would read: "There exists an $\epsilon > 0$ such that, for all $\delta > 0$, $|f(x) - f(y)| \geq \epsilon$ for some y with $|x - y| < \delta$."

Exercises

1. Discuss whether or not the given open sentences are equivalent with respect to the given replacement set R.

   (a) $(x - 1)(x + 2) = 0$ and $(x - 1)(x + 2)(3x + 10) = 0$
       i. R is the set of natural numbers.

      ii. R is the set of integers.

     iii. R is the set of rational numbers.

  (b) $(x - 2)(5x + 3) = 0$ and $(x - 2)(5x + 3)(x^2 + 1) = 0$

      i. R is the set of natural numbers.

     ii. R is the set of integers.

    iii. R is the set of rational numbers.

    iv. R is the set of real numbers.

     v. R is the set of complex numbers.

2. Write each of the following sentences in symbolic form using quantifiers. The replacement set is given in parentheses.

  (a) Every prime number is odd. (The set of natural numbers.)

  (b) Every square of a real number is positive. (The set of real numbers.)

  (c) Every differentiable function is continuous. (The set of all real valued functions defined on the set of real numbers.)

  (d) Some triangles are equilateral. (The set of all triangles.)

  (e) Some rational numbers have a multiplicative inverse. (The set of real numbers.)

  (f) Some dogs are vicious. (The set of all animals.)

  (g) All men over six feet tall play basketball. (The set of all people.)

3. For each answer obtained in Exercise 2, write the negation in symbolic form. Translate each negation into everyday language.

4. Read a recent local newspaper and find a statement where a sentence containing a quantifier has been incorrectly negated.

5. Write down a statement that you have recently heard where a sentence containing a quantifier has been incorrectly negated.

6. Show that $(\exists!x)p(x)$ is equivalent to each of the following:

  (a) $(\exists x)\big[p(x) \wedge (\forall y)(p(y) \Rightarrow (x = y))\big]$

  (b) $(\exists x)\big[(\forall y)[p(y) \Leftrightarrow (x = y)]\big]$

7. Suppose a_n is a real number for each natural number n. State the negation of "There is a number $M > 0$ such that $|a_n| \leq M$ for all n."

8. Suppose a_n is a real number for each natural number n and that L is a real number. State the negation of "For all $\epsilon > 0$, there is a number N such that $|a_n - L| < \epsilon$ whenever $n > N$."

1.4 Methods of Mathematical Proof

The reader has likely encountered many proofs in elementary mathematics textbooks. However, the proofs that we encounter in more advanced mathematics courses are more involved because the subject is usually more abstract. Our goal in this text is not only to introduce the reader to some topological concepts, but also to increase the reader's ability to write mathematical proofs correctly. In the present section we introduce the different types of mathematical proofs. We discuss the validity of each type of proof based on the concepts introduced in the previous sections; in addition, we give several examples to illustrate the different styles of writing a proof.

In mathematics, the truth of a statement is not accepted until a proof has been provided, or we can provide a counterexample to show the statement is false.

Counterexample

Consider the statement "For the positive integer n, the integer $n^2 - n + 41$ is prime." It turns out that the statement is true if we replace n successively by $1, 2, 3, \ldots, 40$. If we replace n by 41, however, then we get 41^2, which is obviously not prime. This last situation is a counterexample showing that the statement is false. Note that *one* counterexample is sufficient to prove that the statement is not true. However, the truth of the statement for 40 values of n did not prove the statement was true.

Direct proof

Most of the theorems in mathematics are conditional propositions of the form $p \Rightarrow q$, where p and q are called the *hypothesis* and *conclusion*, respectively. A direct proof consists of finding propositions $r_1, r_2, r_3, \ldots, r_n$ such that

$$p \Rightarrow r_1, \quad r_1 \Rightarrow r_2, \quad r_2 \Rightarrow r_3, \quad \ldots, \quad r_{n-1} \Rightarrow r_n, \quad r_n \Rightarrow q,$$

and inferring that $p \Rightarrow q$ using the Law of Syllogism. The truth of each of these propositions may stem from a definition, an axiom, or a theorem previously proved. The art of providing a proof requires knowing the basic axioms, definitions, and previously proved theorems.

Example 1.4.1. Prove the following theorem: If f is a function that is continuous on the closed interval $[a, b]$, differentiable on the open interval (a, b), and $f'(x) > 0$ for all x in (a, b), then f is strictly increasing on $[a, b]$. (See Figure 1.4.a.)

Solution. We want to prove:

$$\Big(f \text{ continuous on } [a, b] \land f \text{ differentiable on } (a, b) \land f'(x) > 0 \ \forall x \text{ in } (a, b) \Big)$$
$$\implies \big[a \le x_1 < x_2 \le b \Rightarrow f(x_1) < f(x_2) \big].$$

We proceed as follows:

$$\Big(f \text{ continuous on } [a,b] \;\wedge\; f \text{ differentiable on } (a,b) \;\wedge\; [a \le x_1 < x_2 \le b]\Big)$$
$$(*)$$

$$\Longrightarrow \Big(f \text{ continuous on } [x_1, x_2] \;\wedge\; f \text{ differentiable on } (x_1, x_2)\Big)$$

$$\Longrightarrow (\exists c)\Big[c \text{ is in } (x_1, x_2) \text{ such that } f'(c) = \frac{f(x_2) - f(x_1)}{x_2 - x_1}\Big].$$

This last assertion follows from the Mean Value Theorem. We have not yet made use of the fact that $f'(x) > 0$ for all x in $[a, b]$. Therefore,

$$(*) \;\wedge\; \big(f'(x) > 0 \text{ for all } x \text{ in } [a, b]\big) \;\Longrightarrow\; \frac{f(x_2) - f(x_1)}{x_2 - x_1} > 0,$$

because $f'(c) > 0$. In the previous line we used $(*)$ as shorthand to represent everything that appeared on the line labeled $(*)$. Continuing, we have

$$\frac{f(x_2) - f(x_1)}{x_2 - x_1} > 0 \;\Longrightarrow\; f(x_2) - f(x_1) \text{ and } x_2 - x_1 \text{ have the same sign}$$

$$\Longrightarrow\; f(x_2) - f(x_1) > 0 \;\; \big[(x_1 < x_2) \Leftrightarrow (x_2 - x_1 > 0)\big]$$

$$\Longrightarrow\; f(x_1) < f(x_2).$$

Since $f(x_1) < f(x_2)$ whenever $a \le x_1 < x_2 \le b$, the function f is strictly increasing on $[a, b]$.   Q.E.D.

At the end of the proof in the previous example, we used the symbol Q.E.D. to indicate that the proof was complete. We remind the reader that Q.E.D. abbreviates the Latin phrase *quod erat demonstrandum*, which means "what was to be demonstrated."

Our goal in the previous example was to prove $p \Rightarrow (q \Rightarrow r)$, where p denotes $[f$ is continuous on $[a, b] \wedge f$ is differentiable on $(a, b) \wedge f'(x) > 0 \;\forall x$ in $(a, b)]$, where q denotes $[a \le x_1 < x_2 \le b]$, and r denotes $[f(x_1) < f(x_2)]$. We actually did not prove $p \Rightarrow (q \Rightarrow r)$, instead proving $(p \wedge q) \Rightarrow r$. However, the two propositional forms $p \Rightarrow (q \Rightarrow r)$ and $(p \wedge q) \Rightarrow r$ are equivalent. (See Exercise 4 in Section 1.2.)

Also note that part of the hypothesis of the theorem to be proved was the same as the hypothesis of the Mean Value Theorem. This suggested we might want to use the Mean Value Theorem as a step in our proof.

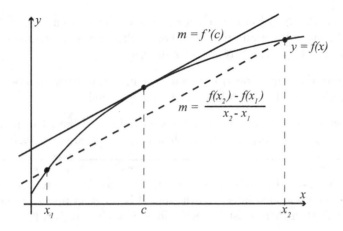

Figure 1.4.a: By the Mean Value Theorem, there is a number $c \in [x_1, x_2]$ such that $f'(c) = \frac{f(x_2) - f(x_1)}{x_2 - x_1}$. Thus, since $f'(c) > 0$ and $x_2 > x_1$, it must be that $f(x_2) > f(x_1)$. (See Example 1.4.1.)

Proof by contradiction

Suppose that we wish to prove $p \Rightarrow q$. If we can prove $(p \wedge \neg q) \Rightarrow r$, where r is a statement known to be false, then we have given a proof by contradiction. The statement r is known as a contradiction. An example of a contradiction is $s \wedge \neg s$, where s is any proposition. The validity of this technique is based on the fact that $p \Rightarrow q$ and $(p \wedge \neg q) \Rightarrow r$ are equivalent, if r is known to be false. (See Exercise 5 in Section 1.2.)

Example 1.4.2. Prove that if the replacement set for the variable x is the set of real numbers, then the following is true:

$$(x = 0) \Leftrightarrow \big(|x| < \epsilon \text{ for every } \epsilon > 0\big).$$

Solution. Certainly, $(x = 0) \Rightarrow (|x| = 0) \Rightarrow \big(|x| < \epsilon \text{ for every } \epsilon > 0\big)$. We now prove the converse, $\big(|x| < \epsilon \text{ for every } \epsilon > 0\big) \Rightarrow (x = 0)$. Assume to the contrary that $\big(|x| < \epsilon \text{ for every } \epsilon > 0\big)$ and $(x \neq 0)$. We will derive a contradiction. Let $\epsilon = \frac{|x|}{2}$. Since $x \neq 0$, we know that $\epsilon > 0$, and so it follows that $|x| < \frac{|x|}{2}$. Since $|x| > 0$, we may divide both sides of this inequality by $|x|$ to get $1 < \frac{1}{2}$. We know this last statement to be false—it is a contradiction. We must reject the assumption $x \neq 0$ and conclude that $x = 0$.   Q.E.D.

In order to give a more difficult example, we first state some facts from calculus.

1. If a function f is continuous on the closed bounded interval $[a, b]$, then it is a bounded function on $[a, b]$. That is, there are real numbers M and N such that $M \leq f(x) \leq N$ for all x in $[a, b]$.

2. (Axiom of Completeness) If S is a nonempty set of real numbers that has an upper bound, then S has a least upper bound.

3. If f is a function that is continuous on the closed bounded interval $[a, b]$, and $f(x) \neq 0$ for any x in $[a, b]$, then $1/f$ is continuous on $[a, b]$.

Example 1.4.3. Prove the following: If f is a continuous function defined on the closed bounded interval $[a, b]$, and A is the least upper bound of the range of f, then there exists some x_0 in $[a, b]$ such that $f(x_0) = A$.

Solution. We assume the hypotheses (f is continuous on $[a, b]$) and (A is the least upper bound of the range of f), but we also assume our desired conclusion is false; that is, we assume

$$\left[\neg (\exists x_0)(x_0 \text{ is in } [a, b] \text{ and } f(x_0) = A) \right]$$

is true, and we will arrive at a contradiction.

Our assumptions allow us to conclude $(\forall x)(x \text{ in } [a, b] \Rightarrow f(x) < A)$, because A is an upper bound and A is not in the range of f. We proceed as follows:

$(\forall x)(x \text{ in } [a, b] \Rightarrow f(x) < A)$

$\implies g(x) = \dfrac{1}{A - f(x)}$ defines a continuous positive function g on $[a, b]$

$\implies$ There is a positive number B such that $g(x) \leq B$ for all x in $[a, b]$

$\implies \dfrac{1}{A - f(x)} \leq B$ for all x in $[a, b]$

$\implies 1 \leq B(A - f(x))$ for all x in $[a, b]$

$\implies B f(x) \leq AB - 1$ for all x in $[a, b]$

$\implies f(x) \leq A - \dfrac{1}{B}$ for all x in $[a, b]$

$\implies A - \dfrac{1}{B}$ is an upper bound for the range of f

$\implies A$ is not the least upper bound for the range of f.

This last statement follows from the fact that B is a positive number. We have now (A is the least upper bound for the range of f) $\wedge$ (A is not the least upper bound for the range of f). This is a contradiction. We must reject the assumption $\left[\neg(\exists x_0)(x_0 \text{ is in } [a, b] \text{ and } f(x_0) = A)\right]$ and conclude that there exists some x_0 in $[a, b]$ such that $f(x_0) = A$. Q.E.D.

Contrapositive proof

Suppose that we wish to prove $p \Rightarrow q$ but we find that it is easier to prove $\neg q \Rightarrow \neg p$. We may proceed to prove the contrapositive and claim the truth of $p \Rightarrow q$. The validity of this technique stems from the fact that an implication and its contrapositive are equivalent. (See Theorem 1.2.4).

Example 1.4.4. Prove the following: If n is an integer and n^2 is odd, then n is odd.

Solution. We proceed as follows:

$$\neg \,(n \text{ is odd}) \quad \Longrightarrow \quad (n \text{ is even, since } n \text{ is an integer})$$

$$\Longrightarrow \quad (n = 2k, \text{ where } k \text{ is an integer})$$

$$\Longrightarrow \quad (n^2 = 4k^2 = 2(2k^2))$$

$$\Longrightarrow \quad (n^2 \text{ is even, since } 2k^2 \text{ is an integer})$$

$$\Longrightarrow \quad \neg(n^2 \text{ is odd}).$$

Thus, we have proved the contrapositive of the desired implication, and so the desired implication is true. Q.E.D.

Proof by induction

For simplicity, we consider only the following, which is called the *First Principle of Mathematical Induction*: Suppose that the set $\mathbb{N}$ of natural numbers is the replacement set for an open sentence $S(n)$ with variable n. Suppose further that

(i) $S(1)$ is true, and

(ii) the truth of $S(k)$ implies the truth of $S(k + 1)$.

Then the solution set of $S(n)$ is $\mathbb{N}$. In symbols, this is written:

$$\big[S(1) \wedge (\forall k)\big(S(k) \Rightarrow S(k+1)\big)\big] \Rightarrow (\forall n)(S(n)).$$

Example 1.4.5. Prove that the following is true for all natural numbers n:

$$\sum_{i=1}^{n} i^2 = \frac{n(n+1)(2n+1)}{6}.$$

Solution. Let $S(n)$ denote the given statement.

(i) $S(1)$ is true because it is true that $1^2 = \dfrac{(1)(1+1)(2+1)}{6}$.

(ii) Assume that $S(k)$ is true for a given natural number k. That is, assume

$$\sum_{i=1}^{k} i^2 = \frac{k(k+1)(2k+1)}{6}$$

is true. We will show that this implies the truth of $S(k+1)$. Adding $(k+1)^2$ to both sides of the above equation, we get

$$\sum_{i=1}^{k} i^2 + (k+1)^2 = \frac{k(k+1)(2k+1)}{6} + (k+1)^2.$$

With a little algebra, we see

$$\sum_{i=1}^{k+1} i^2 = \frac{k(k+1)(2k+1) + 6(k+1)^2}{6} = \frac{(k+1)[k(2k+1) + 6(k+1)]}{6}$$

$$= \frac{(k+1)(2k^2 + 7k + 6)}{6} = \frac{(k+1)(k+2)(2k+3)}{6}.$$

Therefore,

$$\sum_{i=1}^{k+1} i^2 = \frac{(k+1)[(k+1)+1][2(k+1)+1]}{6}.$$

We have thus shown that the truth of $S(k)$ implies the truth of $S(k+1)$. Hence, $S(n)$ is true for all natural numbers n, by the First Principle of Mathematical Induction.   Q.E.D.

The Pigeonhole Principle

A simple property of finite sets, often referred to as the *Pigeonhole Principle*, is surprisingly useful in many instances. This property is stated as follows: "If a flock of n pigeons comes to roost in a place that has m pigeonholes, where

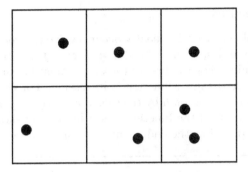

Figure 1.4.b: The Pigeonhole Principle. Seven pigeons can fit into six pigeon-holes only if one pigeonhole contains at least two pigeons.

$n > m$, then at least one pigeonhole will have more than one pigeon." (See Figure 1.4.b.)

We illustrate the usefulness of the Pigeonhole Principle in the next two examples.

Example 1.4.6. Show that the decimal representation of any rational number is a repeating decimal.

Solution. Let r be a (nonzero) rational number. There are positive integers m and n such that $r = \pm\frac{m}{n}$. To obtain the decimal representation of the number r, we must perform the division $m \div n$. At each step of the division, the possible partial remainders are $0, 1, 2, \ldots, n-1$. Thus, after at most $n + 1$ steps, a partial remainder must be either 0 or a repetition of one of the previous partial remainders. From that point on, the decimal representation will be repeating. (See Exercises 26 and 27 at the end of this section.)   Q.E.D.

Example 1.4.7. Let R be a rectangle whose length and width are 5 and 3 units, respectively. Let S be a set of 16 points in R. Show that at least two points in S are at a distance no more than $\sqrt{2}$ units apart.

Solution. Divide the rectangle into 15 congruent squares, each with dimensions 1 unit by 1 unit. Since there are 16 points and only 15 squares, at least one of the squares must contain two (or more) of the points. The distance between these two points cannot exceed the length of the diagonal, which is $\sqrt{2}$ units.   Q.E.D.

Proofs using calculus

Some types of problems that frequently occur involve the verification of propositions about inequalities, such as "$x < y \Rightarrow f(x) < f(y)$" or "$f(x) < g(x)$ for all x in some domain." For the first of these, a standard approach is to use the first derivative to show that a function is increasing. For the other, a natural approach is to define a function h by the rule $h(x) = f(x) - g(x)$, use calculus to find the minimum value of the function h, say $h(x_0)$, and show that $h(x_0) > 0$. (See Exercises 31 and 32 at the end of this section.)

Example 1.4.8. Show that $x < \tan x$ for all numbers x in the interval $(0, \pi/2)$.

Solution. Let $f(x) = \tan x - x$. Then $f'(x) = \sec^2 x - 1$. Certainly, $f'(0) = 0$ and $f'(x) > 0$ for all x in $(0, \pi/2)$. It follows that f is strictly increasing on $[0, \pi/2)$. But $f(0) = 0$, and so $f(x) = \tan x - x > 0$ whenever $0 < x < \pi/2$. Therefore, $x < \tan x$ for all x in $(0, \pi/2)$.   Q.E.D.

Exercises

1. Give a direct proof of the following: If f is a continuous function defined on the closed interval $[a, b]$, and $f'(x) < 0$ for all x in the open interval (a, b), then f is strictly decreasing on $[a, b]$.

2. Prove the following theorem: If $f'(x) = 0$ for all x in the open interval (a, b), then f is constant on (a, b).

3. Prove the following: If f is a strictly increasing function on an interval I, then

   (a) the function f is one-to-one, and

   (b) the inverse of f, denoted f^{-1}, is strictly increasing on $f(I)$, the image of I under f.

4. Prove the following: If f is a strictly decreasing function on an interval I, then

   (a) the function f is one-to-one, and

   (b) the inverse function f^{-1} is strictly decreasing on $f(I)$.

5. Suppose c is a positive real number. Prove the following:

   (a) $|x| < c \Leftrightarrow -c < x < c$, and

   (b) $|x| > c \Leftrightarrow [(x < -c) \vee (x > c)]$.

6. Give a contrapositive proof of the following: If n is an odd integer, then the equation $x^2 + x - n = 0$ has no solutions that are odd integers.

7. Prove the following theorem: If p is a prime number greater than 2, then the equation $x^2 + x - p = 0$ has no integer solutions.

8. Give a contrapositive proof of the following: If n is an integer such that its cube is odd, then n is odd.

9. The *Fundamental Theorem of Arithmetic* asserts that the prime factorization of a natural number is unique, except for the order of the factors. Use this to give a proof by contradiction of the following theorem: If p is a prime number, then $\sqrt{p}$ is irrational.

10. Prove the following: If k is an integer greater than 2, and p is a prime number, then $\sqrt[k]{p}$ is irrational. (*Hint:* See Exercise 8.)

11. Show that there is no set of integers $\{m, n, k\}$ such that $m + n\sqrt{2} + k\sqrt{3} = 0$ except for $m = n = k = 0$.

12. Prove that if x, y, and z are positive integers and $x^2 + y^2 = z^2$, then x and y cannot both be odd.

13. Prove that for every integer m, the number $\dfrac{m}{3} + \dfrac{m^2}{2} + \dfrac{m^3}{6}$ is an integer.

14. Prove that the product of any four consecutive integers plus 1 is a perfect square.

15. Use mathematical induction to prove that, for all natural numbers n,

$$\sum_{i=1}^{n} i = \frac{n(n+1)}{2}.$$

16. Use mathematical induction to prove that, for all natural numbers n,

$$\sum_{i=1}^{n} i^3 = \left[\frac{n(n+1)}{2}\right]^2.$$

17. Use mathematical induction to prove that, for all natural numbers n,

$$x^n - 1 = (x - 1)(x^{n-1} + x^{n-2} + \cdots + x + 1).$$

(*Hint:* $x^{k+1} - 1 = x(x^k - 1) + (x - 1)$.)

18. Use mathematical induction to prove the following: If n is any natural number, then

$$1 + \frac{1}{2} + \frac{1}{4} + \cdots + \frac{1}{2^n} < 2.$$

19. Use mathematical induction to prove that, for all natural numbers n,

$$\sum_{k=1}^{n} k(k!) = (n+1)! - 1.$$

20. Let $S(n)$ denote the statement "$\displaystyle\sum_{i=1}^{n} 2i = n^2 + n + 1$."

    (a) Prove that the truth of $S(k)$ implies the truth of $S(k+1)$.

    (b) There are some positive integers n that are not solutions of the open sentence $S(n)$. Does this contradict the induction principle?

    (c) Are there any positive integers that are solutions of $S(n)$?

21. (a) Let $S(n)$ denote the statement

$$\sum_{i=1}^{n} i = \frac{n(n+1)}{2} + (n-1)(n-2)(n-3)(n-4)(n-5)(n-6)(n-7).$$

    Verify that the first seven natural numbers are solutions of the statement and explain why no other natural number is a solution.

    (b) Write an open sentence with variable n and replacement set being the set of natural numbers and with the solution set the first 10,000 natural numbers.

22. Observe that

$$1 = 1$$
$$1 - 4 = -(1+2)$$
$$1 - 4 + 9 = 1 + 2 + 3$$
$$1 - 4 + 9 - 16 = -(1+2+3+4)$$
$$1 - 4 + 9 - 16 + 25 = 1 + 2 + 3 + 4 + 5.$$

Based on this pattern, conjecture an open sentence with solution set being the set of natural numbers. Use induction to prove your conjecture is correct.

23. Repeat Exercise 22 using the following equalities:

$$1 = 1$$
$$1 - 3 = -2$$
$$1 - 3 + 5 = 3$$
$$1 - 3 + 5 - 7 = -4.$$

24. Repeat Exercise 22 using the following equalities:

$$\left(1 - \frac{1}{2}\right) = \frac{1}{2}$$

$$\left(1 - \frac{1}{2}\right)\left(1 - \frac{1}{3}\right) = \frac{1}{3}$$

$$\left(1 - \frac{1}{2}\right)\left(1 - \frac{1}{3}\right)\left(1 - \frac{1}{4}\right) = \frac{1}{4}$$

$$\left(1 - \frac{1}{2}\right)\left(1 - \frac{1}{3}\right)\left(1 - \frac{1}{4}\right)\left(1 - \frac{1}{5}\right) = \frac{1}{5}.$$

25. Do Exercise 24 without using induction.

26. Find a decimal representation of the rational number $\frac{10}{7}$ by performing the division $10 \div 7$ (by hand). What are the partial remainders before a repetition occurs?

27. Find a decimal representation of the rational number $\frac{13}{4}$ by performing the division $13 \div 4$ (by hand). What are the partial remainders before a zero occurs?

28. Let R be a rectangle with dimensions 4×6 (in centimeters). Suppose R contains a set that has 27 points. Show that at least two of these points are no more than $\sqrt{2}$ centimeters apart.

29. Let R be a three-dimensional rectangular region that has length, width, and height measuring 7, 5, and 3 inches, respectively. Suppose R contains a set that has 106 points. Show that at least two of these points are no more than $\sqrt{3}$ inches apart.

30. Let R be an equilateral triangle with sides having length 6 units. Suppose R contains a set that has 10 points. Show that at least two of these points are no more than 2 units apart.

31. Prove that $\sin x < x$ for all $x > 0$.

32. Prove that if $0 < x < y < \frac{\pi}{2}$, then

$$\frac{\sin x}{x} > \frac{\sin y}{y} \quad \text{and} \quad \frac{\tan x}{x} < \frac{\tan y}{y}.$$

(*Hint:* Use Exercise 31 and the result of Example 1.4.8.)

33. The lengths of the sides of a triangle are in arithmetic progression. That is, the lengths are a, $a + d$, and $a + 2d$, where a and d are positive real numbers. Prove that the radius of the inscribed circle is one third the length of one of the altitudes.

34. Show that there are infinitely many prime numbers.

1.5   Introduction to Elementary Set Theory

A *set* is a collection of objects. A set must be *well defined* so that there is no confusion as to which objects belong to the set. For example, the set of all good students attending Seattle University is not well defined. There may not be agreement on whether a given student is considered to be a good student. However, the set of all students who have a G.P.A. of at least 3.5 and who are currently attending Seattle University is a well defined set.

The usual convention is to use capital letters to denote sets and lowercase letters to represent members of sets. Frequently, a script letter is used to denote a collection of sets.

Notation. The symbol $\in$ indicates membership in a set. Thus, $a \in A$ is read "a is a member (or *element*) of A." On the other hand, $a \notin B$ is read "a is not a member (or element) of the set B."

There are two techniques to describe sets. The first, called the *Roster Technique*, is where the names of the elements are written within braces and separated by commas. For example, the set of the first five counting numbers is represented as $\{1, 2, 3, 4, 5\}$. This works well for a set having only a finite number of elements; however, some sets have infinitely many elements. For some such sets, we may use a variation of the Roster Technique. For example, the set of all natural numbers (or counting numbers) may be represented by $\{1, 2, 3, 4, \ldots\}$. The symbol "$\ldots$" indicates that the elements will continue to exhibit the demonstrated pattern.

For many sets with a large number of elements, it is convenient to use the *Open Sentence Technique*. For this technique, we use the symbol $\{x \mid p(x)\}$ to represent the solution set of the open sentence $p(x)$. In words, we say $\{x \mid p(x)\}$ is "The set of all x such that $p(x)$." The replacement set is usually understood from context; otherwise it must be specified.

Example 1.5.1. Let $S = \{x \mid (x \text{ is a whole number}) \wedge (5 < x < 100)\}$. Which of the following statements are true and which are false?

     (a) $8 \in S$ (b) $3\pi \in S$ (c) $150 \in S$ (d) $3 \notin S$ (e) $5 \in S$

Solution.   (a) $8 \in S$ is true because 8 is a whole number that is larger than 5 and less than 100.

   (b) $3\pi \in S$ is false because 3π is not a whole number.

   (c) $150 \in S$ is false because 150 is not less than 100.

   (d) $3 \notin S$ is true because 3 is less than 5, and thus is not an element of S.

   (e) $5 \in S$ is false because 5 is not greater than 5.

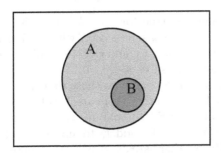

Figure 1.5.a: In this Venn diagram, the disk representing B is contained in the disk representing A. From this we see that B is a subset of A.

In everyday life, we are familiar with the idea of a subcollection of a collection. For example, the collection of all dogs is a subcollection of the collection of all animals. We now describe a similar relationship between sets.

Definitions 1.5.2. Let A and B be two sets. If $(\forall x)(x \in B \Rightarrow x \in A)$, then we say that B is a *subset* of A and we write $B \subseteq A$. We say that A is *equal* to B, and write $A = B$, if and only if both $A \subseteq B$ and $B \subseteq A$; that is, if A and B contain precisely the same elements. If $B \subseteq A$ and $A \neq B$, then we say that B is a *proper subset* of A and we write $B \subset A$.

A convenient way to visualize relationships between sets is to draw a *Venn diagram*. In a Venn diagram, sets are represented by geometric shapes, often disks. We use the geometric relationship between the shapes to illustrate the relationship between the sets. In Figure 1.5.a, we use a Venn diagram to illustrate the notion of one set being a subset of another.

It is convenient to consider an *empty set*, which we denote $\emptyset$. We define the empty set by $\emptyset = \{x \mid x \neq x\}$. It is clear that the open sentence $x \neq x$ has no solution.

Theorem 1.5.3. *If A is a set, then $\emptyset \subseteq A$.*

Proof. The implication $x \in \emptyset \Rightarrow x \in A$ is true because the antecedent $x \in \emptyset$ is false. Thus $\emptyset \subseteq A$, by definition.   Q.E.D.

Theorem 1.5.4. *There is one and only one empty set.*

Proof. Suppose that $\emptyset_1$ and $\emptyset_2$ are both empty sets. It follows from Theorem 1.5.3 that $\emptyset_1 \subseteq \emptyset_2$, because $\emptyset_1$ is empty. However, $\emptyset_2 \subseteq \emptyset_1$ for the same reason. Thus $\emptyset_1 = \emptyset_2$.   Q.E.D.

Theorem 1.5.5. *If a set S has n elements, then it has 2^n subsets.*

Proof. Certainly, the statement is true when $n = 0$, since the only subset of $\emptyset$ is $\emptyset$ and $2^0 = 1$. The statement is also true when $n = 1$, because the only subsets of a set with one element are $\emptyset$ and the set S itself, and because $2^1 = 2$.

Assume the statement is true for $n = k$. That is, assume a set with k elements has 2^k subsets. Now consider a set S with $k+1$ elements, where $k \geq 1$. Remove from S one element, say x_0, and call the resulting set T. Clearly, T has k elements, and hence has 2^k subsets.

Among all the subsets of S, there are those that have x_0 as a member, and those that do not. We have seen that there are 2^k subsets that do not have x_0 as a member. For each subset of S that contains x_0 as a member, if we remove x_0, we get a subset of T. We may also add x_0 to any subset of T to get a subset of S that contains x_0. Hence, there are exactly 2^k such subsets. Thus, the number of subsets of S must be $2^k + 2^k = 2(2^k) = 2^{k+1}$. Therefore, by the Principle of Mathematical Induction, the statement is true for all $n \in \mathbb{N}$.   Q.E.D.

For an alternate proof of Theorem 1.5.5, see Exercise 5 at the end of this section.

Definitions 1.5.6. Let n be a positive integer. The symbol $n!$ is called n *factorial* and represents the product $n(n-1)(n-2)\cdots(3)(2)(1)$. It is convenient to define $0! = 1$. (See Example 1.5.7, Parts (a) and (f).)

If n and k are positive integers with $n \geq k$, then the symbol $\binom{n}{k}$, called a *binomial coefficient*, denotes the number $\frac{n!}{k!(n-k)!}$.

It can be shown that if a set has n elements, then the number of subsets containing exactly k elements is $\binom{n}{k}$. That means that there are $\binom{n}{k}$ ways to select k elements from a collection of n elements. For this reason, the symbol $\binom{n}{k}$ is sometimes read as "n choose k." Of course, we must have $0 \leq k \leq n$.

Example 1.5.7. Suppose a set has 5 elements. How many subsets of exactly k elements does it have if

  (a) $k = 0$, (b) $k = 1$, (c) $k = 2$, (d) $k = 3$, (e) $k = 4$, (f) $k = 5$.

Use (a)–(f) to verify that the total number of subsets is $2^5 = 32$.

Solution. (a) The only subset with zero elements is the empty set $\emptyset$. This is consistent with $\binom{5}{0} = \frac{5!}{0!(5-0)!} = 1$. (We use here the fact that $0! = 1$.)

  (b) If the set has five elements, it is evident there are five subsets containing exactly one element, consistent with the fact that

$$\binom{5}{1} = \frac{5!}{1!(5-1)!} = \frac{5!}{4!} = 5.$$

  (c) There are ten subsets containing exactly two elements. This can be seen by writing them out directly or by observing that

$$\binom{5}{2} = \frac{5!}{2!(5-2)!} = \frac{5!}{2!3!} = 10.$$

(d) The number of subsets containing exactly three elements is again ten, because to each subset A of three elements there corresponds a set containing exactly two elements (those that are not in A). This is consistent with the fact that

$$\binom{5}{3} = \frac{5!}{3!(5-3)!} = \frac{5!}{3!2!} = 10.$$

(e) The number of ways to choose four elements is the same as the number of ways to exclude one element. Therefore, the number of sets containing exactly four elements is the same as the number of sets containing exactly one, which is already known to be five. Again, this is consistent with the fact that

$$\binom{5}{4} = \frac{5!}{4!(5-4)!} = \frac{5!}{4!} = 5.$$

(f) Certainly, there is only one subset containing five elements, and that is the set itself. Observe that

$$\binom{5}{5} = \frac{5!}{5!(5-5)!} = \frac{5!}{5!} = 1.$$

(Here again we use $0! = 1$.)

Finally, notice that the total number of subsets is

$$1 + 5 + 10 + 10 + 5 + 1 = 32 = 2^5.$$

Definition 1.5.8. Let A be a set. The set of all subsets of A is called the *power set* of A and is denoted $\mathscr{P}(A)$. (Some texts use the symbol 2^A to denote the power set of A.)

Example 1.5.9. Find $\mathscr{P}(A)$ if $A = \{a, b, c\}$.

Solution. $\mathscr{P}(A) = \{\emptyset, \{a\}, \{b\}, \{c\}, \{a, b\}, \{a, c\}, \{b, c\}, A\}$.

In any discussion, we have a set U such that every set we consider in that discussion is a subset of the set U. This set U is called the *universe* or the *universal set*. For example, in algebra, the universe can be the set of all real numbers, or the set of all complex numbers. In a sociology course, the universe could be the set of all people, etc. In a Venn diagram, the universe is represented by the large rectangle that contains all of the geometric shapes.

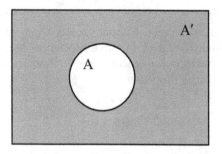

Figure 1.5.b: This Venn diagram represents the set A (white) together with its complement A' (shaded).

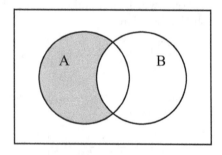

Figure 1.5.c: The shaded region represents $A \backslash B$ and includes the elements in A that are not also in B.

Definition 1.5.10. If U is the universe and $A \subseteq U$, then the *complement* of A is the set $\{x \mid (x \in U) \wedge (x \notin A)\}$. The complement of A is denoted A'. (See Figure 1.5.b.)

For example, if the universe is the set of all real numbers and A is the set of all rational numbers, then A' is the set of all irrational numbers.

Definition 1.5.11. If A and B are sets, then the *difference of A and B*, or the *complement of B with respect to A*, is the set $\{x \mid (x \in A) \wedge (x \notin B)\}$ and is denoted $A \backslash B$. (See Figure 1.5.c.) Often $A \backslash B$ is read "A minus B" or "A without B".

Observe that if U is the universe and $A \subseteq U$, then $A' = U \backslash A$.

Example 1.5.12. Find $A \backslash B$ in each of the following cases.

(a) $A = \{1, 2, 3, 4, 5\}$ and $B = \{2, 4, 6, 8\}$.

(b) $A = \{1, 2, 3, 4, 5\}$ and $B = \{0, 1, 2, 3, 4, 5, 6, 7, 8\}$.

(c) $A = \{1, 2, 3, 4, 5\}$ and $B = \{7, 8, 9, 10, 11\}$.

Solution. (a) $A \backslash B = \{1, 2, 3, 4, 5\} \backslash \{2, 4, 6, 8\} = \{1, 3, 5\}$. Note that $A \backslash B \subseteq A$ and that we have kept only those elements of A that do not belong to B.

(b) $A \backslash B = \{1, 2, 3, 4, 5\} \backslash \{0, 1, 2, 3, 4, 5, 6, 7, 8\} = \emptyset$. Observe that every element of A belongs to B, and so we had to eliminate all elements of A. Therefore, we are left with the empty set. Again, $A \backslash B \subseteq A$ because the empty set is a subset of any set.

(c) $A \backslash B = \{1, 2, 3, 4, 5\} \backslash \{7, 8, 9, 10, 11\} = \{1, 2, 3, 4, 5\}$. In this case, no element of A belongs to B, and so we did not have to eliminate any member of A. Thus, we end up with the entire set A. Again, $A \backslash B \subseteq A$ because any set is a subset of itself.

Definitions 1.5.13. Let A and B be sets. The *union* of A and B, written $A \cup B$, is the set $\{x \mid (x \in A) \vee (x \in B)\}$. The *intersection* of A and B, written $A \cap B$, is the set $\{x \mid (x \in A) \wedge (x \in B)\}$. If $A \cap B = \emptyset$, we say A and B are *disjoint*. (See Figure 1.5.d.)

Example 1.5.14. Consider the sets $A = \{1, 3, 5, 7, 9\}$, $B = \{0, 3, 7, 10, 13\}$, and $C = \{2, 4, 14\}$. Find:

(a) $A \cup B$

(b) $A \cap B$

(c) $B \cup C$

(d) $B \cap C$

(e) $A \cup C$

(f) $A \cap C$

(g) $A \cap (B \cup C)$ and $(A \cap B) \cup (A \cap C)$

(h) $A \cup (B \cap C)$ and $(A \cup B) \cap (A \cup C)$

Solution.

(a) $A \cup B = \{1, 3, 5, 7, 9\} \cup \{0, 3, 7, 10, 13\} = \{0, 1, 3, 5, 7, 9, 10, 13\}$.

(b) $A \cap B = \{1, 3, 5, 7, 9\} \cap \{0, 3, 7, 10, 13\} = \{3, 7\}$.

(c) $B \cup C = \{0, 3, 7, 10, 13\} \cup \{2, 4, 14\} = \{0, 2, 3, 4, 7, 10, 13, 14\}$.

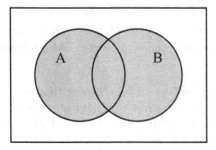 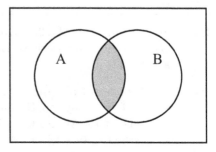

Figure 1.5.d: The shaded area in the left Venn diagram represents the union $A \cup B$ and the shaded area in the right Venn diagram represents the intersection $A \cap B$.

(d) $B \cap C = \{0, 3, 7, 10, 13\} \cap \{2, 4, 14\} = \emptyset$.

(e) $A \cup C = \{1, 3, 5, 7, 9\} \cup \{2, 4, 14\} = \{1, 2, 3, 4, 5, 7, 9, 14\}$.

(f) $A \cap C = \{1, 3, 5, 7, 9\} \cap \{2, 4, 14\} = \emptyset$.

(g) $A \cap (B \cup C) = \{1, 3, 5, 7, 9\} \cap \{0, 2, 3, 4, 7, 10, 13, 14\} = \{3, 7\}$ and $(A \cap B) \cup (A \cap C) = \{3, 7\} \cup \emptyset = \{3, 7\}$. Observe that both are the same.

(h) $A \cup (B \cap C) = \{1, 3, 5, 7, 9\} \cup \emptyset = \{1, 3, 5, 7, 9\}$ and $(A \cup B) \cap (A \cup C) = \{0, 1, 3, 5, 7, 9, 10, 13\} \cap \{1, 2, 3, 4, 5, 7, 9, 14\} = \{1, 3, 5, 7, 9\}$. Again, observe that both are the same.

Theorem 1.5.15. *Let A, B, and C be sets. The following are true:*

(a) $(A')' = A$

(b) $(A \cup B)' = A' \cap B'$ .... (De Morgan's Law)

(c) $(A \cap B)' = A' \cup B'$ .... (De Morgan's Law)

(d) $A \cup (B \cap C) = (A \cup B) \cap (A \cup C)$ .... (Distributive Law)

(e) $A \cap (B \cup C) = (A \cap B) \cup (A \cap C)$ .... (Distributive Law)

Proof. (a) Let U be the universe and suppose $x \in U$. Then

$$x \in (A')' \implies (x \in U) \wedge (x \notin A') \implies (x \in U) \wedge \neg(x \in A')$$

$$\implies (x \in U) \wedge \neg(x \notin A) \implies x \in A.$$

Since the choice of x was arbitrary, we have shown $(\forall x)\big(x \in (A')' \Rightarrow x \in A\big)$, and therefore $(A')' \subseteq A$. Conversely,

$$x \in A \implies (x \in U) \wedge (x \in A) \implies (x \in U) \wedge \neg(x \notin A)$$

$$\implies (x \in U) \wedge \neg(x \in A') \implies (x \in U) \wedge (x \notin A')$$

$$\implies x \in (A')'.$$

Thus $A \subseteq (A')'$, and so the sets coincide.

(b) Again, we let U be the universe and suppose $x \in U$. Then

$$x \in (A \cup B)'$$

$$\implies (x \in U) \wedge (x \notin A \cup B) \qquad \implies (x \in U) \wedge \neg(x \in A \vee x \in B)$$

$$\implies (x \in U) \wedge [\neg(x \in A) \wedge \neg(x \in B)] \implies (x \in U) \wedge (x \notin A \wedge x \notin B)$$

$$\implies (x \in U) \wedge (x \in A' \wedge x \in B') \qquad \implies x \in A' \cap B'.$$

We have shown that $(A \cup B)' \subseteq A' \cap B'$. The steps are reversible, and so we can replace each $\Rightarrow$ by $\Leftrightarrow$. Thus, it is also true that $A' \cap B' \subseteq (A \cup B)'$, and so the sets are the same.

(c) The proof is similar to that of Part (b) and is left as an exercise. (See Figure 1.5.e for an illustration of Parts (b) and (c) using Venn diagrams.)

(d) Let U be the universe and suppose $x \in U$. Then

$$x \in A \cup (B \cap C)$$

$$\Longleftrightarrow (x \in A) \vee (x \in B \cap C) \Longleftrightarrow (x \in A) \vee (x \in B \wedge x \in C)$$

$$\Longleftrightarrow (x \in A \vee x \in B) \wedge (x \in A \vee x \in C) \Longleftrightarrow (x \in A \cup B) \wedge (x \in A \cup C)$$

$$\Longleftrightarrow x \in (A \cup B) \cap (A \cup C).$$

It follows that $A \cup (B \cap C) = (A \cup B) \cap (A \cup C)$.

(e) The proof is similar to that of Part (d) and is left as an exercise. (See Figure 1.5.f for an illustration of this property using Venn diagrams.)   Q.E.D.

The notions of union and intersection of sets may be generalized as follows:

Definitions 1.5.16. Let $\mathscr{R}$ be a collection of sets. The *union* of the members of $\mathscr{R}$, written $\bigcup_{A \in \mathscr{R}} A$, is the set $\{x \mid x \in B \text{ for some } B \in \mathscr{R}\}$. The *intersection* of the members of $\mathscr{R}$, written $\bigcap_{A \in \mathscr{R}} A$, is the set $\{x \mid x \in B \text{ for every } B \in \mathscr{R}\}$.

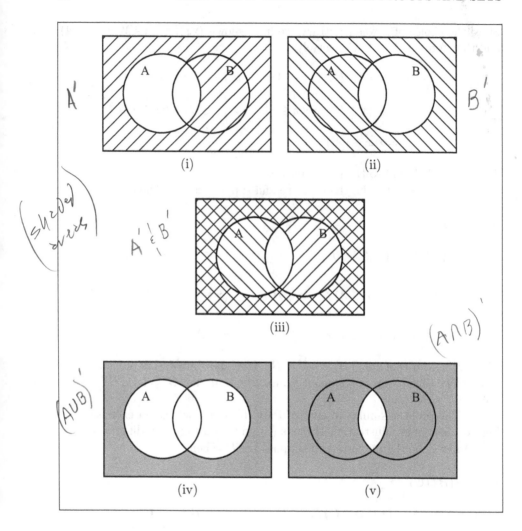

A′

B′

(shaded areas)

A′ & B′

(A∩B)′

(A∪B)′

Figure 1.5.e: De Morgan's Laws: Diagram (i) represents A' and Diagram (ii) represents B'. In Diagram (iii), both A' and B' are shown together. Diagram (iv) shows $(A \cup B)'$, which is the same as the region in (iii) that has both types of shading. (Part (b) of Theorem 1.5.15.) Diagram (v) shows $(A \cap B)'$, which is the same as the region in (iii) having either type of shading. (Part (c) of Theorem 1.5.15.)

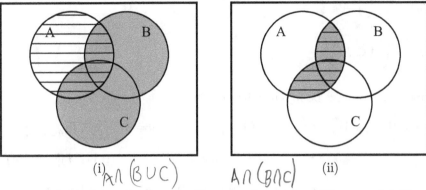

(i) $A \cap (B \cup C)$ $A \cap (B \cap C)$ (ii)

Figure 1.5.f: A Venn diagram to illustrate Part (e) of Theorem 1.5.15 and
Exercise 13.

Example 1.5.17. Let $\mathscr{R} = \{A, B, C, D, E\}$, where $A = \{1, 3, 4, 6, 7\}$,
$B = \{0, 3, 9\}$, $C = \{-3, 0, 3, 5, 9, 10\}$, $D = \{-2, 0, 1, 3, 5\}$, and $E = \{0, 3, 7, 9, 13\}$. Find $\bigcup_{S \in \mathscr{R}} S$ and $\bigcap_{S \in \mathscr{R}} S$.

Solution. $\displaystyle\bigcup_{S \in \mathscr{R}} S = \{-3, -2, 0, 1, 3, 4, 5, 6, 7, 9, 10, 13\}$ and $\displaystyle\bigcap_{S \in \mathscr{R}} S = \{3\}$.

Theorem 1.5.18. *Let C be a set, $\mathscr{R}$ a collection of sets, $\mathscr{T} \subseteq \mathscr{R}$, and $B \in \mathscr{R}$. The following are true:*

(a) $\displaystyle\left(\bigcap_{A \in \mathscr{R}} A \right)' = \bigcup_{A \in \mathscr{R}} A'$ .... *(De Morgan's Law)*

(b) $\displaystyle\left(\bigcup_{A \in \mathscr{R}} A \right)' = \bigcap_{A \in \mathscr{R}} A'$ .... *(De Morgan's Law)*

(c) $\displaystyle C \cup \left(\bigcap_{A \in \mathscr{R}} A \right) = \bigcap_{A \in \mathscr{R}} (C \cup A)$ .... *(Distributive Law)*

(d) $\displaystyle C \cap \left(\bigcup_{A \in \mathscr{R}} A \right) = \bigcup_{A \in \mathscr{R}} (C \cap A)$ .... *(Distributive Law)*

(e) $\displaystyle\bigcup_{A \in \mathscr{T}} A \subseteq \bigcup_{A \in \mathscr{R}} A$

(f) $\displaystyle\bigcap_{A \in \mathscr{R}} A \subseteq \bigcap_{A \in \mathscr{T}} A$

(g) $B \subseteq \bigcup\limits_{A \in \mathscr{R}} A$

(h) $\bigcap\limits_{A \in \mathscr{R}} A \subseteq B$

Proof. (a) We let x be in our universe. Then

$$x \in \left(\bigcap_{A \in \mathscr{R}} A\right)' \iff x \notin \bigcap_{A \in \mathscr{R}} A \iff \neg\left[x \in \bigcap_{A \in \mathscr{R}} A\right]$$

$$\iff \neg(\forall A \in \mathscr{R})(x \in A) \iff (\exists A \in \mathscr{R})[\neg(x \in A)]$$

$$\iff (\exists A \in \mathscr{R})(x \in A') \iff x \in \bigcup_{A \in \mathscr{R}} A'.$$

(b) The proof is similar to (a). Let x be in the universe. Then

$$x \in \left(\bigcup_{A \in \mathscr{R}} A\right)' \iff x \notin \bigcup_{A \in \mathscr{R}} A \iff \neg\left[x \in \bigcup_{A \in \mathscr{R}} A\right]$$

$$\iff \neg(\exists A \in \mathscr{R})(x \in A) \iff (\forall A \in \mathscr{R})[\neg(x \in A)]$$

$$\iff (\forall A \in \mathscr{R})(x \in A') \iff x \in \bigcap_{A \in \mathscr{R}} A'.$$

The proofs of (c)–(h) are left as exercises.   Q.E.D.

If we consider $\emptyset$ as the empty collection of sets in a situation where the universe is a set U, then we have the following startling result.

Theorem 1.5.19. *If U is the universe and $\emptyset$ is the empty collection of sets, then*

$$(a)\ \bigcap_{A \in \emptyset} A = U \quad and \quad (b)\ \bigcup_{A \in \emptyset} A = \emptyset.$$

Proof. (a) If $\mathscr{R}$ is any collection of sets, then

$$\bigcap_{A \in \mathscr{R}} A = \{x \mid x \in B \text{ for every } B \in \mathscr{R}\} = \{x \mid B \in \mathscr{R} \Rightarrow x \in B\}.$$

Thus,

$$\bigcap_{A \in \emptyset} A = \{x \mid B \in \emptyset \Rightarrow x \in B\} = U,$$

because the implication $B \in \emptyset \Rightarrow x \in B$ is true for every x in the universe (since $B \in \emptyset$ is false).

(b) If $\mathscr{R}$ is any collection of sets, then

$$\bigcup_{A \in \mathscr{R}} A = \{x \mid x \in B \text{ for some } B \in \mathscr{R}\} = \{x \mid (\exists B \in \mathscr{R})(x \in B)\}.$$

Thus,

$$\bigcup_{A \in \emptyset} A = \{x \mid (\exists B \in \emptyset)(x \in B)\} = \emptyset,$$

since the statement $(\exists B \in \emptyset)(x \in B)$ is false for all $x \in U$ by virtue of the fact that $\exists B \in \emptyset$ is necessarily false. Q.E.D.

Definitions 1.5.20. A collection $\mathscr{R}$ of sets is called an *indexed collection* if there exists a set I, called an *index set*, such that for each $\alpha \in I$ there corresponds a set A_α in $\mathscr{R}$, and every set in $\mathscr{R}$ is given as A_α for some $\alpha \in I$.

When a collection $\mathscr{R}$ of sets is an indexed collection, the union and intersection of the members of $\mathscr{R}$ can be denoted $\bigcup_{\alpha \in I} A_\alpha$ and $\bigcap_{\alpha \in I} A_\alpha$, respectively. Furthermore, if $I = \{1, 2, \ldots, n\}$, then we may denote the union and intersection by $\bigcup_{i=1}^{n} A_i$ and $\bigcap_{i=1}^{n} A_i$, respectively. (If $I = \mathbb{N}$, then we replace n with the symbol ∞.)

Exercises

1. Let $A = \{t \mid (t \text{ is a whole number}) \wedge (2 \leq t \leq 195)\}$. Identify each of the following statements as either true or false:

   (a) $9 \in A$ (b) $\frac{11}{3} \in A$ (c) $\sqrt[3]{27} \in A$ (d) $\pi \notin A$ (e) $195 \in A$

2. Let $B = \{x \mid (x \text{ is a rational number}) \wedge (-50 < x < 50)\}$. Identify each of the following statements as either true or false:

   (a) $-\frac{20}{3} \in B$ (b) $\frac{11}{4} \in B$ (c) $\sqrt{2} \in B$ (d) $\pi \notin B$ (e) $-50 \in B$

3. Let $A = \{1, 2\}$. List all subsets of A; that is, find $\mathscr{P}(A)$.

4. Let $B = \{1, 2, 3, 4\}$. List all subsets of B; that is, find $\mathscr{P}(B)$. (*Hint:* First list the subset with no elements, then all subsets with one element, then all subsets with two elements, etc.)

5. The following proposition is known as the *Binomial Theorem*: If a and b are any real numbers and n is a positive integer, then

$$(a + b)^n = \sum_{k=0}^{n} \binom{n}{k} a^{n-k} b^k.$$

   Use the Binomial Theorem to prove Theorem 1.5.5.

6. Let the universe be the set of integers. Let A be the set of odd numbers. What is A'?

7. Let the universe be the set of real numbers. Let C be the set of irrational numbers. What is C'?

8. Let the universe be the set of natural numbers. Let A be the set of prime numbers. List five members of A'.

9. Let $A = \{1, 3, 5, 7, 9\}$ and $B = \{0, 1, 9, 10, 13\}$. Find $A \backslash B$ and $B \backslash A$.

10. Let $A = \{a, b, c, d, e, f\}$, $B = \{a, d, f, g, k\}$, and $C = \{b, d, f, k, l, n\}$. Find the following:

   (a) $A \cup B$ (c) $B \cup C$ (e) $A \cap (B \cup C)$ (g) $(A \cap B) \cup (A \cap C)$
   (b) $A \cap B$ (d) $B \cap C$ (f) $A \cup (B \cap C)$ (h) $(A \cup B) \cap (A \cup C)$.

11. Repeat Exercise 10 for $A = \{-5, -3, 0, 2, 5, 8\}$, $B = \{-8, -3, 2, 8, 13\}$, and $C = \{-9, -3, 9, 12, 15\}$.

12. Prove Part (c) of Theorem 1.5.15. (See Figure 1.5.e.)

13. Prove Part (e) of Theorem 1.5.15. (See Figure 1.5.f.)

14. Let $\mathscr{R} = \{A, B, C, D, E\}$, where $A = \{1, 3, 5, 7, 9\}$, $B = \{0, 1, 2, 7, 8, 9, 11\}$, $C = \{-1, 1, 4, 8, 13\}$, $D = \{1, 3, 9, 12\}$, and $E = \{-2, 0, 1, 8, 15\}$. Find:

   (a) $\displaystyle\bigcup_{S \in \mathscr{R}} S$

   (b) $\displaystyle\bigcap_{S \in \mathscr{R}} S$

15. Let $\mathbb{R}^+$ be the set of positive real numbers and for each $c \in \mathbb{R}^+$, let A_c be the open interval $(-c, c)$. Find the following:

   (a) $\displaystyle\bigcup_{c \in \mathbb{R}^+} A_c$

   (b) $\displaystyle\bigcap_{c \in \mathbb{R}^+} A_c$

16. Prove Part (c) of Theorem 1.5.18.

17. Prove Part (d) of Theorem 1.5.18.

18. Prove Part (e) of Theorem 1.5.18.

19. Prove Part (f) of Theorem 1.5.18.

20. Prove Part (g) of Theorem 1.5.18.

21. Prove Part (h) of Theorem 1.5.18.

22. Let U be the universe and for each $A \subseteq U$ define the *characteristic function* of A, denoted χ_A, as follows:

$$\chi_A(x) = \begin{cases} 1 & \text{if } x \in A, \\ 0 & \text{if } x \notin A. \end{cases}$$

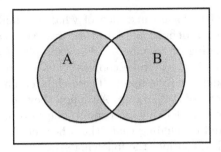

Figure 1.5.g: The shaded area in this Venn diagram represents the symmetric difference $A \triangle B$.

Prove the following, where A and B are subsets of the same universe U:

(a) $(A = B) \Leftrightarrow (\chi_A = \chi_B)$

(b) $(A \subseteq B) \Leftrightarrow (\chi_A \leq \chi_B)$

(c) $\chi_{A'} = 1 - \chi_A$

(d) $\chi_{A \cap B} = \chi_A \cdot \chi_B$

(e) $\chi_{A \cup B} = \chi_A + \chi_B - \chi_A \cdot \chi_B$

(f) $(\chi_A)^2 = \chi_A$

23. If A and B are sets in the same universe, then their *symmetric difference* is the set $A \triangle B = (A \backslash B) \cup (B \backslash A)$. (See Figure 1.5.g.) Find $A \triangle B$ if $A = \{1, 3, 5, 7, 9, 11\}$ and $B = \{-1, 0, 3, 7, 11, 13, 15\}$.

24. If A and B are subsets of the same universe, show $\chi_{A \triangle B} = |\chi_A - \chi_B|$. (See Exercises 22 and 23.)

25. Use Exercises 22 and 24 to prove the following: If A, B, and C are subsets of the same universe, then

(a) $A' \triangle B' = A \triangle B$

(b) $A \cap (B \triangle C) = (A \cap B) \triangle (A \cap C)$

(c) $A \triangle C \subset (A \triangle B) \cup (B \triangle C)$

(d) $A \triangle (B \triangle C) = (A \triangle B) \triangle C$

1.6 Cardinality

In this section, we are assuming that the student remembers the meaning of the terms used in elementary calculus concerning functions. We will use, without compunction, the terms *function, one-to-one correspondence, into, onto, domain, range,* etc. (See Section 3.1 for a quick review of definitions.)

Although many of us have some idea of what we think a number is, it is difficult to give a formal definition. All of us, at a very early age, have some conceptual understanding of "numbers." For example, if a child is given two cookies, and his sister is given three cookies, chances are that the child who received only two cookies will be upset. These children know that one of them is getting the better end of the bargain. Although they do not realize what they are doing, they are actually looking for a one-to-one correspondence between the two sets of cookies and concluding that "the other set" has an "extra" cookie. To expand on this idea, consider the following sets:

$$\{\{a,b,c\},\ \{\#,\%,*\},\ \{\Pi,\Sigma,\Delta\},\ \{2,6,12\},\ \{\triangle,\square,\Diamond\}\}.$$

If we were to add another element to this collection, we would likely select another set that also has three distinct members. The collection of all sets that can be put in a one-to-one correspondence with any of the above sets can be chosen to represent the number "three."

We will extend the concepts of elementary arithmetic to numbers that deal with infinite sets, as opposed to numbers that deal with finite sets. Infinite sets have some peculiar properties. We illustrate one of them in the next example.

Example 1.6.1. Define a one-to-one correspondence between the set $\mathbb{N}$ of positive integers and a proper subset of $\mathbb{N}$.

Solution. For each $n \in \mathbb{N}$, let $f(n) = 2n$. The function f is a mapping from the set $\mathbb{N}$ of positive integers to the set $\mathbb{E}$ of even positive integers. Clearly $\mathbb{E}$ is a proper subset of $\mathbb{N}$ because, for example, $3 \in \mathbb{N}$ but $3 \notin \mathbb{E}$. The function f is evidently one-to-one. Also, f is onto, because if $k \in \mathbb{E}$, then $k/2 \in \mathbb{N}$ and $f(k/2) = k$. Therefore there is a one-to-one correspondence between $\mathbb{N}$ and its proper subset $\mathbb{E}$.

This behavior cannot occur with finite sets, and indeed it is this property that characterizes infinite sets.

Definition 1.6.2. A set S is said to be *infinite* if and only if there is a proper subset T of S and a one-to-one correspondence between S and T.

Recall that if A and B are sets, then the set of *ordered pairs* (a,b), where $a \in A$ and $b \in B$, is denoted $A \times B$ and is called the *Cartesian product* of A and B.

Definition 1.6.3. A *relation* on a set A is a subset of $A \times A$. If $R \subseteq A \times A$ and $(a,b) \in R$, then we say a *is in relation to* b and we write aRb.

Example 1.6.4. Describe the *less than* relation ($<$) on the set $B = \{1, 2, 3, 4, 5\}$ as a set of ordered pairs.

Solution. Let R be the following subset of $B \times B$:

$$\{(1,2), (1,3), (1,4), (1,5), (2,3), (2,4), (2,5), (3,4), (3,5), (4,5)\}.$$

Since $R \subset B \times B$, it is a relation on B. Note that $(x, y) \in R$ if and only if $x < y$, where x and y are in B. Therefore, the subset R of $B \times B$ describes the *less than* relation $<$ on B.

A very important type of relation on a set is the so-called *equivalence relation*.

Definitions 1.6.5. A relation R on a set S is said to be:

(a) *reflexive* if $(a, a) \in R$ for all $a \in S$,

(b) *symmetric* if $(a, b) \in R \Rightarrow (b, a) \in R$, and

(c) *transitive* if $[(a, b) \in R] \wedge [(b, c) \in R] \Rightarrow (a, c) \in R$.

Any relation that has all three of these properties is called an *equivalence relation*. (See Figure 1.6.a.)

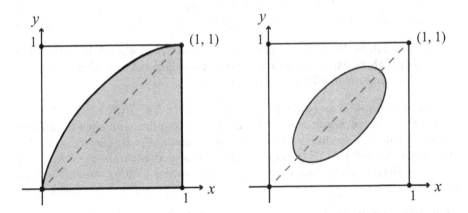

Figure 1.6.a: These two regions illustrate relations on subsets of the interval $[0, 1]$. The first relation is a reflexive relation on $[0, 1]$ because it contains the diagonal; however, it is not symmetric, because it is not symmetric with respect to the diagonal. The second relation is symmetric on a proper subset of $[0, 1]$, because it is symmetric with respect to the diagonal. Since the relation does not contain the entire diagonal, it is not reflexive. (So neither of these is an equivalence relation.)

On the set $\mathbb{R}$ of real numbers, the *equality* relation ($=$) is an equivalence relation. The *less than* relation ($<$) is neither reflexive nor symmetric, but it is transitive. The *less than or equal to* relation ($\leq$) is reflexive and transitive, but not symmetric.

Definitions 1.6.6. If R is an equivalence relation on a set S, and $x \in S$, then the set $\{y \mid yRx\}$ is called an *equivalence class* and is denoted $[x]$. The element x is called a *representative* of the equivalence class $[x]$.

Theorem 1.6.7. *If R is an equivalence relation on a set S, then any two equivalence classes are either equal or disjoint.*

Proof. Let x and y be in S. If $[x] \cap [y] = \emptyset$, then there is nothing to show. Assume $[x] \cap [y] \neq \emptyset$. We will show that $[x] = [y]$. We proceed:

$$[x] \cap [y] \neq \emptyset \implies (\exists z)(z \in [x] \cap [y]).$$

Furthermore,

$$z \in [x] \cap [y] \implies (z \in [x]) \wedge (z \in [y]) \implies zRx \wedge zRy \implies xRz \wedge zRy.$$

(The last implication follows from the symmetric property.) Now,

$$w \in [x] \implies wRx \implies wRy \implies w \in [y].$$

(The middle implication follows from $xRz \wedge zRy$ and two applications of the transitive property.) We have shown that $w \in [x] \Rightarrow w \in [y]$, and hence $[x] \subseteq [y]$. A similar argument shows that $[y] \subseteq [x]$, and so $[x] = [y]$.   Q.E.D.

Definition 1.6.8. We say that two subsets A and B of a set S have the same *cardinality* if there exists a one-to-one correspondence between them. If A and B have the same cardinality, we write $A \sim B$.

The relation $\sim$ is an equivalence relation on the set $\mathscr{P}(S)$ of all subsets of S, and two subsets A and B of S are in the same equivalence class if and only if there is a one-to-one function with domain A and range B. (See Exercise 1 at the end of this section). The requirement that A and B are both subsets of S is not restrictive. Given any two sets, we may always assume they are subsets of a common set.

Definition 1.6.9. If A is a set, then the *cardinal number* of A is the equivalence class (with respect to the equivalence relation $\sim$ from Definition 1.6.8) containing A and is written $|A|$.

For finite sets, the cardinal number refers to the number of elements that are in the finite set. In the literature, the symbol $\aleph_0$ (read "aleph-naught") is often used to denote the cardinal number of the set of positive integers $\mathbb{N}$; that is, $|\mathbb{N}| = \aleph_0$. (The symbol $\aleph$ is the first letter of the Hebrew alphabet.)

Example 1.6.10. Show that any two open intervals have the same cardinality.

Solution. Let (a, b) and (c, d) be any two open intervals. (We may assume they are not the same interval.) We will define a one-to-one function with domain (a, b) and range (c, d). There are many such functions, but the simplest is to define a linear function. The equation of the line through the points (a, c) and (b, d) is $y = mx + k$ where $m = \frac{d-c}{b-a}$ and $k = \frac{bc-ad}{b-a}$. The function $L(x) = mx + k$ is the desired one-to-one and onto function.

Definitions 1.6.11. A set S is called *countably infinite* (or *countable*) if it can be put in one-to-one correspondence with $\mathbb{N}$; that is, if $|S| = \aleph_0$. If S is neither countable nor finite, then S is called *uncountably infinite* (or *uncountable*).

If a set S is countably infinite, then there is a one-to-one function with domain $\mathbb{N}$, the set of natural numbers, and range S. In practice, one demonstrates this one-to-one correspondence by labeling the elements of the set S by means of a subscript, as in: $x_1, x_2, x_3, \ldots$. The one-to-one function, then, is given by $f(n) = x_n$ for each $n \in \mathbb{N}$.

Example 1.6.12. Show that the open interval $(0, 1)$ is uncountable.

Solution. Assume that $(0, 1)$ is a countable set. Then the elements of $(0, 1)$ can be put in one-to-one correspondence with the natural numbers. This means we can label the elements of $(0, 1)$ as $x_1, x_2, x_3, \ldots$ (as described above). Each number can be expressed as an infinite decimal. Therefore, we may write the elements of $(0, 1)$ as follows:

$$x_1 = 0.d_{11}d_{12}d_{13}\cdots$$
$$x_2 = 0.d_{21}d_{22}d_{23}\cdots$$
$$x_3 = 0.d_{31}d_{32}d_{33}\cdots$$
$$\vdots$$

We continue until we have such a representation for each x_n, where $n \in \mathbb{N}$.

Remark: In order to avoid ambiguity, we agree that for real numbers that have two different representations, we always use the one that terminates. For example, $0.4200000\ldots$ and $0.4199999\ldots$ both represent the number $21/50$, but we agree to never use the second representation.

We will now construct a number between 0 and 1 which is different from x_n for each $n \in \mathbb{N}$. Let $y = 0.c_1c_2c_3\ldots$, where

$$c_n = \begin{cases} d_{nn} - 1 & \text{if } d_{nn} \neq 0, \\ 1 & \text{if } d_{nn} = 0. \end{cases}$$

For each $n \in \mathbb{N}$, the number y is different than x_n because it differs from x_n at the nth decimal place. This is a contradiction, and so we conclude that the interval $(0,1)$ is not countable.

The method of proof used in the previous solution is called *Cantor's diagonal argument* (after Georg Cantor).

Definition 1.6.13. If A and B are two sets and there is a one-to-one function from A to B, then we write $|A| \leq |B|$.

Given two sets A and B, it may be that $|A| = |B|$, but it is difficult to exhibit a one-to-one and onto function between the two sets. The following theorem is useful in dealing with this situation.

Theorem 1.6.14 (Schröder-Bernstein Theorem). *Let A and B be subsets of a common set. Suppose there exists a one-to-one function f from A to B and a one-to-one function g from B to A. Then there exists a function h that is a one-to-one function from A onto B. Therefore, if $|A| \leq |B|$ and $|B| \leq |A|$, then $|A| = |B|$.*

Proof. Without loss of generality, assume A and B are disjoint subsets of the same set. (See Exercise 8 at the end of this section for a justification of this assumption.) Suppose f is a one-to-one function from A to B and g is a one-to-one function from B to A. We will partition each of A and B into three disjoint subsets that are defined in terms of "ancestors." We say that x is an *ancestor* of y if y can be obtained from x by successive applications of f and g (or of g and f). The three subsets of A are called A_E, A_O, and A_∞ and are defined as follows:

$$A_E = \{x \mid x \in A \text{ and } x \text{ has an even number of ancestors}\},$$
$$A_O = \{x \mid x \in A \text{ and } x \text{ has an odd number of ancestors}\}, \text{ and}$$
$$A_\infty = \{x \mid x \in A \text{ and } x \text{ has infinitely many ancestors}\}.$$

The three subsets of B are B_E, B_O, and B_∞ and are defined similarly.

Observe that f maps A_E onto B_O, and f also maps A_∞ onto B_∞. Further, g^{-1} maps A_O onto B_E. So it follows that we can define a function h by

$$h(x) = \begin{cases} f(x) & \text{if } x \in A_E \cup A_\infty, \\ g^{-1}(x) & \text{if } x \in A_O. \end{cases}$$

The function h is one-to-one from A onto B. Q.E.D.

Often, the Schröder-Bernstein Theorem provides a convenient way to prove that two sets have the same cardinality.

Example 1.6.15. Let $\mathbb{N}$ denote the set of natural numbers. The Cartesian product $\mathbb{N} \times \mathbb{N}$ has the same cardinality as $\mathbb{N}$.

Solution. Define a function $f : \mathbb{N} \to \mathbb{N} \times \mathbb{N}$ by $f(n) = (n, n)$. This function is certainly one-to-one, but not onto. Now define a function $g : \mathbb{N} \times \mathbb{N} \to \mathbb{N}$ by $g(m, n) = 2^m 3^n$. This function is also one-to-one (by the Fundamental Theorem of Arithmetic), but not onto. By the Schröder-Bernstein Theorem, there exists a function $h : \mathbb{N} \to \mathbb{N} \times \mathbb{N}$ that is one-to-one and onto. Thus, we conclude that $|\mathbb{N}| = |\mathbb{N} \times \mathbb{N}|$, and so the set $\mathbb{N} \times \mathbb{N}$ is countable.

In the preceding argument, we used the Schröder-Bernstein Theorem to show the existence of a one-to-one and onto function $h : \mathbb{N} \to \mathbb{N} \times \mathbb{N}$, but at no point did we explicitly construct such a function. Although we did not choose to do it in our solution, it is possible to construct a one-to-one and onto function between $\mathbb{N}$ and $\mathbb{N} \times \mathbb{N}$. (See Exercises 13 and 14 at the end of this section.)

A question that is frequently asked is: "How many infinite cardinal numbers are there?" Or one may ask, "Is there a largest cardinal number?" The following theorem provides an answer: Given any cardinal number, there is always one that is larger.

Theorem 1.6.16 (Cantor's Cardinality Theorem). *If X is any nonempty set, then $|X| < |\mathscr{P}(X)|$.*

Proof. Define a function $f : X \to \mathscr{P}(X)$ by $f(x) = \{x\}$. Clearly f is a one-to-one function (which is certainly not onto). Therefore, $|X| \leq |\mathscr{P}(X)|$. It remains to show that $|X| \neq |\mathscr{P}(X)|$.

Assume there exists a function $g : X \to \mathscr{P}(X)$ that is one-to-one and onto. Define a set A as follows:

$$A = \{ x \mid (x \in X) \wedge (x \notin g(x)) \}.$$

Since $A \subseteq X$, we know $A \in \mathscr{P}(X)$. Therefore, since g is onto $\mathscr{P}(X)$, there is some element $a \in X$ such that $g(a) = A$. Either $a \in A$ or $a \notin A$.

By the definition of the set A, if $a \in A$, then $a \notin g(a) = A$. Thus we have obtained both $a \in A$ and $a \notin A$, which is a contradiction. Suppose instead that $a \notin A$. Then $a \notin g(a)$, and so it follows that $a \in A$. Again we have $a \in A$ and $a \notin A$, which is a contradiction. Therefore, we reject the assumption that there is a one-to-one function from X onto $\mathscr{P}(X)$, and so $|X| < |\mathscr{P}(X)|$. Q.E.D.

Exercises

1. Prove that the relation $\sim$ of Definition 1.6.8 is an equivalence relation.

2. Let $\mathbb{R}$ be the set of real numbers and let (a, b) be a nonempty open bounded interval such that $a < b$. Show that $\mathbb{R} \sim (a, b)$. (*Hint:* First show that $\mathbb{R} \sim (-\frac{\pi}{2}, \frac{\pi}{2})$.)

3. Let S be the closed interval $[0, 1]$. Define a relation R on S that is both symmetric and transitive, but not reflexive.

4. Let S be the closed interval $[0, 1]$. Define a relation R on S that is both reflexive and symmetric, but not transitive.

5. Let S be the closed interval $[0, 1]$. Define a relation R on S that is both reflexive and transitive, but not symmetric.

6. Let $\mathbb{N}$ be the set of natural numbers. Define a relation R on $\mathbb{N} \times \mathbb{N}$ as follows: $(a, b)R(c, d)$ if and only if $ad = bc$. Prove that R is an equivalence relation.

7. Let S be the set of all continuous functions on the closed bounded interval $[a, b]$, where $a < b$. Define a relation $\sim$ on S as follows:

$$ f \sim g \text{ if and only if } \int_a^b \big(f(x) - g(x)\big)\,dx = 0. $$

Is $\sim$ an equivalence relation? Justify your answer.

8. Let A and B be subsets of X. Define two subsets of the product $X \times \{0, 1\}$ as follows:
$$ C = A \times \{0\} \quad \text{and} \quad D = B \times \{1\}. $$

   (a) Show that $C \cap D = \emptyset$.

   (b) Show that $A \sim B$ if and only if $C \sim D$.

   This result allows us, in any proof about the cardinality of two sets, to assume those two sets are disjoint.

9. Let $\mathbb{N}$ be the set of natural numbers and let $\mathbb{Z}$ be the set of all integers. Show that $|\mathbb{N}| = |\mathbb{Z}|$.

10. Define a one-to-one function from the closed interval $[0, 1]$ *onto* the open interval $(0, 1)$.

11. Use the Schröder-Bernstein Theorem to show that the set $\mathbb{Q}$ of rational numbers is countable.

12. Let $\mathbb{R}$ be the set of all real numbers and let $\mathbb{Q}$ be the set of all rational numbers. Then $\mathbb{R}\backslash\mathbb{Q}$ is the set of all irrational numbers. Use the Schröder-Bernstein Theorem to show there is a function $f : \mathbb{R} \to \mathbb{R}\backslash\mathbb{Q}$ that is both one-to-one and onto.

13. Define a "zig-zag" function $h : \mathbb{N} \to \mathbb{N} \times \mathbb{N}$ so that $h(1) = (1,1), h(2) = (1,2), h(3) = (2,1), h(4) = (3,1), h(5) = (2,2), h(6) = (1,3)$, and so on. Show that h is a one-to-one and onto function. Conclude that $\mathbb{N}$ and $\mathbb{N} \times \mathbb{N}$ have the same cardinality. (*Hint:* See Exercise 14.)

14. Show that the function defined by $f(i,j) = i + \frac{(i+j-2)(i+j-1)}{2}$ is a one-to-one and onto function from

$$
\begin{array}{ccccc}
(1,1) & (2,1) & (3,1) & (4,1) & (5,1) & \cdots \\
(1,2) & (2,2) & (3,2) & (4,2) & (5,2) & \cdots \\
(1,3) & (2,3) & (3,3) & (4,3) & (5,3) & \cdots \\
(1,4) & (2,4) & (3,4) & (4,4) & (5,4) & \cdots \\
\vdots & \vdots & \vdots & \vdots & \vdots
\end{array}
$$

to $\{1,2,3,4,5,\ldots\}$. Conclude that $\mathbb{N}$ and $\mathbb{N} \times \mathbb{N}$ have the same cardinality.

1.7 Cardinal Arithmetic

We introduced cardinal numbers in the preceding section. In this section, we will briefly discuss how the operations of addition, multiplication, and exponentiation of finite cardinals can be extended to infinite cardinals.

Consider the following sets:

$$A = \{a,b,c\}, \quad B = \{d,e\}, \quad \text{and } A \cup B = \{a,b,c,d,e\}.$$

Note that $|A| = 3$, $|B| = 2$, and $|A \cup B| = 5$, and that $3 + 2 = 5$. However, if we consider the sets

$$A = \{a,b,c\}, \quad C = \{c,d\}, \quad \text{and } A \cup C = \{a,b,c,d\},$$

then $|A| = 3$, $|C| = 2$, $|A \cup C| = 4$, but $3 + 2 \neq 4$. The difference between these two situations is that A and B are disjoint, whereas A and C are not.

Definition 1.7.1. If $|A|$ and $|B|$ are cardinal numbers, then the *sum* of the two is defined by

$$|A| + |B| = |A \cup B|,$$

where A and B are assumed to be disjoint. (If the sets are not disjoint, replace A by $A \times \{0\}$ and replace B by $B \times \{1\}$.)

It is easy to see that the definition does not depend on the choice of representatives A and B in $|A|$ and $|B|$. (See Exercise 1 at the end of this section.)

Example 1.7.2. Recall the notation $\aleph_0 = |\mathbb{N}|$, where $\mathbb{N}$ is the set of natural numbers. Show that $\aleph_0 + \aleph_0 = \aleph_0$.

Solution. Let A be the set of even natural numbers and let B be the set of odd natural numbers. Both A and B are countably infinite sets. To see this, observe that $f(n) = 2n$ is a one-to-one function from $\mathbb{N}$ onto A, and $g(n) = 2n - 1$ is a one-to-one function from $\mathbb{N}$ onto B. Thus, $|A| = |\mathbb{N}| = \aleph_0$ and $|B| = |\mathbb{N}| = \aleph_0$. Note that A and B are disjoint. Therefore,

$$\aleph_0 + \aleph_0 = |A| + |B| = |A \cup B| = |\mathbb{N}| = \aleph_0,$$

which is the desired result.

Many properties of addition for finite cardinal numbers (for example, commutativity and associativity) are valid also for infinite cardinal numbers. Verification of these properties is left for the exercises.

Now consider the sets

$$A = \{a, b, c\}, \; B = \{d, e\}, \; \text{and } A \times B = \{(a, d), (a, e), (b, d), (b, e), (c, d), (c, e)\}.$$

Observe that $|A| = 3$, $|B| = 2$, $|A \times B| = 6$, and $3 \times 2 = 6$. This motivates us to give the following definition.

Definition 1.7.3. If $|A|$ and $|B|$ are cardinal numbers, then their *product* $|A|\,|B|$ is defined by $|A|\,|B| = |A \times B|$.

The definition of $|A|\,|B|$ does not depend on the choices of A and B picked as representatives from each equivalence class. (See Exercise 2 at the end of this section.)

Consider the two sets $A = \{x, y\}$ and $B = \{a, b, c\}$. How many functions $f : B \to A$ are there? All possible functions are displayed in the following table:

f	$f(a)$	$f(b)$	$f(c)$
f_1	x	x	x
f_2	x	x	y
f_3	x	y	x
f_4	x	y	y
f_5	y	x	x
f_6	y	x	y
f_7	y	y	x
f_8	y	y	y

We see there are 8 functions with domain $\{a, b, c\}$ and range contained in $\{x, y\}$. For example, the function f_3 is defined so that $f_3(a) = x$, $f_3(b) = y$, and $f_3(c) = x$.

Observe that $|A| = 2$, $|B| = 3$, and $2^3 = 8$. This motivates us to define exponentiation in cardinal arithmetic in the following way.

Definition 1.7.4. If $|A|$ and $|B|$ are cardinal numbers, the *exponential* $|A|^{|B|}$ is defined to be $|A|^{|B|} = |\{f \mid f \text{ is a function from } B \text{ to } A\}|$.

Once again, the definition does not depend on the choices of A and B. (See Exercise 3 at the end of this section.) We comment that, because of Definition 1.7.4, the set of all functions from B to A is sometimes denoted in the literature by the symbol A^B. (See Exercise 4 at the end of this section for a connection between A^B and 2^B.)

In this section, we have illustrated how the concepts of elementary addition, multiplication, and exponentiation can be extended to infinite cardinal numbers. (However, there is no notion of "subtraction" for infinite cardinal numbers.) The familiar laws of arithmetic for addition, multiplication, and exponentiation still hold in this setting of infinite cardinal numbers. For illustration, we state these laws and prove two of them. The proofs of the remaining laws are left to the exercises.

Theorem 1.7.5 (Laws of Cardinal Arithmetic). *Let $|A|$, $|B|$, and $|C|$ be cardinal numbers. The following properties hold:*

(a) $|A| + |B| = |B| + |A|$ .... *(Commutative Law of Addition)*

(b) $(|A| + |B|) + |C| = |A| + (|B| + |C|)$ ...... *(Associative Law of Addition)*

(c) $|A|\,|B| = |B|\,|A|$ .... *(Commutative Law of Multiplication)*

(d) $|A|\,(|B|\,|C|) = (|A|\,|B|)\,|C|$ .... *(Associative Law of Multiplication)*

(e) $|A|\,(|B| + |C|) = |A|\,|B| + |A|\,|C|$ .... *(Distributive Law)*

(f) $(|A|\,|B|)^{|C|} = |A|^{|C|}\,|B|^{|C|}$ .... *(Exponentiation of a Product)*

(g) $|A|^{|B|}\,|A|^{|C|} = |A|^{|B|+|C|}$ .... *(Multiplication of Exponentials)*

(h) $(|A|^{|B|})^{|C|} = |A|^{|B|\,|C|}$ .... *(Exponentiation of a Power)*

Proof. (c) By definition of the product, $|A|\,|B| = |A \times B|$ and $|B|\,|A| = |B \times A|$. We need only show that $A \times B \sim B \times A$. That is, we need to define a one-to-one function from $A \times B$ onto $B \times A$. To that end, define $f : A \times B \to B \times A$ by

$$f(x, y) = (y, x)$$

for all $(x, y) \in A \times B$. It is easy to show that this function has the desired properties. We conclude that $A \times B \sim B \times A$, and so $|A|\,|B| = |B|\,|A|$.

(g) Recall that, for any sets X and Y, we defined the exponential $|X|^{|Y|}$ as the cardinal number $|X|^{|Y|} = |\{f \mid f \text{ is a function from } Y \text{ to } X\}|$. For ease of notation, in the current proof we will write this as $|X|^{|Y|} = |\{f : Y \to X\}|$. Consequently, we have

$$|A|^{|B|} = |\{f : B \to A\}| \text{ and } |A|^{|C|} = |\{g : C \to A\}|.$$

Therefore, by the definition of multiplication for infinite cardinal numbers,

$$|A|^{|B|}|A|^{|C|} = |\{f : B \to A\} \times \{g : C \to A\}|.$$

Without loss of generality, we may assume that B and C are disjoint. Therefore $|B| + |C| = |B \cup C|$, and consequently,

$$|A|^{|B|+|C|} = |A|^{|B \cup C|} = |\{h : B \cup C \to A\}|.$$

The proof will be complete if we can find a one-to-one and onto function ϕ from $\{f : B \to A\} \times \{g : C \to A\}$ to $\{h : B \cup C \to A\}$.

Define such a function ϕ as follows: For each $f : B \to A$ and $g : C \to A$, let $\phi(f,g)$ be the function from $B \cup C$ to A given by the rule

$$\phi(f,g)(x) = \begin{cases} f(x) & \text{if } x \in B, \\ g(x) & \text{if } x \in C. \end{cases}$$

The function ϕ is well defined because B and C were assumed to be disjoint.

We next show that ϕ is one-to-one. Suppose that $\phi(f_1, g_1) = \phi(f_2, g_2)$, where f_1 and f_2 are functions from B to A, and g_1 and g_2 are functions from C to A. Then, for all $x \in B \cup C$, we have $\phi(f_1, g_1)(x) = \phi(f_2, g_2)(x)$. We now see, from the definition of ϕ, that if $x \in B$, then $f_1(x) = f_2(x)$; and if $x \in C$, then $g_1(x) = g_2(x)$. Therefore, $(f_1, g_1) = (f_2, g_2)$, and so ϕ is a one-to-one function.

We now show that ϕ is onto. Let $h : B \cup C \to A$ be a function. Recalling that B and C are disjoint, define for all $x \in B \cup C$ a function $f : B \to A$ by $f(x) = h(x)$ if $x \in B$, and a function $g : C \to A$ by $g(x) = h(x)$ if $x \in C$. Then $\phi(f,g) = h$, and so ϕ is onto.

We have shown that $\{f : B \to A\} \times \{g : C \to A\} \sim \{h : B \cup C \to A\}$, and consequently $|\{f : B \to A\} \times \{g : C \to A\}| = |\{h : B \cup C \to A\}|$. Therefore,

$$|A|^{|B|}|A|^{|C|} = |A|^{|B \cup C|} = |A|^{|B|+|C|}.$$

Q.E.D.

Exercises

1. Show that if $C \in |A|$ and $D \in |B|$, where $A \cap B = \emptyset$ and $C \cap D = \emptyset$, then $C \cup D \sim A \cup B$. That is, show $|A| + |B| = |C| + |D|$ if $|A| = |C|$ and $|B| = |D|$.

2. Show that if $C \in |A|$ and $D \in |B|$, then $C \times D \sim A \times B$. That is, show $|A| \, |B| = |C| \, |D|$ whenever $|A| = |C|$ and $|B| = |D|$.

3. Show that if $C \in |A|$ and $D \in |B|$, then $\{f : B \to A\} \sim \{g : D \to C\}$. That is, show $|A|^{|B|} = |C|^{|D|}$ whenever $|A| = |C|$ and $|B| = |D|$.

4. Suppose B is a nonempty set. Let 2^B be the collection of all subsets of B and let $\{0,1\}^B$ be the collection of all functions from B to $\{0,1\}$. (See Definition 1.5.8 and the comments after Definition 1.7.4.) Show that there is a one-to-one correspondence between 2^B and $\{0,1\}^B$. (*Hint:* See Exercise 22 in Section 1.5.)

5. Show that $\aleph_0 + \aleph_0 + \aleph_0 = \aleph_0$.

6. Let $\mathbb{R}$ be the set of real numbers and let $\mathbb{Q}$ be the set of rational numbers.

   (a) If you know $\mathbb{Q}$ is countable, can you conclude that $\mathbb{R}\backslash\mathbb{Q}$ is uncountable?

   (b) If you know $\mathbb{R}\backslash\mathbb{Q}$ is uncountable, can you conclude that $\mathbb{Q}$ is countable?

   (c) Find two disjoint subsets of $\mathbb{R}$ such that both have the same cardinality as $\mathbb{R}$.

   (d) Is it possible to define subtraction for cardinal numbers $|A|$ and $|B|$ by letting $|A| - |B| = |A\backslash B|$?

7. Let $|A|$, $|B|$, and $|C|$ be cardinal numbers. Prove the Laws of Cardinal Arithmetic from Theorem 1.7.5 that were not proved within the section. (Recall that we proved (c) and (g).)

   (a) $|A| + |B| = |B| + |A|$ .... *(Commutative Law of Addition)*
   (b) $(|A| + |B|) + |C| = |A| + (|B| + |C|)$ .. *(Associative Law of Addition)*
   (d) $|A|\,(|B|\,|C|) = (|A|\,|B|)\,|C|$ ...... *(Associative Law of Multiplication)*
   (e) $|A|\,(|B| + |C|) = |A|\,|B| + |A|\,|C|$ .... *(Distributive Law)*
   (f) $(|A|\,|B|)^{|C|} = |A|^{|C|}\,|B|^{|C|}$ .... *(Exponentiation of a Product)*
   (h) $(|A|^{|B|})^{|C|} = |A|^{|B||C|}$ .... *(Exponentiation of a Power)*

Chapter 2

Topological Spaces

2.1 Introduction

One of the basic notions in topology is that of an open set. Our goal is to identify among all the subsets of a universe those that we wish to consider open. We will base our definition on some of the properties of open sets that we encountered in calculus. Thus, we first review some of these properties.

Recall that if a and b are real numbers, the distance between them is $|a - b|$. If ϵ is a positive number, and x is any real number, the ϵ-neighborhood of x is the set of all real numbers that are within ϵ of x and is denoted $N_\epsilon(x)$. That is,

$$N_\epsilon(x) = \{y \mid |y - x| < \epsilon\}.$$

Definition 2.1.1. A set S of real numbers is said to be *open* provided that for any $x \in S$, there exists some $\epsilon > 0$ such that $N_\epsilon(x) \subseteq S$.

Example 2.1.2. Show that any open interval (a, b) is an open set, where a and b are real numbers such that $a < b$.

Solution. Let $x \in (a, b)$. We must find some $\epsilon > 0$ such that $N_\epsilon(x) \subseteq (a, b)$. Let $\epsilon = \min\{x - a, b - x\}$. If $y \in N_\epsilon(x)$, then $|y - x| < \epsilon$, which implies both $|y - x| < x - a$ and $|y - x| < b - x$. Thus,

$$-(x - a) < y - x < x - a \quad \text{and} \quad -(b - x) < y - x < b - x.$$

It follows that $a < y < b$, and so $y \in (a, b)$. Therefore, $N_\epsilon(x) \subseteq (a, b)$.

Example 2.1.3. Show the set $\mathbb{Q}$ of rational numbers is not an open set.

Solution. Let $x \in \mathbb{Q}$. By the density of irrational numbers, there are irrational numbers in $N_\epsilon(x)$ for any positive number ϵ. Thus, there does not exist $\epsilon > 0$ such that $N_\epsilon(x) \subseteq \mathbb{Q}$, and so $\mathbb{Q}$ is not an open set.

Example 2.1.4. Show that the closed interval $[a, b]$ is not an open set, where a and b are real numbers such that $a < b$.

Solution. Note that $a \in [a, b]$. For any positive number ϵ, there is a real number y such that $a - \epsilon < y < a$. Therefore, $y \in N_\epsilon(a)$, but $y \notin [a, b]$. Consequently, $[a, b]$ is not an open set.

The following two theorems are introduced to justify the definition of open sets that we will give in the next section.

Theorem 2.1.5. *If $\mathscr{U}$ is any collection of open sets, then $\bigcup_{U \in \mathscr{U}} U$ is an open set.*

Proof. Let $x \in \bigcup_{U \in \mathscr{U}} U$. Then there exists some $V \in \mathscr{U}$ such that $x \in V$. Since V is open, there exists some $\epsilon > 0$ such that $N_\epsilon(x) \subseteq V$. It follows that $N_\epsilon(x) \subseteq \bigcup_{U \in \mathscr{U}} U$, and so $\bigcup_{U \in \mathscr{U}} U$ is an open set.   Q.E.D.

Theorem 2.1.6. *If $\{U_1, \ldots, U_n\}$ is a finite collection of open sets, then $\bigcap_{i=1}^{n} U_i$ is an open set.*

Proof. Let $x \in \bigcap_{i=1}^{n} U_i$. Then $x \in U_i$ for each $i \in \{1, \ldots, n\}$. Since each U_i is open, there exists some $\epsilon_i > 0$ such that $N_{\epsilon_i}(x) \subseteq U_i$. Let $\epsilon = \min\{\epsilon_1, \ldots, \epsilon_n\}$. We claim that $N_\epsilon(x) \subseteq \bigcap_{i=1}^{n} U_i$. To show this, let $y \in N_\epsilon(x)$. Then $|y - x| < \epsilon$, from which it follows that $|y - x| < \epsilon_i$ for each $i \in \{1, \ldots, n\}$. Thus $y \in N_{\epsilon_i}(x)$, and so $y \in U_i$, for each $i \in \{1, \ldots, n\}$. It follows that $y \in \bigcap_{i=1}^{n} U_i$. Therefore, we have shown that $N_\epsilon(x) \subseteq \bigcap_{i=1}^{n} U_i$.   Q.E.D.

We should observe that if $\mathscr{U}$ is an infinite collection of open sets, then $\bigcap_{U \in \mathscr{U}} U$ is not necessarily an open set. (See Exercise 3.)

Exercises

1. Let a be a real number. Show that the set $\{a\}$ is not an open set.

2. Show that the set $(a, b]$ is not an open set, where a and b are real numbers such that $a < b$.

3. For each natural number n, let $U_n = \left(-\frac{1}{n}, \frac{1}{n}\right)$. Show that $\bigcap_{n=1}^{\infty} U_n = \{0\}$ and conclude that $\bigcap_{n=1}^{\infty} U_n$ is not an open set.

4. Show that the set of all real numbers is an open set.

5. Show that the empty set is an open set. (*Hint:* Recall that the implication $p \Rightarrow q$ is true whenever p is false, regardless of the truth value of q.)

6. Show that the set of all irrational numbers is not an open set.

2.2 Topologies

We are now ready to describe how to identify subsets of a universe that we wish to call open. We will base our definition on the properties of open sets in $\mathbb{R}$ that we saw in Section 2.1. Given the definition of an open set in $\mathbb{R}$ (as stated in Definition 2.1.1), we saw that the following are true:

(i) The empty set $\emptyset$ is open. (Section 2.1, Exercise 5.)

(ii) The universe $\mathbb{R}$ is open. (Section 2.1, Exercise 4.)

(iii) The union of an arbitrary collection of open sets is open. (Theorem 2.1.5.)

(iv) The intersection of finitely many open sets is an open set. (Theorem 2.1.6.)

We want to model our definition of an *open set* on the definition of an open set in $\mathbb{R}$. Thus, we give the following definition.

Definition 2.2.1. Let X be a set. A collection $\mathscr{T}$ of subsets of X is called a *topology* on X if the following conditions are satisfied:

(i) $\emptyset \in \mathscr{T}$,

(ii) $X \in \mathscr{T}$,

(iii) If $\mathscr{U}$ is *any* subcollection of elements in $\mathscr{T}$, then $\bigcup_{U \in \mathscr{U}} U \in \mathscr{T}$, and

(iv) If $\mathscr{U}$ is a *finite* subcollection of elements in $\mathscr{T}$, then $\bigcap_{U \in \mathscr{U}} U \in \mathscr{T}$.

The ordered pair $(X, \mathscr{T})$ is called a *topological space*. Each member of $\mathscr{T}$ is said to be $\mathscr{T}$-*open*, or simply *open*, if no confusion arises.

Example 2.2.2. Let X be any nonempty set and let $\mathscr{T} = \{\emptyset, X\}$. It is clear that $\mathscr{T}$ satisfies the four conditions in Definition 2.2.1. Therefore, $\mathscr{T}$ is a topology on X. It is the *smallest* topology on X, in the sense that if $\mathscr{S}$ is any topology on X, then $\mathscr{T} \subseteq \mathscr{S}$. The topology $\mathscr{T}$ is called the *indiscrete topology*.

Example 2.2.3. Let X be a nonempty set and let $\mathscr{T}$ be the collection of all subsets of X. That is, $\mathscr{T} = \mathscr{P}(X)$. Then $\mathscr{T}$ satisfies the four conditions in Definition 2.2.1 and is therefore a topology on X. It is the *largest* topology on X, in the sense that if $\mathscr{S}$ is any topology on X, then $\mathscr{S} \subseteq \mathscr{T}$. The topology $\mathscr{T}$ is called the *discrete topology*.

Example 2.2.4. Let $\mathbb{R}$ be the set of real numbers and let $\mathscr{T}$ be the collection of sets that are open in the sense of Definition 2.1.1. Then $\mathscr{T}$ is a topology by the results of Section 2.1. (See Exercises 4 and 5 and Theorems 2.1.5 and 2.1.6.) The topology $\mathscr{T}$ is called the *standard topology on* $\mathbb{R}$.

Example 2.2.5. Let X be a nonempty set and let

$$\mathscr{T} = \{\emptyset\} \cup \{U \mid U \subseteq X \text{ and } U' \text{ is finite}\}.$$

(Recall that U' is the complement of U in X.) Show that $\mathscr{T}$ is a topology on X. (This topology is called the *cofinite topology*.)

Solution. The empty set $\emptyset$ is in $\mathscr{T}$, by definition. Also, $X \in \mathscr{T}$ because $X' = \emptyset$ is finite. Now let $\{U_1, \ldots, U_n\}$ be a finite subcollection of sets in $\mathscr{T}$. If the intersection is empty, so that $\bigcap_{i=1}^{n} U_i = \emptyset$, then $\bigcap_{i=1}^{n} U_i \in \mathscr{T}$. Suppose that $\bigcap_{i=1}^{n} U_i \neq \emptyset$. Then, for each $i \in \{1, \ldots, n\}$, the set $U_i \in \mathscr{T}$ is nonempty, and so it follows that U_i' is finite. Therefore, the set $\left(\bigcap_{i=1}^{n} U_i\right)' = \bigcup_{i=1}^{n} U_i'$ is finite, and hence $\bigcap_{i=1}^{n} U_i \in \mathscr{T}$.

Finally, we let $\mathscr{U}$ be an arbitrary subcollection of sets in $\mathscr{T}$. If $\bigcup_{U \in \mathscr{U}} U = \emptyset$, then $\bigcup_{U \in \mathscr{U}} U \in \mathscr{T}$. Suppose that $\bigcup_{U \in \mathscr{U}} U \neq \emptyset$. Then there exists some $U_0 \in \mathscr{U}$ such that $U_0 \in \mathscr{T}$ is nonempty. Therefore, U_0' is finite. Observe that

$$\left(\bigcup_{U \in \mathscr{U}} U\right)' = \bigcap_{U \in \mathscr{U}} U' \subseteq U_0'.$$

Thus, $\left(\bigcup_{U \in \mathscr{U}} U\right)'$ is finite, and consequently $\bigcup_{U \in \mathscr{U}} U \in \mathscr{T}$.

We have therefore verified the four conditions necessary to be a topology, and so $\mathscr{T}$ is a topology on X.

We now introduce the following.

Definition 2.2.6. Let $(X, \mathscr{T})$ be a topological space. A subset C of X is said to be $\mathscr{T}$-*closed*, or simply *closed*, if and only if C' is open.

Theorem 2.2.7. *Let $(X, \mathscr{T})$ be a topological space. If $\{C_1, \ldots, C_n\}$ is a finite collection of closed sets, then $\bigcup_{i=1}^n C_i$ is closed.*

Proof. Observe that $\left(\bigcup_{i=1}^n C_i\right)' = \bigcap_{i=1}^n C_i'$ is open, because it is the intersection of finitely many open sets. Therefore, $\bigcup_{i=1}^n C_i$ is closed, because it has an open complement. Q.E.D.

It should be noted that the union of an arbitrary collection of closed sets is not necessarily a closed set. (See Exercise 7 at the end of this section.)

Theorem 2.2.8. *Let $(X, \mathscr{T})$ be a topological space. If $\mathscr{C}$ is an arbitrary collection of closed sets, then $\bigcap_{C \in \mathscr{C}} C$ is a closed set.*

Proof. Notice that $\left(\bigcap_{C \in \mathscr{C}} C\right)' = \bigcup_{C \in \mathscr{C}} C'$ is open, because it is the union of open sets. Therefore, $\bigcap_{C \in \mathscr{C}} C$ is a closed set, because it has an open complement. Q.E.D.

Whether or not a set is open or closed in a universe depends on the topology assigned to the universe. For example, if the universe is $\mathbb{R}$, the set of real numbers, then the interval $(0, 1)$ is open in both the standard and discrete topologies, but it is not open in the indiscrete or cofinite topologies. As another example, the set $\{1\}$ is open in the discrete topology on $\mathbb{R}$, but it is not open in the standard, indiscrete, or cofinite topologies. On the other hand $\{1\}$ is closed in the standard, discrete, and the cofinite topologies, but not in the indiscrete topology.

It is important to realize that "closed" is not the negation of "open." In any topological space $(X, \mathscr{T})$, it is necessarily the case that X and $\emptyset$ are each both open and closed. Furthermore, if $\mathscr{T}$ is the discrete topology on X, then every subset of X is both open and closed. Also, there may exist subsets that are neither open nor closed. When $\mathbb{R}$ is equipped with the standard topology, for example, the interval $[0, 1)$ is neither open nor closed. (See Exercise 2 at the end of this section.)

Example 2.2.9. Let $X = \{1, 2, 3, 4, 5\}$ and $\mathscr{T} = \{\emptyset, X, \{1, 2, 3\}, \{4, 5\}\}$.

(i) Show that $\mathscr{T}$ is a topology on X.

(ii) Show that each open set is also closed.

(iii) Give an example of a subset of X that is neither open nor closed.

Solution. (i) It is clear that X and $\emptyset$ are both in $\mathscr{T}$. We must check that for each subcollection $\mathscr{U}$ of $\mathscr{T}$, both $\bigcup_{U \in \mathscr{U}} U \in \mathscr{T}$ and $\bigcap_{U \in \mathscr{U}} U \in \mathscr{T}$.

Since $\mathscr{T}$ has 4 members, there are $2^4 = 16$ choices for $\mathscr{U}$. One such choice for $\mathscr{U}$ is $\mathscr{U}_1 = \{\{1,2,3\},\{4,5\}\}$. In this case, $\bigcup_{U \in \mathscr{U}_1} U = X$ and $\bigcap_{U \in \mathscr{U}_1} U = \emptyset$, both of which are in $\mathscr{T}$. The remaining 15 cases can be done in a similar fashion and we leave them as an exercise for the interested reader.

(ii) The open subsets of X are X, $\emptyset$, $\{1,2,3\}$, and $\{4,5\}$. Note that X is closed because $X' = \emptyset$ is open. Similarly, $\emptyset$, $\{1,2,3\}$, and $\{4,5\}$ are closed because $\emptyset' = X$, $\{1,2,3\}' = \{4,5\}$, and $\{4,5\}' = \{1,2,3\}$ are open.

(iii) Let $A = \{1,3\}$. Then A is not open because $A \notin \mathscr{T}$, and A is not closed because $A' = \{2,4,5\} \notin \mathscr{T}$. Note that X has five elements, and so has 32 subsets. Among these subsets, we have identified 4 that are open and found that those same 4 subsets are also the only closed sets. Thus, there are 28 subsets of X that are neither open nor closed.

Exercises

1. Let $(\mathbb{R}, \mathscr{T})$ be the set of real numbers with the cofinite topology. (See Example 2.2.5.)

   (a) Show that any finite subset of $\mathbb{R}$ is closed.

   (b) Give an example of a subset of $\mathbb{R}$ that is neither open nor closed in the cofinite topology.

2. Show that $[0,1)$ is neither open nor closed in $(\mathbb{R}, \mathscr{T})$, where $\mathscr{T}$ is the standard topology on $\mathbb{R}$.

3. Show that $\{1\}$ is closed in the set of real numbers $\mathbb{R}$ equipped with the standard topology.

4. Let $(\mathbb{R}, \mathscr{T})$ be the set of real numbers with $\mathscr{T}$ the standard topology. Is the set $[0,\infty)$ open, closed, or neither? (Justify your answer.)

5. Let $X = \{1,2,3,4,5\}$ and $\mathscr{T} = \{X, \emptyset, \{1,2,3\}, \{4,5\}\}$. (This is the topological space from Example 2.2.9.) Let $A = \{1,3\}$ and find the subset B of X that has the following properties:

   (a) $A \subseteq B$,

   (b) B is closed,

   (c) If C is closed and $A \subseteq C$, then $B \subseteq C$.

   That is, find the "smallest" closed set containing A.

6. Let $X = \{1,2,3,4,5\}$ and $\mathscr{T} = \{X, \emptyset, \{1,2,3\}, \{4,5\}\}$. (This is the topological space from Example 2.2.9.) Let $A = \{1,2,3,4\}$ and find the subset

U of X that has the following properties:

(a) $U \subseteq A$,

(b) U is open,

(c) If V is open and $V \subseteq A$, then $V \subseteq U$.

That is, find the "largest" open set contained in A.

7. Let $\mathbb{R}$ be the set of real numbers and let $\mathscr{T}$ be the standard topology on $\mathbb{R}$. For each positive integer n, let

$$A_n = \left[-1 + \frac{1}{n+1}, \ 1 - \frac{1}{n+1} \right].$$

(a) Show that A_n is closed for $n \in \mathbb{N}$.

(b) Show that $\bigcup_{n=1}^{\infty} A_n = (-1, 1)$.

(c) Show that $(-1, 1)$ is not closed.

8. Let $X = \{a, b, c, d, e, f\}$ and $\mathscr{T} = \{\emptyset, X, \{a, c, f\}, \{b, d, e, f\}, \{f\}\}$.

(a) Show that $\mathscr{T}$ is a topology on X.

(b) List all the closed subsets of X.

(c) Identify all subsets of X that are both open and closed.

9. Let $(X, \mathscr{T})$ be the topological space of Exercise 8. Let $A = \{b, e\}$ and find the "smallest" closed subset of X containing A. That is, find the subset B of X that has the following properties:

(a) $A \subseteq B$,

(b) B is closed,

(c) If C is closed and $A \subseteq C$, then $B \subseteq C$.

10. Let $(X, \mathscr{T})$ be the topological space of Exercise 8. Let $A = \{a, b, c, f\}$ and find the "largest" open set contained in A. That is, find the subset U of X that has the following properties:

(a) $U \subseteq A$,

(b) U is open,

(c) If V is open and $V \subseteq A$, then $V \subseteq U$.

11. Let X be the set of real numbers and let $I = [-1, 1]$. Let

$$\mathscr{T} = \{U \mid I \subseteq U \subseteq X\} \cup \{\emptyset\}.$$

Is $\mathscr{T}$ a topology for X?

2.3   Bases

In the preceding section, we introduced the concept of a topology. We gave some very simple examples of topologies; however, topologies are often very large collections of sets, even if the universe itself is small. For example, if X has five elements, then its power set $\mathscr{P}(X)$ has 32 members. A topology for X is a subcollection of members from $\mathscr{P}(X)$, and there are $2^{32} = 4,294,967,296$ such subcollections. Of course, not all of these are topologies; nevertheless, it is tedious to describe a topology and to verify that a subcollection of $\mathscr{P}(X)$ satisfies the four conditions required of a topology. It is often easier to describe a subset of a topology that characterizes the topology itself. We will show how this is done in this section. We begin with the following definition.

Definition 2.3.1. Suppose that $\mathscr{T}$ and $\mathscr{R}$ are collections of sets such that $A \in \mathscr{T}$ if and only if A is a union of members of $\mathscr{R}$. That is,

$$\mathscr{T} = \left\{ A \ \middle| \ A = \bigcup_{V \in \mathscr{U}} V, \text{ where } \mathscr{U} \subseteq \mathscr{R} \right\}.$$

In such a circumstance, we say $\mathscr{T}$ is *generated* by $\mathscr{R}$, or that $\mathscr{R}$ *generates* $\mathscr{T}$.

Example 2.3.2. Let $V_1 = \{a,b,c\}$, $V_2 = \{a,d\}$, and $V_3 = \{b,e,f\}$. Assuming $\mathscr{R} = \{V_1, V_2, V_3\}$, find the collection $\mathscr{T}$ generated by $\mathscr{R}$.

Solution. Since $\mathscr{R}$ has 3 members, it has $2^3 = 8$ subsets, each of which is a collection of sets. These subsets are $\mathscr{U}_1 = \emptyset$, $\mathscr{U}_2 = \{V_1\}$, $\mathscr{U}_3 = \{V_2\}$, $\mathscr{U}_4 = \{V_3\}$, $\mathscr{U}_5 = \{V_1, V_2\}$, $\mathscr{U}_6 = \{V_1, V_3\}$, $\mathscr{U}_7 = \{V_2, V_3\}$, and $\mathscr{U}_8 = \mathscr{R}$. We must compute $\bigcup_{V \in \mathscr{U}_n} V$ for each $n \in \{1, \ldots, 8\}$. Thus,

$n = 1$: $\bigcup_{V \in \mathscr{U}_1} V = \emptyset$,

$n = 2$: $\bigcup_{V \in \mathscr{U}_2} V = V_1 = \{a,b,c\}$,

$n = 3$: $\bigcup_{V \in \mathscr{U}_3} V = V_2 = \{a,d\}$,

$n = 4$: $\bigcup_{V \in \mathscr{U}_4} V = V_3 = \{b,e,f\}$,

$n = 5$: $\bigcup_{V \in \mathscr{U}_5} V = V_1 \cup V_2 = \{a,b,c,d\}$,

$n = 6$: $\bigcup_{V \in \mathscr{U}_6} V = V_1 \cup V_3 = \{a,b,c,e,f\}$,

$n = 7$: $\bigcup_{V \in \mathscr{U}_7} V = V_2 \cup V_3 = \{a,b,d,e,f\}$,

$n = 8$: $\bigcup_{V \in \mathscr{U}_8} V = V_1 \cup V_2 \cup V_3 = \{a,b,c,d,e,f\}$.

Therefore, the collection $\mathscr{T}$ of sets generated by $\mathscr{R}$ is $\mathscr{T} = \{\ \emptyset, \{a,b,c\}, \{a,d\}, \{b,e,f\}, \{a,b,c,d\}, \{a,b,c,e,f\}, \{a,b,d,e,f\}, \{a,b,c,d,e,f\}\ \}$.

Observe that if a collection of sets $\mathscr{R}$ generates $\mathscr{T}$, then $\mathscr{R} \subseteq \mathscr{T}$.

Definition 2.3.3. Let $(X, \mathscr{T})$ be a topological space. If $\mathscr{R}$ is a subcollection of $\mathscr{T}$ such that $\mathscr{R}$ generates $\mathscr{T}$, then $\mathscr{R}$ is said to be a *base* for the topology $\mathscr{T}$.

It is desirable to find an answer to the following question: If X is a nonempty set and $\mathscr{R}$ is a collection of subsets of X, then under what conditions can we be sure that $\mathscr{R}$ is a base for some topology for X? The answer will be provided in Theorem 2.3.8, below. We first prove a lemma.

Lemma 2.3.4. *Let A be a set and $\mathscr{R}$ a collection of sets. If for each $x \in A$ there exists a set $B_x \in \mathscr{R}$ such that $x \in B_x$ and $B_x \subseteq A$, then there exists a subcollection $\mathscr{B}$ of $\mathscr{R}$ such that $A = \bigcup_{C \in \mathscr{B}} C$.*

Proof. By assumption, for each $x \in A$, there exists a set $B_x \in \mathscr{R}$ such that $x \in B_x$ and $B_x \subseteq A$. Let $\mathscr{B} = \{B_x \mid x \in A\}$. Since $B_x \in \mathscr{R}$ for each $x \in A$, it follows that $\mathscr{B} \subseteq \mathscr{R}$. It remains to show that $A = \bigcup_{C \in \mathscr{B}} C$.

Let $y \in A$. By the definition of $\mathscr{B}$, there exists a set $B_y \in \mathscr{B}$ such that $y \in B_y$. Certainly, since $B_y \in \mathscr{B}$, it must be the case that $B_y \subseteq \bigcup_{C \in \mathscr{B}} C$. Therefore, for each $y \in A$, we have established that $y \in \bigcup_{C \in \mathscr{B}} C$, and consequently we have the inclusion $A \subseteq \bigcup_{C \in \mathscr{B}} C$.

Conversely, suppose $z \in \bigcup_{C \in \mathscr{B}} C$. Then $z \in D$ for some $D \in \mathscr{B}$. But each member of $\mathscr{B}$ is a subset of A. Therefore, $z \in A$, and so $\bigcup_{C \in \mathscr{B}} C \subseteq A$.

Having demonstrated mutual inclusion, it follows that $A = \bigcup_{C \in \mathscr{B}} C$. Q.E.D.

Example 2.3.5. Let $A = \{5, 7, 9, 11, 13, 15\}$ be a set and let $\mathscr{R}$ be a collection of sets, say $\mathscr{R} = \{C, D, E, F, G, H, I, J, K, L, M\}$, where

$$C = \{5, 13, 15\}, \quad D = \{1, 7\}, \quad E = \{7, 9, 13\}, \quad F = \{2, 5, 6\},$$

$$G = \{5, 11, 13\}, \quad H = \{5, 7, 9, 15\}, \quad I = \{2, 5, 7, 15\},$$

$$J = \{3, 5, 9\}, \quad K = \{0, 3, 7\}, \quad L = \{7, 11, 15\}, \quad M = \{5\}.$$

Find a subcollection $\mathscr{B}$ of $\mathscr{R}$ such that $A = \bigcup_{S \in \mathscr{B}} S$.

Solution. For each $n \in A$, we will identify a set $B_n \in \mathscr{R}$ such that $n \in B_n$ and $B_n \subseteq A$. Let

$$B_5 = C, \quad B_7 = E, \quad B_9 = E, \quad B_{11} = G, \quad B_{13} = G, \quad B_{15} = C.$$

If $\mathscr{B} = \{B_5, B_7, B_9, B_{11}, B_{13}, B_{15}\}$, then

$$\bigcup_{S \in \mathscr{B}} S = B_5 \cup B_7 \cup B_9 \cup B_{11} \cup B_{13} \cup B_{15} = C \cup E \cup G$$

$$= \{5, 13, 15\} \cup \{7, 9, 13\} \cup \{5, 11, 13\} = \{5, 7, 9, 11, 13, 15\}$$

$$= A.$$

Observe that the sets B_n, for $n \in A$, are not distinct and that the choice of each B_n is not unique. (For example, we could have chosen $B_5 = H$.)

Example 2.3.6. Let $\mathbb{R}$ be the set of real numbers and let $\mathcal{T}$ be the standard topology on $\mathbb{R}$. Let $\mathcal{R}$ be the collection of all open intervals. Show that $\mathcal{R}$ is a base for $\mathcal{T}$.

Solution. If U is an open interval, then $U \in \mathcal{T}$. (See Example 2.1.2.) Thus, $\mathcal{R} \subseteq \mathcal{T}$. Now let $V \in \mathcal{T}$ and let $x \in V$. Since V is an open set in the standard topology on $\mathbb{R}$, there is an $\epsilon > 0$ such that $(x - \epsilon, x + \epsilon) \subseteq V$. (See Definition 2.1.1.) We have $x \in (x - \epsilon, x + \epsilon)$ and $(x - \epsilon, x + \epsilon) \in \mathcal{R}$. The choice of $x \in V$ was arbitrary, and so, by Lemma 2.3.4, there is a subcollection $\mathcal{B}$ of $\mathcal{R}$ such that $V = \bigcup_{I \in \mathcal{B}} I$.

Further, because $\mathcal{T}$ is a topology and $\mathcal{R} \subseteq \mathcal{T}$, any union of members of $\mathcal{R}$ is an element of $\mathcal{T}$. It follows that $\mathcal{R}$ generates the topology $\mathcal{T}$ and is therefore a base for $\mathcal{T}$.

In defining a topology, we required that the intersection of finitely many open sets is an open set. In general, it is sufficient to show that the intersection of any two open sets is an open set. This claim is justified by the following lemma:

Lemma 2.3.7. *Let $\mathcal{T}$ be a collection of sets with the property that the intersection of any two members of $\mathcal{T}$ is in $\mathcal{T}$. If $\{U_1, \ldots, U_n\}$ is a finite subcollection of sets from $\mathcal{T}$, then the intersection $\bigcap_{i=1}^{n} U_i$ is in $\mathcal{T}$.*

The proof is a straightforward application of induction and is left as an exercise. (See Exercise 8 at the end of this section.)

Theorem 2.3.8. *Let X be a nonempty set and let $\mathcal{R} \subseteq \mathcal{P}(X)$. The collection of sets $\mathcal{R}$ is a base for a topology $\mathcal{T}$ on X if and only if the following two conditions are satisfied:*

(1) $X = \bigcup_{V \in \mathcal{R}} V$.

(2) If U and V are members of $\mathcal{R}$ and $x \in U \cap V$, then there exists a $W \in \mathcal{R}$ such that $x \in W$ and $W \subseteq U \cap V$.

Proof. First, suppose conditions (1) and (2) are satisfied and let $\mathcal{T}$ be generated by $\mathcal{R}$. We want to show that $\mathcal{T}$ is a topology. By (1), we know that $X \in \mathcal{T}$. Also, $\emptyset \subseteq \mathcal{R}$, and so $\bigcup_{V \in \emptyset} V = \emptyset \in \mathcal{T}$. (See Theorem 1.5.19(b).)

Now suppose that $\mathcal{B}$ is a finite subcollection of members in $\mathcal{T}$. We will show that $\bigcap_{V \in \mathcal{B}} V \in \mathcal{T}$. By Theorem 1.5.19(a), if $\mathcal{B} = \emptyset$, then $\bigcap_{V \in \mathcal{B}} V = X$ and $X \in \mathcal{T}$, by (1). Assume instead that $\mathcal{B} \neq \emptyset$. By Lemma 2.3.7, it suffices to show that the intersection $\bigcap_{S \in \mathcal{B}} S \in \mathcal{T}$ whenever $\mathcal{B}$ has two members. Suppose U and V are in $\mathcal{T}$ and let $W = U \cap V$. If $x \in W$, then $x \in U$ and $x \in V$. Since U and V are in $\mathcal{T}$, and $\mathcal{T}$ is generated by $\mathcal{R}$, it follows that U and V are each the union of members of $\mathcal{R}$. Consequently, there are sets U_x

and V_x in $\mathscr{R}$ such that $x \in U_x \subseteq U$ and $x \in V_x \subseteq V$. It follows that $x \in U_x \cap V_x$ and $U_x \cap V_x \subseteq W$. Thus, by (2), there is a set W_x in $\mathscr{R}$ such that $x \in W_x$ and $W_x \subseteq U_x \cap V_x \subseteq W$. We have demonstrated that for each $x \in W$, there is a $W_x \in \mathscr{R}$ such that $x \in W_x$ and $W_x \subseteq W$. Therefore, by Lemma 2.3.4, there exists a subcollection $\mathscr{S}$ of $\mathscr{R}$ such that $W = \bigcup_{S \in \mathscr{S}} S$. Since $\mathscr{T}$ is generated by $\mathscr{R}$, we conclude that $W \in \mathscr{T}$, and so the intersection of two sets in $\mathscr{T}$ is again in $\mathscr{T}$.

Finally, let $\mathscr{B}$ be an arbitrary subcollection of sets from $\mathscr{T}$. Each member of $\mathscr{B}$ is a union of members of $\mathscr{R}$, and so $\bigcup_{V \in \mathscr{B}} V$ is itself a union of members of $\mathscr{R}$. Therefore $\bigcup_{V \in \mathscr{B}} V \in \mathscr{T}$. It follows that $\mathscr{T}$ is a topology.

Conversely, we now assume $\mathscr{R}$ is a base for the topology $\mathscr{T}$. We will show that the conditions (1) and (2) are satisfied. Since $X \in \mathscr{T}$ and $\mathscr{T}$ is generated by $\mathscr{R}$, it follows that $X = \bigcup_{U \in \mathscr{R}} U$. Now suppose $x \in U \cap V$, where U and V are in $\mathscr{R}$. Since $\mathscr{R} \subseteq \mathscr{T}$, it follows that U and V are elements of $\mathscr{T}$. Because $\mathscr{T}$ is a topology, it follows that $U \cap V \in \mathscr{T}$. Therefore, $U \cap V$ is a union of members of $\mathscr{R}$, and so there exists some $W \in \mathscr{R}$ such that $x \in W$ and $W \subseteq U \cap V$. Q.E.D.

At this point, one may wonder if it is possible for two different bases to generate the same topology, and if so, under what circumstances. An answer to these questions is provided by the following theorem.

Theorem 2.3.9. *Let X be a nonempty set and suppose $\mathscr{T}_1$ and $\mathscr{T}_2$ are two topologies on X that are generated by $\mathscr{R}_1$ and $\mathscr{R}_2$, respectively. The topologies $\mathscr{T}_1$ and $\mathscr{T}_2$ coincide if and only if the following two conditions are satisfied:*

(1) If $U_1 \in \mathscr{R}_1$ and $x \in U_1$, then there exists a set $U_2 \in \mathscr{R}_2$ such that $x \in U_2$ and $U_2 \subseteq U_1$.

(2) If $V_2 \in \mathscr{R}_2$ and $y \in V_2$, then there exists a set $V_1 \in \mathscr{R}_1$ such that $y \in V_1$ and $V_1 \subseteq V_2$.

Proof. Suppose $\mathscr{T}_1 = \mathscr{T}_2$ and let $U_1 \in \mathscr{R}_1$ and $x \in U_1$. Since $\mathscr{T}_1$ is generated by $\mathscr{R}_1$, it follows that $U_1 \in \mathscr{R}_1 \subseteq \mathscr{T}_1$. By assumption, then, we have that $U_1 \in \mathscr{T}_2$. Therefore, U_1 is a union of members of $\mathscr{R}_2$, and so there exists some $U_2 \in \mathscr{R}_2$ such that $x \in U_2$ and $U_2 \subseteq U_1$. The proof of the second condition is proved similarly, and the details are left to the interested reader.

Now suppose the conditions (1) and (2) are satisfied. (See Figure 2.3.a.) Let $U_1 \in \mathscr{T}_1$. We will show that $U_1 \in \mathscr{T}_2$. Suppose $x \in U_1$. Since U_1 is a union of members of $\mathscr{R}_1$, there exists a set $V_x \in \mathscr{R}_1$ such that $x \in V_x$ and $V_x \subseteq U_1$. By (1), there exists a set $W_x \in \mathscr{R}_2$ such that $x \in W_x$ and $W_x \subseteq V_x \subseteq U_1$. We have shown that for each $x \in U_1$, there exists a set $W_x \in \mathscr{R}_2$ such that $x \in W_x$ and $W_x \subseteq U_1$. Therefore, the set U_1 is a union of members from $\mathscr{R}_2$, by Lemma 2.3.4. Consequently, we have $U_1 \in \mathscr{T}_2$, and so $\mathscr{T}_1 \subseteq \mathscr{T}_2$. The reverse inclusion is proved with a similar argument and is left to the reader.   Q.E.D.

It is perhaps surprising that two disjoint bases can generate the same topology. For examples of this phenomenon, see Exercises 6 and 7 at the end of this section.

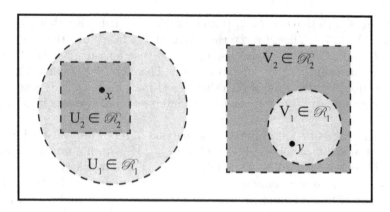

Figure 2.3.a: The topology $\mathscr{R}_1$ is generated by the open disks, while the topology $\mathscr{R}_2$ is generated by the open squares. Since for any point x in an open disk there is an open square subset of that disk containing x, and since for any point y in an open square there is an open disk subset of that square containing y, the topologies $\mathscr{R}_1$ and $\mathscr{R}_2$ are equivalent.

A base is given for a topology in order to describe that topology using a smaller collection of open sets. It is possible to describe a topology using a collection of open sets even smaller than a base, as we shall now see.

Suppose that X is a nonempty set and $\mathscr{S}$ is a nonempty collection of subsets of X. Let $\mathscr{R}$ be the collection of all sets that are the intersection of finite subcollections of $\mathscr{S}$. That is,

$$\mathscr{R} = \left\{ U \ \middle| \ U = \bigcap_{S \in \mathscr{C}} S, \text{ where } \mathscr{C} \subseteq \mathscr{S} \text{ and } |\mathscr{C}| < \infty \right\}.$$

Observe that $X \in \mathscr{R}$, because $\emptyset \subseteq \mathscr{S}$, $\emptyset$ is finite, and $\bigcap_{S \in \emptyset} S = X$. Also observe that the intersection of any two members in $\mathscr{R}$ is again in $\mathscr{R}$. Therefore, we can use Theorem 2.3.8 to show that $\mathscr{R}$ is a base for a topology on X. (See Exercise 9 at the end of this section.)

The preceding paragraph suggests the following definition.

Definition 2.3.10. Let X be a nonempty set with topology $\mathscr{T}$. If $\mathscr{S}$ is a nonempty collection of subsets of X such that

$$\left\{ U \ \middle| \ U = \bigcap_{S \in \mathscr{C}} S, \text{ where } \mathscr{C} \subseteq \mathscr{S} \text{ and } |\mathscr{C}| < \infty \right\}$$

is a base for the topology $\mathscr{T}$, then $\mathscr{S}$ is called at *subbase* for $\mathscr{T}$.

Exercises

1. Let $\mathbb{R}$ be the set of real numbers and let $\mathscr{B}$ be the collection of all intervals of the form $[a, b)$, where $a < b$. Show that $\mathscr{B}$ is a base for a topology on $\mathbb{R}$. (Recall that $[a, b) = \{x \mid a \leq x < b\}$.)

2. Let $\mathbb{R}$ be the set of real numbers and let $\mathscr{B}$ be the collection of all intervals of the form $[a, \infty)$. Show that $\mathscr{B}$ is a base for a topology on $\mathbb{R}$. (Recall that $[a, \infty) = \{x \mid x \geq a\}$.)

3. Let $\mathscr{T}_1$ be the topology on $\mathbb{R}$ from Exercise 1 and let $\mathscr{T}_2$ be the topology on $\mathbb{R}$ from Exercise 2. Are the topologies $\mathscr{T}_1$ and $\mathscr{T}_2$ the same? Prove that they are the same or demonstrate they are not by exhibiting a set which is open in one topology and not in the other.

4. Let $X = \{a, b, c, d, e, f, g\}$ and $\mathscr{B} = \{\{a, c, e, f\}, \{b, c, d, f, g\}, \{c\}, \{f\}\}$. Show that $\mathscr{B}$ is a base for a topology on X and find that topology.

5. Let $X = \{1, 2, 3, 4, 5, 6, 7, 8, 9\}$ and

$$\mathscr{B} = \{\{1, 3, 5, 9\}, \{5, 7, 8, 9\}, \{2, 4, 6, 7, 8\}, \{5, 9\}, \{7\}, \{8\}\}.$$

Show that $\mathscr{B}$ is a base for a topology on X and find that topology.

6. Let $\mathbb{R}$ be the set of real numbers. Let $\mathscr{R}_1$ be the collection of all open intervals with rational endpoints, let $\mathscr{R}_2$ be the collection of all open intervals with irrational endpoints, and let $\mathscr{R}_3$ be the collection of all open intervals. Show that $\mathscr{R}_1$, $\mathscr{R}_2$, and $\mathscr{R}_3$ are all bases that generate the same topology. (Notice that $\mathscr{R}_1 \cap \mathscr{R}_2 = \emptyset$.)

7. Let X be the Cartesian coordinate plane. Let $\mathscr{R}_1$ be the collection of all interiors of circles, and let $\mathscr{R}_2$ be the collection of all interiors of squares that have sides parallel to the coordinate axes. Show that $\mathscr{R}_1$ and $\mathscr{R}_2$ are bases that generate the same topology despite the fact that $\mathscr{R}_1 \cap \mathscr{R}_2 = \emptyset$.

8. Prove Lemma 2.3.7.

9. Suppose X is a nonempty set and $\mathscr{S}$ is a nonempty collection of subsets of X. Let $\mathscr{B}$ be the collection of all sets that are the intersection of finite subcollections of $\mathscr{S}$. That is,

$$\mathscr{B} = \left\{U \mid U = \bigcap_{S \in \mathscr{C}} S, \text{ where } \mathscr{C} \subseteq \mathscr{S} \text{ and } |\mathscr{C}| < \infty\right\}.$$

Prove that $\mathscr{B}$ is a base for a topology on X.

10. Show that the collection $\mathscr{S}$ of all intervals of form (a, ∞) and $(-\infty, b)$, where a and b are real numbers, is a subbase for the standard topology on $\mathbb{R}$.

2.4 Subspaces

Suppose that $(X, \mathscr{T})$ is a topological space and let Y be a nonempty subset of X. It is possible to define a topology on Y that is related to $\mathscr{T}$. We describe the topology in the following theorem.

Theorem 2.4.1. *Suppose $(X, \mathscr{T})$ is a topological space and Y is a nonempty subset of X. If $\mathscr{R} = \{Y \cap V \mid V \in \mathscr{T}\}$, then $\mathscr{R}$ is a topology on Y.*

Proof. Observe that $X \in \mathscr{T}$, and so $Y = Y \cap X$ is in $\mathscr{R}$. Similarly, $\emptyset \in \mathscr{T}$, and so $\emptyset = Y \cap \emptyset$ is in $\mathscr{R}$.

Let $\{W_1, W_2, \dots, W_n\}$ be a finite subcollection of $\mathscr{R}$. By the definition of $\mathscr{R}$, there exists a subcollection $\{V_1, V_2, \dots, V_n\}$ of $\mathscr{T}$ such that $W_k = Y \cap V_k$ for each number $k \in \{1, 2, \dots, n\}$. It follows that

$$\bigcap_{k=1}^{n} W_k = \bigcap_{k=1}^{n} (Y \cap V_k) = Y \cap \left(\bigcap_{k=1}^{n} V_k \right).$$

Because $\bigcap_{k=1}^{n} V_k \in \mathscr{T}$, it follows that $\bigcap_{k=1}^{n} W_k \in \mathscr{R}$.

Finally, let $\mathscr{B}$ be an arbitrary subcollection of $\mathscr{R}$. By the definition of $\mathscr{R}$, there exists a subcollection $\mathscr{C}$ of $\mathscr{T}$ such that $W \in \mathscr{B}$ if and only if there exists a set $V \in \mathscr{C}$ such that $W = Y \cap V$. Consequently,

$$\bigcup_{W \in \mathscr{B}} W = \bigcup_{V \in \mathscr{C}} (Y \cap V) = Y \cap \left(\bigcup_{V \in \mathscr{C}} V \right).$$

Since $\bigcup_{V \in \mathscr{C}} V$ is in $\mathscr{T}$, it follows that $\bigcup_{W \in \mathscr{B}} W$ is in $\mathscr{R}$.

Having verified all the conditions necessary to define a topology, we conclude that $\mathscr{R}$ is a topology on Y.   Q.E.D.

Definition 2.4.2. Let $(X, \mathscr{T})$ be a topological space. If Y is a nonempty subset of X and $\mathscr{R} = \{Y \cap V \mid V \in \mathscr{T}\}$, then $\mathscr{R}$ is called the *topology on Y induced by $\mathscr{T}$*, or the *induced topology*, when there is no risk of confusion. The space $(Y, \mathscr{R})$ is called a *subspace* of $(X, \mathscr{T})$. For this reason, $\mathscr{R}$ is also called the *subspace topology* on Y (when there is no risk of confusion).

Example 2.4.3. Let $\mathbb{R}$ be the set of real numbers with $\mathscr{T}$ the standard topology. Let Y be the set of integers. Show that the induced topology on Y is the discrete topology.

Solution. We first show that the single-point set $\{a\}$ is open in the induced topology for each $a \in Y$. Observe that, for any $a \in Y$, we have that $\{a\} = Y \cap \left(a - \frac{1}{2}, a + \frac{1}{2} \right)$. Since the interval $\left(a - \frac{1}{2}, a + \frac{1}{2} \right)$ is open in the standard topology on $\mathbb{R}$, it follows that $\{a\}$ is open in the induced topology. Now we observe that any subset of Y is the union of single-point sets, and so is open in the induced topology. Therefore, the induced topology is the discrete topology $\mathscr{P}(Y)$.

We remark that a set may be open in the subspace topology even though it is not open in the original topology. (See Exercise 3 at the end of this section.)

Theorem 2.4.4. *Let* $(Y, \mathscr{R})$ *be a subspace of a topological space* $(X, \mathscr{T})$. *A set* $A \subseteq Y$ *is* $\mathscr{R}$-*closed if and only if* $A = Y \cap B$ *for some* $\mathscr{T}$-*closed subset* B *of* X.

Proof. Suppose first that A is $\mathscr{R}$-closed. It follows that $Y \setminus A$ is $\mathscr{R}$-open. Therefore, there is some $\mathscr{T}$-open set U such that $Y \setminus A = Y \cap U$. Let $B = X \setminus U$. Then, B is $\mathscr{T}$-closed and

$$Y \cap B = Y \cap (X \setminus U) = (Y \cap X) \setminus (Y \cap U) = Y \setminus (Y \setminus A) = A.$$

(See Exercises 1 and 2 at the end of this section.)

Conversely, suppose $A = Y \cap B$ for some $\mathscr{T}$-closed subset B of X. Let $V = X \setminus B$. Then V is $\mathscr{T}$-open and so $Y \cap V$ is $\mathscr{R}$-open. Thus, $Y \setminus (Y \cap V)$ is $\mathscr{R}$-closed; but

$$\begin{aligned} Y \setminus (Y \cap V) &= Y \setminus [Y \cap (X \setminus B)] = Y \setminus [(Y \cap X) \setminus (Y \cap B)] \\ &= Y \setminus (Y \setminus A) = A. \end{aligned}$$

Therefore, A is $\mathscr{R}$-closed.   Q.E.D.

Exercises

1. Let X be a nonempty set and let $A \subseteq X$. Prove that $X \setminus (X \setminus A) = A$.

2. Let X be a nonempty set with subsets A, B, and C. Prove that
$$A \cap (B \setminus C) = (A \cap B) \setminus (A \cap C).$$

3. Give an example of a topological space $(X, \mathscr{T})$ and a subspace $(Y, \mathscr{R})$ where some subset of Y is $\mathscr{R}$-open but fails to be $\mathscr{T}$-open.

4. Let $(X, \mathscr{T})$ be a topological space and let Y be a nonempty $\mathscr{T}$-open subset of X. Let $\mathscr{R}$ be the topology on Y induced by $\mathscr{T}$. Show that a subset of Y is $\mathscr{R}$-open if and only if it is $\mathscr{T}$-open.

5. Let $(X, \mathscr{T})$ be a topological space and let Y be a nonempty $\mathscr{T}$-closed subset of X. Let $\mathscr{R}$ be the topology on Y induced by $\mathscr{T}$. Show that a subset of Y is $\mathscr{R}$-closed if and only if it is $\mathscr{T}$-closed.

6. Let $(X, \mathscr{T})$ be a topological space with subspace $(Y, \mathscr{R})$. Suppose that Z is a nonempty subset of Y. If $\mathscr{B}_1$ is the topology on Z induced by $\mathscr{T}$, and $\mathscr{B}_2$ is the topology on Z induced by $\mathscr{R}$, prove that $\mathscr{B}_1 = \mathscr{B}_2$.

7. Let $X = \{1, 2, 3, 4, 5\}$ and $\mathscr{T} = \{\emptyset, X, \{1, 2, 3\}, \{4, 5\}\}$. (This is the topological space from Example 2.2.9.) Let $Y = \{1, 3, 4\}$. What is the subspace topology on Y?

8. Let $X = \{a, b, c, d, e, f\}$ and $\mathscr{T} = \{\emptyset, X, \{a, c, f\}, \{b, d, e, f\}, \{f\}\}$. (This is the topology from Exercise 8 in Section 2.2.) Let $Y = \{a, b, c, f\}$.

Figure 2.5.a: The interior of A is the union of all open sets contained in A. The open sets that constitute A° are illustrated in the third image.

(a) What is the subspace topology $\mathscr{R}$ on Y induced by $\mathscr{T}$?

(b) List all of the $\mathscr{R}$-closed subsets of Y.

(c) For each $\mathscr{R}$-closed subset A of Y, find a $\mathscr{T}$-closed subset B of X such that $A = Y \cap B$.

9. Let $(X, \mathscr{T})$ be a topological space and suppose $\mathscr{B}$ is a base for the topology $\mathscr{T}$. Let Y be a nonempty subset of X and let $\mathscr{B}_Y = \{Y \cap U \mid U \in \mathscr{B}\}$. Show that $\mathscr{B}_Y$ is a base for the subspace topology on Y.

2.5 Interior, Closure, and Boundary

Let $(X, \mathscr{T})$ be a topological space and $A \subseteq X$. In the present section, we consider three sets that are closely related to A. The first of these sets is the largest open set contained in A; the second is the smallest closed set containing A. (See Exercises 5, 6, 9 and 10 in Section 2.2.)

Definition 2.5.1. Let $(X, \mathscr{T})$ be a topological space and $A \subseteq X$. The *interior* of A is the set $A^\circ = \bigcup_{U \in \mathscr{R}} U$, where $\mathscr{R} = \{U \mid U \in \mathscr{T} \text{ and } U \subseteq A\}$. That is, A° is the union of all open sets that are subsets of A.

Theorem 2.5.2. *Let $(X, \mathscr{T})$ be a topological space and $A \subseteq X$. The set A° is the unique subset of X satisfying the following three properties:*

(i) A° is open.

(ii) $A^\circ \subseteq A$.

(iii) If V is open and $V \subseteq A$, then $V \subseteq A^\circ$.

That is, A° is the largest open subset of A.

Proof. Let $\mathscr{R} = \{U \mid U \in \mathscr{T} \text{ and } U \subseteq A\}$, so that $A^\circ = \bigcup_{U \in \mathscr{R}} U$. We first observe that A° is open as the union of open sets. To prove *(ii)*, suppose that $x \in A^\circ$. Since $A^\circ = \bigcup_{U \in \mathscr{R}} U$, it follows that $x \in U_x$ for some $U_x \in \mathscr{R}$; however, each member of $\mathscr{R}$ is a subset of A, by assumption. Therefore $x \in A$, and so $A^\circ \subseteq A$. We next prove *(iii)*. If V is an open set contained in A, then $V \in \mathscr{R}$, by the definition of $\mathscr{R}$. Consequently $V \subseteq \bigcup_{U \in \mathscr{R}} U = A^\circ$.

It remains to verify that A° is the unique subset of A satisfying *(i)–(iii)*. Assume that E also satisfies *(i)–(iii)*. Then E is open and contained in A. It follows that $E \subseteq A^\circ$, because A° is the union of all such sets. Finally, we observe that A° is an open subset of A, and so, since E satisfies *(iii)*, we conclude that $A^\circ \subseteq E$. Therefore, $A^\circ = E$, as required. Q.E.D.

Example 2.5.3. Let $X = \{a, b, c, d, e, f\}$ be a set and define a topology on X by $\mathscr{T} = \{\emptyset, X, \{a, b, c, d\}, \{b, d\}, \{b, d, e, f\}\}$. Let $A = \{a, b\}$ and $B = \{a, b, c, d, f\}$. Find A° and B°.

Solution. The only open set contained in A is $\emptyset$, and so $A^\circ = \emptyset$. The only open sets contained in B are $\{a, b, c, d\}$ and $\{b, d\}$. Thus,

$$B^\circ = \{a, b, c, d\} \cup \{b, d\} = \{a, b, c, d\}.$$

Definition 2.5.4. Let $(X, \mathscr{T})$ be a topological space and $A \subseteq X$. The *closure* of A is the set $\overline{A} = \bigcap_{C \in \mathscr{R}} C$, where $\mathscr{R} = \{C \mid X \setminus C \in \mathscr{T} \text{ and } A \subseteq C\}$. That is, $\overline{A}$ is the intersection of all closed sets that contain A as a subset.

Theorem 2.5.5. *Let $(X, \mathscr{T})$ be a topological space and $A \subseteq X$. The set $\overline{A}$ is the unique subset of X satisfying the following three properties:*

(i) $\overline{A}$ *is closed.*

(ii) $A \subseteq \overline{A}$.

(iii) *If B is closed and $A \subseteq B$, then $\overline{A} \subseteq B$.*

That is, $\overline{A}$ is the smallest closed set containing A.

The proof of Theorem 2.5.5 is left as an exercise. (See Exercise 2 at the end of this section.)

Example 2.5.6. Consider the set $X = \{a, b, c, d, e, f\}$ with the topology $\mathscr{T} = \{\emptyset, X, \{a, b, c, d\}, \{b, d\}, \{b, d, e, f\}\}$. (This is the topological space from Example 2.5.3.) Suppose $A = \{a, b\}$ and $B = \{a, e, f\}$. Find $\overline{A}$ and $\overline{B}$.

Solution. The only closed set containing A is X, and consequently $\overline{A} = X$. The only closed sets containing B are $\{a, c, e, f\}$ and X. Therefore,

$$\overline{B} = \{a, c, e, f\} \cap X = \{a, c, e, f\}.$$

Before stating the next theorem, we recall that if E is a subset of a universe U, then E' is the *complement* of E in U. That is, $E' = U \setminus E$. (See Definition 1.5.10.)

Theorem 2.5.7. *If $(X, \mathscr{T})$ is a topological space and $A \subseteq X$, then $A^\circ = (\overline{A'})'$.*

Proof. We start by observing that $A^\circ \subseteq A$, and hence $A' \subseteq (A^\circ)'$. Since $(A^\circ)'$ is a closed set containing A', it must be the case that $\overline{A'} \subseteq (A^\circ)'$. Consequently, $A^\circ \subseteq (\overline{A'})'$

To show the reverse inclusion, note that $A' \subseteq \overline{A'}$. Thus, $(\overline{A'})' \subseteq A$. Since $(\overline{A'})'$ is an open set contained within A, it follows that $(\overline{A'})' \subseteq A^\circ$. We have shown both $A^\circ \subseteq (\overline{A'})'$ and $(\overline{A'})' \subseteq A^\circ$. Therefore, $A^\circ = (\overline{A'})'$.   Q.E.D.

Example 2.5.8. Once again consider the set $X = \{a, b, c, d, e, f\}$ with the topology $\mathscr{T} = \{\emptyset, X, \{a, b, c, d\}, \{b, d\}, \{b, d, e, f\}\}$. (This is the topological space from Example 2.5.3.) Let $A = \{a, b, d, e, f\}$ and find $A^\circ, A', \overline{A'}$ and $(\overline{A'})'$. Verify the equality in Theorem 2.5.7 for this example.

Solution. The open sets contained in A are $\emptyset$, $\{b, d\}$, and $\{b, d, e, f\}$. Consequently,

$$A^\circ = \emptyset \cup \{b, d\} \cup \{b, d, e, f\} = \{b, d, e, f\}.$$

Observe that $A' = \{c\}$. The closed sets that contain $\{c\}$ are X, $\{a, c, e, f\}$, and $\{a, c\}$. Thus,

$$\overline{A'} = X \cap \{a, c, e, f\} \cap \{a, c\} = \{a, c\}.$$

It follows that $(\overline{A'})' = \{a, c\}' = X \setminus \{a, c\} = \{b, d, e, f\}$, and so $(\overline{A'})' = A^\circ$.

A concept closely related to that of the interior and closure of a set is the *boundary* of a set. Intuitively, a boundary point of a set ought to be a point that is somehow "close" to both the set and its complement. However, we have not defined a notion of "closeness" in an abstract topological space. To that end, we give the following definition.

Definition 2.5.9. Let $(X, \mathscr{T})$ be a topological space and let $x \in X$. If N is a set in $\mathscr{T}$ such that $x \in N$, then N is called a $\mathscr{T}$*-neighborhood* (or simply a *neighborhood*) of x in X.

Frequently, in the literature, a neighborhood of a point x is defined to be any set that contains an open set to which x belongs. By this definition, both sets N and A in Figure 2.5.b are neighborhoods of the point x. (See Exercise 7 at the end of this section.) The definition we have given would then be called an *open neighborhood*. For our purposes, it is convenient (and sufficient) to consider only neighborhoods that are open, and so we will build this into the definition.

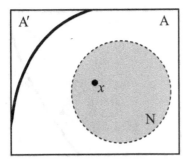

Figure 2.5.b: A neighborhood of the point x is any open set N containing x.

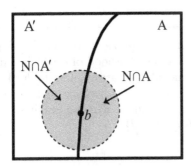

Figure 2.5.c: The point b is a boundary point of A.

Definitions 2.5.10. Let $(X, \mathscr{T})$ be a topological space and suppose $A \subseteq X$. A point $b \in X$ is called a *boundary point* of A if and only if each neighborhood of b intersects both A and A'. That is, if N is a neighborhood of b, then $N \cap A \neq \emptyset$ and $N \cap A' \neq \emptyset$.

The collection of all boundary points of A is called the *boundary* of A and is denoted ∂A.

Example 2.5.11. Let $(X, \mathscr{T})$ be the topological space that is given by $X = \{a, b, c, d\}$ and $\mathscr{T} = \{\emptyset, X, \{a, b, c\}, \{a, d\}, \{a\}\}$. Let $A = \{a, c, d\}$ and $B = \{b, d\}$. Find the boundaries ∂A and ∂B.

Solution. Observe that $A' = \{b\}$. It follows that $a \notin \partial A$, because $\{a\}$ is a neighborhood of a which is disjoint from A'. The neighborhoods of b are X and $\{a, b, c\}$. Each of these sets intersect both $A = \{a, c, d\}$ and $A' = \{b\}$. Thus, $b \in \partial A$. The neighborhoods of c are the same as those of b, and so $c \in \partial A$ for the same reasons. On the other hand, $\{a, d\}$ is a neighborhood of d which is disjoint from A'. Therefore, d is not a boundary point of A, and so $\partial A = \{b, c\}$.

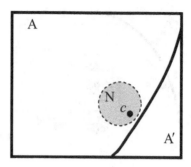

Figure 2.5.d: The point c is an interior point of A.

The complement of $B = \{b, d\}$ is the set $B' = \{a, c\}$. We see that $a \notin \partial B$, because $\{a\}$ is a neighborhood of a which is disjoint from B. The neighborhoods of b (and also c) are X and $\{a, b, c\}$. Each of these sets intersects B and B', and so b (as well as c) is a boundary point. The neighborhoods of d are X and $\{a, d\}$. Again, these sets intersect both B and B'. Therefore $\partial B = \{b, c, d\}$.

Definition 2.5.12. Let $(X, \mathscr{T})$ be a topological space and suppose $A \subseteq X$. A point $c \in A$ is said to be an *interior point* of A provided there exists a set $N \in \mathscr{T}$ such that $c \in N$ and $N \subseteq A$. That is, provided there is a neighborhood of c contained in A.

Notice that an interior point cannot be a boundary point.

Example 2.5.13. Let $(X, \mathscr{T})$ be the topological space from Example 2.5.11, where $X = \{a, b, c, d\}$ and $\mathscr{T} = \{\emptyset, X, \{a, b, c\}, \{a, d\}, \{a\}\}$. Let $A = \{a, c, d\}$ and $B = \{b, d\}$. Find each interior point of the sets A and B, and also compute A° and B°.

Solution. The points a and d are both interior points of A because each is in the open set $\{a, d\} \subseteq A$. The point b cannot be interior to A because $b \notin A$. We next observe that c is not an interior point of A because $\{a, b, c\}$ and X are the only neighborhoods of c, and neither is contained in A.

Because neither a nor c is an element of B, neither can be an interior point of B. The neighborhoods of b are X and $\{a, b, c\}$, but neither of these is a subset of B, and so b is not an interior point of B. Similarly, the neighborhoods of d are X and $\{a, d\}$. Again, neither of these is a subset of B, and so d is not an interior point of B. It follows that B has no interior points.

We now compute the sets A° and B°. The open subsets of A are $\emptyset$, $\{a\}$, and $\{a, d\}$. Thus, $A^\circ = \{a, d\}$. The only open set contained in B is the empty set, and so $B^\circ = \emptyset$.

In the above example, the sets of interior points of A and B coïncide with the sets A° and B°, respectively. This is no coincidence, as the next theorem shows.

Theorem 2.5.14. *Let* $(X, \mathscr{T})$ *be a topological space and suppose* $A \subseteq X$. *The set* A° *is precisely the set of interior points of* A.

Proof. Suppose x is an interior point of A. Then there exists an open set N such that $x \in N$ and $N \subseteq A$. Since A° is the union of all open sets which are subsets of A, it follows that $N \subseteq A^\circ$, and hence $x \in A^\circ$.

Conversely, suppose $x \in A^\circ$. It follows that A° is a neighborhood of x that is a subset of A. Therefore, x is an interior point of A. Q.E.D.

Example 2.5.15. Let $\mathbb{R}$ be the set of real numbers with the standard topology. Let a and b be real numbers such that $a < b$. For each of the intervals (a, b), $[a, b)$, $(a, b]$, and $[a, b]$, show that each point in (a, b) is an interior point. Further, show that each of the four intervals has boundary $\{a, b\}$ and closure $[a, b]$.

Solution. Consider the interval (a, b) where $a < b$. We begin by identifying the interior points. Suppose x is a real number such that $a < x < b$. Let

$$\epsilon = \frac{1}{2} \min\{x - a, b - x\}.$$

Then $(x - \epsilon, x + \epsilon)$ is a neighborhood of the point x and $(x - \epsilon, x + \epsilon) \subseteq (a, b)$. Therefore, x is an interior point of (a, b).

Let N be any neighborhood of a. Because N is an open set in $\mathbb{R}$, there is an interval (c, d) such that $a \in (c, d)$ and $(c, d) \subseteq N$. By choosing a smaller interval, if necessary, we may assume $d < b$. It follows that $c < a < d < b$, and so $(a, d) = (c, d) \cap (a, b) \subseteq N \cap (a, b)$. Next, observe that $(a, b)' = (-\infty, a] \cup [b, \infty)$. Thus, $(c, a) = (c, d) \cap (-\infty, a) \subseteq N \cap (a, b)'$. We have shown both $N \cap (a, b) \neq \emptyset$ and $N \cap (a, b)' \neq \emptyset$. Therefore, $a \in \partial(a, b)$.

A similar argument shows that $b \in \partial(a, b)$. We note that boundary points cannot be interior points, and so we have also shown that (a, b) is the set of all interior points of (a, b).

We have demonstrated that both a and b are in $\partial(a, b)$. Certainly, no interior point can be in the boundary. We now show that $x \notin [a, b]$ implies x is not in the boundary. Suppose $x \notin [a, b]$. Then $\min\{|x - a|, |x - b|\} > 0$.

Choose some $\epsilon > 0$ such that $\epsilon < \frac{1}{2}\min\{|x - a|, |x - b|\}$. It follows that $(x-\epsilon, x+\epsilon)$ is a neighborhood of x and $(x-\epsilon, x+\epsilon)\cap(a, b) = \emptyset$. Consequently, if x is not in $[a, b]$, then it cannot be a boundary point of (a, b). It follows that $\partial(a, b) = \{a, b\}$, as required.

To find the closure of (a, b), observe that $[a, b]$ is the smallest closed set that contains (a, b). Therefore, $[a, b] = \overline{(a, b)}$.

The arguments for the other three intervals are similar and are left to the reader.

There are several relations between interior, boundary, and closure that can be easily discovered for the special case of the real numbers with the standard topology. The next theorem describes these relations for a general topological space. In the proof of the theorem, we will make use of some of the symbols introduced in Chapter 1. Recall that the symbols $\Rightarrow$, $\Leftrightarrow$, $\wedge$, and $\vee$ signify *implies* (if ..., then), *is equivalent to* (if and only if), *and*, and *or*, respectively.

Theorem 2.5.16. *Let $(X, \mathscr{T})$ be a topological space and let $A \subseteq X$. Then*

(i) $\partial A = \overline{A} \cap \overline{A'} = \overline{A} \setminus A°$,

(ii) $(\partial A)' = (A')° \cup A°$,

(iii) $\overline{A} = A° \cup \partial A$,

(iv) $A° = A \setminus \partial A$.

Proof. (i) Suppose $x \notin \overline{A} \cap \overline{A'}$. Then, by Theorems 1.5.15(c) and 2.5.7,

$$x \in (\overline{A} \cap \overline{A'})' = (\overline{A})' \cup (\overline{A'})' = (\overline{A})' \cup A°.$$

If $x \in (\overline{A})'$, then $x \notin \partial A$, because $(\overline{A})' \subseteq A'$ is a neighborhood of x and $(\overline{A})' \cap A = \emptyset$. If $x \in A°$, then $x \notin \partial A$, because $A°$ is a neighborhood of x and $A° \cap A' = \emptyset$. Either way, x is not in the boundary of A. We have shown

$$x \notin \overline{A} \cap \overline{A'} \implies x \notin \partial A,$$

which is equivalent to the contrapositive statement

$$x \in \partial A \implies x \in \overline{A} \cap \overline{A'}.$$

Conversely, suppose $x \notin \partial A$. Then there is a neighborhood U of x such that either $U \cap A = \emptyset$ or $U \cap A' = \emptyset$. We have:

$$(U \cap A = \emptyset) \vee (U \cap A' = \emptyset) \implies (U \subseteq A') \vee (U \subseteq A).$$

Keeping in mind that U' is closed, we find

$$(U \subseteq A') \vee (U \subseteq A) \implies (A \subseteq U') \vee (A' \subseteq U') \implies (\overline{A} \subseteq U') \vee (\overline{A'} \subseteq U').$$

Thus,

$$x \notin \partial A \implies (U \subseteq (\overline{A})') \vee (U \subseteq (\overline{A'})').$$

It follows that if $x \notin \partial A$, then there exists a neighborhood U of x which is either disjoint from $\overline{A}$ or disjoint from $\overline{A'}$. That is, x cannot be in both $\overline{A}$ and $\overline{A'}$. Consequently,

$$x \notin \partial A \implies x \notin \overline{A} \cap \overline{A'},$$

or equivalently,

$$x \in \overline{A} \cap \overline{A'} \implies x \in \partial A.$$

We have shown

$$\left(x \in \partial A \implies x \in \overline{A} \cap \overline{A'}\right) \wedge \left(x \in \overline{A} \cap \overline{A'} \implies x \in \partial A\right),$$

or

$$x \in \overline{A} \cap \overline{A'} \iff x \in \partial A,$$

and so $\overline{A} \cap \overline{A'} = \partial A$, as required.

To complete the proof of (i), we observe that

$$\overline{A} \setminus A^\circ = \overline{A} \cap (A^\circ)' = \overline{A} \cap \left((\overline{A'})'\right)' = \overline{A} \cap \overline{A'}.$$

(See Theorems 2.5.7 and 1.5.15(a).)

(ii) We make use of (i), along with Theorems 1.5.15(c) and 2.5.7, to conclude

$$(\partial A)' = \left(\overline{A} \cap \overline{A'}\right)' = (\overline{A})' \cup (\overline{A'})' = \left(\overline{(A')'}\right)' \cup A^\circ = (A')^\circ \cup A^\circ.$$

(iii) Similarly,

$$A^\circ \cup \partial A = A^\circ \cup \left(\overline{A} \setminus A^\circ\right) = A^\circ \cup \left(\overline{A} \cap (A^\circ)'\right)$$
$$= \left(A^\circ \cup \overline{A}\right) \cap \left(A^\circ \cup (A^\circ)'\right) = \overline{A} \cap X = \overline{A}.$$

(iv) Using (ii) and arguing as in the previous two proofs, we have

$$A \setminus \partial A = A \cap (\partial A)' = A \cap \left((A')^\circ \cup A^\circ\right)$$
$$= \left(A \cap (A')^\circ\right) \cup (A \cap A^\circ) = \emptyset \cup A^\circ = A^\circ.$$

Q.E.D.

Suppose $(X, \mathscr{T})$ is a topological space and $(Y, \mathscr{R})$ is a subspace. Let $A \subseteq Y$. The interior, closure, and boundary of A are defined in terms of open sets, and therefore depend on the choice of topology $\mathscr{T}$ or $\mathscr{R}$. The relationships between these sets are discussed in the exercises.

Exercises

1. For each of the following topological spaces, a subspace A is given. Find $A^\circ, \overline{A}, \partial A, \overline{A'}, (A')^\circ$ and verify the equalities of Theorem 2.5.16.

   (a) Let $(X, \mathscr{T})$ be the topological space of Example 2.5.3 and let A be the set $\{a, b, d\}$.

   (b) Let $(X, \mathscr{T})$ be the topological space of Example 2.5.11 and let A be the set $\{a, d\}$.

   (c) Let $(X, \mathscr{T})$ be the set of real numbers with the standard topology and let A be the set of rational numbers.

   (d) Let $(X, \mathscr{T})$ be the set of real numbers with the standard topology and let A be the interval $[0, 2)$.

2. Prove Theorem 2.5.5.

3. Let $(X, \mathscr{T})$ be a topological space and let $A \subseteq X$. Prove that A is an open set if and only if $A = A^\circ$.

4. Let $(X, \mathscr{T})$ be a topological space and let $A \subseteq X$. Prove that A is a closed set if and only if $A = \overline{A}$.

5. Let X be a nonempty set and let $\mathscr{T}$ be the indiscrete topology on X. Let A be a proper subset of X. Find $A^\circ, \overline{A}$, and ∂A.

6. Let X be a nonempty set and let $\mathscr{T}$ be the discrete topology on X. Let A be a proper subset of X. Find $A^\circ, \overline{A}$, and ∂A.

7. In some of the literature, a neighborhood of a point in a topological space $(X, \mathscr{T})$ is defined as follows: The set N is a *neighborhood* of the point $x \in X$ if and only if $x \in N$ and there exists some $U \in \mathscr{T}$ such that $x \in U$ and $U \subseteq N$. Under this definition, a neighborhood of a point need not be an open set, but must contain an open set that contains the point. Using this definition of neighborhood, prove that a set is open if and only if it is a neighborhood of each of its points.

8. Let $(X, \mathscr{T})$ be a topological space and $A \subseteq X$. Prove that $\overline{\overline{A}} = \overline{A}$.

9. Let $(X, \mathscr{T})$ be a topological space and suppose that A and B are subsets of X such that $A \subseteq B$. Prove that $A^\circ \subseteq B^\circ$ and $\overline{A} \subseteq \overline{B}$.

10. Let $(X, \mathscr{T})$ be a topological space and let A and B be subsets of X.

    (a) Prove that $\overline{A \cup B} = \overline{A} \cup \overline{B}$.

    (b) Give an example to show that $\overline{A \cap B} = \overline{A} \cap \overline{B}$ is not necessarily true.

11. Let $(X, \mathcal{T})$ be a topological space and let $A \subseteq X$. A point $x \in X$ (that is not necessarily in A) is called an *accumulation point* (or *limit point*) of A if and only if each neighborhood of x contains one or more points of A that are distinct from x. That is, if $U \in \mathcal{T}$ and $x \in U$, then $A \cap (U \setminus \{x\})$ is nonempty. The set of all accumulation points of A is called the *derived set* of A and is denoted $\delta(A)$. Prove the following:

    (a) The set A is closed if and only if $\delta(A) \subseteq A$.
    (b) $A \cup \delta(A)$ is a closed set.
    (c) $A \setminus \delta(A') = A^\circ$.

12. Let $\mathbb{R}$ be the set of real numbers and let $\mathcal{T}$ be the discrete topology. For each set A, find $\delta(A)$ and verify that $A \setminus \delta(A') = A^\circ$. (See Exercise 11.)

    (a) Let $A = [1, 2]$.
    (b) Let $A = \mathbb{Q}$ be the set of rational numbers.

13. Repeat Exercise 12, but let $\mathcal{T}$ be the standard topology on $\mathbb{R}$.

14. Let $(X, \mathcal{T})$ be a topological space and let $(Y, \mathcal{R})$ be a subspace. Let $A \subseteq Y$ and $a \in Y$. Prove that a is an $\mathcal{R}$-accumulation point of A if and only if it is a $\mathcal{T}$-accumulation point of A. (See Exercise 11.)

15. Let $(X, \mathcal{T})$ be a topological space with $(Y, \mathcal{R})$ a subspace. If $A \subseteq Y$, prove that the $\mathcal{R}$-closure of A is equal to $\overline{A} \cap Y$, where $\overline{A}$ denotes the $\mathcal{T}$-closure of A.

16. Let $(X, \mathcal{T})$ be a topological space and let $A \subseteq X$. Prove the following:

    (a) The boundary ∂A is a closed set.
    (b) A is both open and closed if and only if $\partial A = \emptyset$.
    (c) A is open if and only if $\partial A \subseteq A'$.

17. Let $(X, \mathcal{T})$ be a topological space and let $A \subseteq X$. A point $x \in X$ is said to be an *exterior point* of A if there exists a neighborhood N of x such that $N \cap \overline{A} = \emptyset$. The collection of all exterior points of A is called the *exterior* of A and is denoted by $\text{ext}(A)$. Show that $\text{ext}(A) = (A')^\circ = (\overline{A})'$.

18. Let $(X, \mathcal{T})$ be a topological space and let A and B be subsets of X.

    (a) Prove that $(A \cap B)^\circ = A^\circ \cap B^\circ$.
    (b) Give an example to show that $(A \cup B)^\circ = A^\circ \cup B^\circ$ is not necessarily true.

19. Give an example of a topological space $(X, \mathcal{T})$, a subspace $(Y, \mathcal{R})$, and subsets A and B such that:

    (a) The $\mathcal{R}$-closure of $A \cap Y$ is not equal to $\overline{A} \cap Y$, where $\overline{A}$ denotes the $\mathcal{T}$-closure of A.

(b) The $\mathscr{R}$-interior of $B \cap Y$ is not equal to $B° \cap Y$, where $B°$ denotes the $\mathscr{T}$-interior of B.

(Compare to Exercise 15.)

2.6   Hausdorff spaces

One of the theorems of elementary calculus asserts that the limit of a function, if it exists, is unique. Let us briefly demonstrate the proof of this fact. First, we provide a formal definition for the limit of a function.

Definition 2.6.1. Suppose f is a real valued function of a real variable x and suppose that a is a real number. We say the *limit* of $f(x)$ as x *approaches* a is L, and we write

$$\lim_{x \to a} f(x) = L,$$

if for each $\epsilon > 0$, there is a $\delta > 0$ such that $|L - f(x)| < \epsilon$ whenever $0 < |x - a| < \delta$. That is to say,

$$x \in (a - \delta, a + \delta) \setminus \{a\} \implies f(x) \in (L - \epsilon, L + \epsilon).$$

We now prove that this limit, if it exists, is unique.

Theorem 2.6.2. *Let f be a real valued function of a real variable and suppose a is a real number. If $\lim_{x \to a} f(x)$ exist, then this limit is unique.*

Proof. Suppose $\lim_{x \to a} f(x) = L_1$ and $\lim_{x \to a} f(x) = L_2$ and assume $L_1 \neq L_2$. Let $\epsilon = \frac{|L_2 - L_1|}{2}$ and observe that $\epsilon > 0$. Consequently, there exist numbers $\delta_1 > 0$ and $\delta_2 > 0$ such that

$$x \in (a - \delta_1, a + \delta_1) \setminus \{a\} \implies f(x) \in (L_1 - \epsilon, L_1 + \epsilon)$$

and

$$x \in (a - \delta_2, a + \delta_2) \setminus \{a\} \implies f(x) \in (L_2 - \epsilon, L_2 + \epsilon).$$

Now let $\delta = \min\{\delta_1, \delta_2\}$. Given this δ, we have $x \in (a - \delta, a + \delta) \setminus \{a\}$ implies both $x \in (a - \delta_1, a + \delta_1) \setminus \{a\}$ and $x \in (a - \delta_2, a + \delta_2) \setminus \{a\}$, which in turn implies both $f(x) \in (L_1 - \epsilon, L_1 + \epsilon)$ and $f(x) \in (L_2 - \epsilon, L_2 + \epsilon)$. Therefore,

$$f(x) \in (L_1 - \epsilon, L_1 + \epsilon) \cap (L_2 - \epsilon, L_2 + \epsilon) = \emptyset.$$

This is a contradiction, and so we reject the assumption that $L_1 \neq L_2$. Q.E.D.

The crucial part of this proof was our ability to select two disjoint neighborhoods $(L_1 - \epsilon, L_1 + \epsilon)$ and $(L_2 - \epsilon, L_2 + \epsilon)$ of the two distinct points L_1 and L_2. For other topological spaces, we may not be able to find disjoint neighborhoods of distinct points. For example, if $X = \{a, b, c, d\}$ and $\mathscr{T} = \{\emptyset, X\}$, then, although $a \neq b$, there do not exist disjoint neighborhoods of a and b.

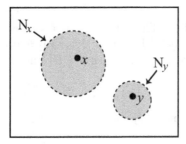

Figure 2.6.a: In a Hausdorff space, any two distinct points x and y can be separated by disjoint neighborhoods N_x and N_y.

Definition 2.6.3. A topological space $(X, \mathcal{T})$ is called a *Hausdorff space* if and only if for each pair of distinct points x and y in X, there exist neighborhoods N_x and N_y such that $x \in N_x$ and $y \in N_y$ and $N_x \cap N_y = \emptyset$. (Hausdorff spaces are sometimes called T_2-*spaces*.)

The set of real numbers with the standard topology is a Hausdorff space. If x and y are real numbers such that $x \neq y$, and if $\epsilon = \frac{1}{2}|x - y|$, then $\epsilon > 0$, and so $N_x = (x - \epsilon, x + \epsilon)$ and $N_y = (y - \epsilon, y + \epsilon)$ are disjoint neighborhoods of x and y (respectively). If $(X, \mathcal{T})$ is a discrete space, then it is necessarily a Hausdorff space. To see this, observe that, if x and y are distinct elements in X, then $N_x = \{x\}$ and $N_y = \{y\}$ are disjoint neighborhoods of x and y. We have remarked that a space with the indiscrete topology is not a Hausdorff space. We now consider another example.

Example 2.6.4. Let $X = \{1, 2, 3, 4, 5\}$ be a set and provide X with the topology $\mathcal{T} = \{\emptyset, X, \{1, 2, 3, 4\}, \{3, 4, 5\}, \{3, 4\}\}$. Show that $(X, \mathcal{T})$ is not a Hausdorff space.

Solution. The elements 1 and 2 are distinct and there do not exist disjoint neighborhoods of 1 and 2. To see this, note that the only open sets containing 1 are X and $\{1, 2, 3, 4\}$, and these both contain 2.

Example 2.6.5. Let $\mathbb{N}$ be the set of natural numbers and let $\mathcal{T}$ denote the cofinite topology on $\mathbb{N}$. (See Example 2.2.5 for the definition of the cofinite topology.) Show that $(\mathbb{N}, \mathcal{T})$ is not a Hausdorff space.

Solution. Consider the distinct elements 1 and 2 of $\mathbb{N}$. Let U_1 be any neighborhood of 1. Since $U_1 \in \mathcal{T}$, it must have a complement with a finite number of elements, say $U_1' = \{a_1, a_2, a_3, \dots, a_k\}$. Now let U_2 be any neighborhood of 2. The set U_2 must also have a finite complement, say $U_2' = \{b_1, b_2, \ldots, b_\ell\}$. Let $n \in \mathbb{N}$ be any natural number such that $n > \max\{a_1, \ldots, a_k, b_1, \ldots, b_\ell\}$. Then $n \notin U_1'$ and $n \notin U_2'$. Consequently,

$$n \in (U_1' \cup U_2')' = U_1 \cap U_2,$$

and so U_1 and U_2 are not disjoint.

The remainder of this section illustrates the usefulness of the Hausdorff property. We begin by generalizing some familiar notions from elementary calculus.

Definitions 2.6.6. Let $(X, \mathscr{T})$ be a topological space. A *sequence* in X is any ordered collection of points $\{x_1, x_2, x_3, \ldots\}$, each of which is a member of X. We denote the sequence by $\{x_n\}_{n=1}^{\infty}$. We say the sequence $\{x_n\}_{n=1}^{\infty}$ *converges* to x if and only if for each neighborhood N_x of x, there exists a positive integer M such that $x_n \in N_x$ whenever $n \geq M$. If the sequence $\{x_n\}_{n=1}^{\infty}$ converges to x, then we say x is a *limit* of the sequence $\{x_n\}_{n=1}^{\infty}$ and we write $\lim_{n \to \infty} x_n = x$.

Remark 2.6.7. In the definition of a sequence, it is not necessary for the first index to be 1. Indeed, it is not necessary that all indices be positive. We require only that the indices consist of consecutive integers listed in increasing order. (So the first term of a sequence may be, for example, x_0 or x_3, or even x_{-1}.)

We now state and prove a theorem that generalizes Theorem 2.6.2, the theorem from calculus that we encountered at the beginning of this section.

Theorem 2.6.8. *If a sequence of points converges in a Hausdorff space, then the limit of the sequence is unique.*

Proof. Let $(X, \mathscr{T})$ be a Hausdorff space and let $\{x_n\}_{n=1}^{\infty}$ be a sequence that converges in X. We wish to show that the limit is unique. Suppose to the contrary that the limit is not unique. Specifically, suppose that x and y are elements in X such that $\lim_{n \to \infty} x_n = x$ and $\lim_{n \to \infty} x_n = y$, but that $x \neq y$. Since X is a Hausdorff space, there exist neighborhoods N_x and N_y of x and y (respectively) such that $N_x \cap N_y = \emptyset$.

By assumption, $\lim_{n \to \infty} x_n = x$, and so there exists a positive integer M_1 such that $x_n \in N_x$ whenever $n \geq M_1$. Similarly, $\lim_{n \to \infty} x_n = y$, and so there exists a positive integer M_2 such that $x_n \in N_y$ whenever $n \geq M_2$. Now, let $M = \max\{M_1, M_2\}$ and choose any $k \geq M$. We have that $k \geq M_1$, and so $x_k \in N_x$. Similarly, $k \geq M_2$, and so $x_k \in N_y$. This is a contradiction, because $N_x \cap N_y = \emptyset$. Therefore, we reject the assumption that $x \neq y$ and conclude that the limit of a sequence, if it exists, must be unique.   Q.E.D.

Example 2.6.9. Let $\mathbb{N}$ be the set of natural numbers and let $\mathscr{T}$ denote the cofinite topology on $\mathbb{N}$. Show that limits need not be unique in this space by showing that any $m \in \mathbb{N}$ is a limit of the sequence $\{n\}_{n=1}^{\infty}$.

Solution. Let $m \in \mathbb{N}$ and let N_m be any neighborhood of m. Since $N_m \in \mathscr{T}$, it has a complement with finitely many elements, say $N'_m = \{a_1, a_2, \ldots, a_k\}$. Let

$$M = \max\{a_1, \ldots, a_k\} + 1.$$

Thus, if $n \geq M$, then $n \notin N'_m$. It follows that $n \in N_m$ for all $n \geq M$. Consequently, the sequence $\{n\}_{n=1}^{\infty}$ is eventually in any neighborhood of m, and so $\lim\limits_{n \to \infty} n = m$.

In the next theorem, we show that a subspace of a Hausdorff space is again a Hausdorff space. This shows that a Hausdorff space behaves in a way that agrees with our intuition, which is nice.

Theorem 2.6.10. *A subspace of a Hausdorff space is a Hausdorff space.*

Proof. Suppose $(X, \mathscr{T})$ is a Hausdorff space and $(Y, \mathscr{R})$ is a subspace. Let y_1 and y_2 be distinct points of Y. Then y_1 and y_2 are distinct points of X, and so there exist sets U_1 and U_2 in $\mathscr{T}$ such that $y_1 \in U_1$ and $y_2 \in U_2$ and $U_1 \cap U_2 = \emptyset$. Let $V_1 = U_1 \cap Y$ and $V_2 = U_2 \cap Y$. It follows that V_1 and V_2 are members of $\mathscr{R}$, by the definition of the subspace topology. Since $y_1 \in V_1$ and $y_2 \in V_2$ and $V_1 \cap V_2 \subseteq U_1 \cap U_2 = \emptyset$, we conclude that $(Y, \mathscr{R})$ is a Hausdorff space.   Q.E.D.

Exercises

1. Let $X = \{a, b, c, d, e, f\}$ and $\mathscr{T} = \{\emptyset, X, \{a, c, e\}, \{b, d, f\}\}$. Show that $(X, \mathscr{T})$ is not a Hausdorff space.

2. Let $(X, \mathscr{T})$ be the space from Exercise 1. Give an example of a sequence of points in X that has more than one limit.

3. Let $\mathbb{R}$ be the set of real numbers and let $\mathscr{B}$ be the collection of all intervals of the form $[a, \infty)$, where $a \in \mathbb{R}$.

   (a) Show that $\mathscr{B}$ is a base for some topology $\mathscr{T}$.

   (b) Show that $(X, \mathscr{T})$ is not a Hausdorff space.

4. Let $\mathbb{R}$ be the set of real numbers and let $\mathbb{R} \times \mathbb{R} = \{(x, y) \mid x \in \mathbb{R} \text{ and } y \in \mathbb{R}\}$ be the set of all ordered pairs of real numbers. For each pair of real numbers a and b with $a < b$, define $U_a^b = \{(x, y) \mid a < x < b \text{ and } y \in \mathbb{R}\}$.

(a) Show that $\mathscr{B} = \{U_a^b \mid a \in \mathbb{R}, b \in \mathbb{R}, \text{ and } a < b\}$ is a base for a topology $\mathscr{T}$.

(b) Show that $(X, \mathscr{T})$ is not a Hausdorff space.

5. Let $(X, \mathscr{T})$ be a Hausdorff space. Suppose $A \subseteq X$ and c is an accumulation point of A. (See Exercise 11 in Section 2.5 for the definition of an accumulation point.) If $U \in \mathscr{T}$ is a neighborhood of c, show that $A \cap U$ is infinite.

6. Prove that any finite subset of a Hausdorff space is closed.

2.7 Metric Spaces

In most of the topological spaces that we have previously discussed, the notion of *distance* was not present. There are, however, many topological spaces in which the topology is derived from some notion of distance. In this section, we discuss these types of spaces.

Definitions 2.7.1. Let X be a nonempty set. A *metric* on X is a real valued function $d : X \times X \to [0, \infty)$ that satisfies the following properties: For all x, y, and z in X,

(i) $d(x, y) = 0$ if and only if $x = y$,

(ii) $d(x, y) = d(y, x)$,

(iii) $d(x, y) \leq d(x, z) + d(z, y)$.

The nonnegative number $d(x, y)$ is called the *distance* between x and y in X. The inequality in (iii) is called the *triangle inequality*.

Example 2.7.2. Let $d(x, y) = |x - y|$ for all x and y in $\mathbb{R}$. Show that d is a metric on $\mathbb{R}$. (This metric is known as the *standard metric* on $\mathbb{R}$.)

Solution. (i) For x and y in $\mathbb{R}$, observe that

$$|x - y| = 0 \iff x - y = 0 \iff x = y.$$

Thus, $d(x, y) = 0$ if and only if $x = y$.

(ii) Because $|x - y| = |-(y - x)| = |y - x|$, it follows that $d(x, y) = d(y, x)$.

(iii) For any two real numbers a and b, we know that $|a + b| \leq |a| + |b|$ (which is also known as the triangle inequality). For x, y, and z in $\mathbb{R}$, let $a = x - z$ and $b = z - y$. Then,

$$|(x - z) + (z - y)| \leq |x - z| + |z - y|.$$

Simplifying this, we have $|x - y| \leq |x - z| + |z - y|$, and hence we conclude that $d(x, y) \leq d(x, z) + d(z, y)$.

Example 2.7.3. Let X be a nonempty set. Define $d : X \times X \to \{0,1\}$ by

$$d(x,y) = \begin{cases} 0 & \text{if } x = y, \\ 1 & \text{if } x \neq y, \end{cases}$$

for all x and y in X. Show that d is a metric on X. (This metric is called the *discrete metric* on X. [See Example 2.7.9.])

Solution. Conditions (i) and (ii) are clearly satisfied. The only way the triangle inequality could fail to be satisfied is if there exist elements x, y, and z in X such that

$$d(x,y) = 1 \quad \text{and} \quad d(x,z) + d(z,y) = 0.$$

Notice that $d(x,z) + d(z,y) = 0$ if and only if $d(x,z) = 0$ and $d(z,y) = 0$. This, in turn, can happen if and only if $x = z$ and $z = y$, and hence $x = y$. We have demonstrated that $x = y$, which contradicts our assumption that $d(x,y) = 1$. It follows that we must have $d(x,y) \leq d(x,z) + d(z,y)$ in all cases. Therefore, the function d defines a metric on X.

Example 2.7.4. Let $\mathbb{R}$ be the set of real numbers and let $X = \mathbb{R} \times \mathbb{R}$. Define $d : X \times X \to [0, \infty)$ by

$$d((x_1, x_2), (y_1, y_2)) = \max\{|x_1 - y_1|, |x_2 - y_2|\},$$

where (x_1, x_2) and (y_1, y_2) are elements in X. Show that d is a metric on X.

Solution. We observe

$$
\begin{aligned}
d((x_1, x_2), (y_1, y_2)) = 0 &\iff \max\{|x_1 - y_1|, |x_2 - y_2|\} = 0 \\
&\iff |x_1 - y_1| = 0 \ \text{ and } \ |x_2 - y_2| = 0 \\
&\iff x_1 = y_1 \ \text{ and } \ x_2 = y_2 \\
&\iff (x_1, x_2) = (y_1, y_2).
\end{aligned}
$$

It follows that (i) is satisfied. To show condition (ii) is satisfied, notice that

$$\max\{|x_1 - y_1|, |x_2 - y_2|\} = \max\{|y_1 - x_1|, |y_2 - x_2|\},$$

and so

$$d((x_1, x_2), (y_1, y_2)) = d((y_1, y_2), (x_1, x_2)).$$

We now check condition (iii):

$$
\begin{aligned}
d((x_1, x_2), (y_1, y_2)) &= \max\{|x_1 - y_1|, \ |x_2 - y_2|\} \\
&= \max\left\{ \big|(x_1 - z_1) + (z_1 - y_1)\big|, \ \big|(x_2 - z_2) + (z_2 - y_2)\big| \right\} \\
&\leq \max\left\{ |x_1 - z_1| + |z_1 - y_1|, \ |x_2 - z_2| + |z_2 - y_2| \right\}.
\end{aligned}
$$

Let $A = \max\{|x_1 - z_1|, \; |x_2 - z_2|\}$ and $B = \max\{|z_1 - y_1|, \; |z_2 - y_2|\}$. Then,

$$|x_1 - z_1| \le A, \quad |x_2 - z_2| \le A,$$

and

$$|z_1 - y_1| \le B, \quad |z_2 - y_2| \le B.$$

It follows that

$$|x_1 - z_1| + |z_1 - y_1| \le A + B \quad \text{and} \quad |x_2 - z_2| + |z_2 - y_2| \le A + B.$$

Thus,

$$
\begin{aligned}
d\big((x_1,x_2),(y_1,y_2)\big) &\le \max\big\{|x_1 - z_1| + |z_1 - y_1|, \; |x_2 - z_2| + |z_2 - y_2|\big\} \\
&\le A + B \\
&= \max\{|x_1 - z_1|, \; |x_2 - z_2|\} + \max\{|z_1 - y_1|, \; |z_2 - y_2|\} \\
&= d\big((x_1,x_2),(z_1,z_2)\big) + d\big((z_1,z_2),(y_1,y_2)\big).
\end{aligned}
$$

Having verified all three conditions, we conclude that d is a metric on X.

Suppose that a metric d is defined on a nonempty set X. The metric generates a topology on X in much the same way that the collection of open intervals generates the standard topology on the set of real numbers. To make this statement precise, we provide the following definition.

Definition 2.7.5. Let d be a metric on a nonempty set X. For $\epsilon > 0$ and $c \in X$, the *open ball* with *center c* and *radius ϵ* is the set $\{x \mid x \in X \text{ and } d(x,c) < \epsilon\}$. We denote the open ball with center c and radius ϵ by $B_\epsilon(c)$.

See Figure 2.7.a for an illustration of Definition 2.7.5.

Theorem 2.7.6. *If d is a metric on a nonempty set X, then the collection of all open balls is a base for a topology on X.*

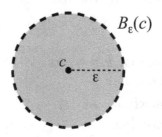

Figure 2.7.a: An open ball in a metric space X with center c and radius ϵ.

Proof. We verify that the two conditions of Theorem 2.3.8 are satisfied. Let $\mathscr{R}$ be the collection of all open balls in X. Each $x \in X$ is the center of at least one open ball, and each open ball is contained in X. Consequently, $X = \bigcup_{B \in \mathscr{R}} B$. This verifies the first condition of Theorem 2.3.8.

Now let $z \in B_\epsilon(x) \cap B_\delta(y)$, where x and y are in X and ϵ and δ are positive numbers. Because z is in both open balls, it must be that $d(z, x) < \epsilon$ and $d(z, y) < \delta$. Let

$$\eta = \min\{\epsilon - d(z, x), \delta - d(z, y)\}.$$

Then $\eta > 0$. We claim that $B_\eta(z) \subseteq B_\epsilon(x) \cap B_\delta(y)$. To see this, suppose $w \in B_\eta(z)$. By the triangle inequality,

$$d(w, x) \leq d(w, z) + d(z, x) < \eta + d(z, x) \leq \epsilon - d(z, x) + d(z, x) = \epsilon,$$

and

$$d(w, y) \leq d(w, z) + d(z, y) < \eta + d(z, y) \leq \delta - d(z, y) + d(z, y) = \delta.$$

Consequently, $w \in B_\epsilon(x) \cap B_\delta(y)$, and so $B_\eta(z) \subseteq B_\epsilon(x) \cap B_\delta(y)$.

We have verified that both conditions of Theorem 2.3.8 are satisfied, and so $\mathscr{R}$ is a base for a topology on X. Q.E.D.

Definitions 2.7.7. If X is a nonempty set and d is a metric on X, then the topology generated by the collection of open balls is called the *metric topology* on X and is denoted by $\mathscr{T}_d$. The ordered pair $(X, \mathscr{T}_d)$ is called a *metric space*. (We sometimes omit the subscript d if no confusion arises as a result.)

Example 2.7.8. In Example 2.7.2, we defined the distance between the real numbers x and y to be $d(x, y) = |x - y|$. Show that the topology generated by this metric is the standard topology on the set of real numbers $\mathbb{R}$. (The metric d is called the *standard metric* on $\mathbb{R}$.)

Solution. If $x \in \mathbb{R}$ and $\epsilon > 0$, then the open ball $B_\epsilon(x)$ is the open interval $(x - \epsilon, x + \epsilon)$. Furthermore, the open interval (a, b), where a and b are real numbers with $a < b$, is the open ball $B_\delta(c)$, where $c = \frac{a+b}{2}$ and $\delta = \frac{b-a}{2}$. Thus, the collection of all open balls in $(\mathbb{R}, \mathscr{T}_d)$ is precisely the collection of all open intervals in $\mathbb{R}$. Therefore, the standard topology on $\mathbb{R}$ is generated by d.

Example 2.7.9. Let X be a nonempty set and let d be the metric of Example 2.7.3. What is the metric topology $\mathscr{T}_d$?

Solution. For each $x \in X$, the open ball of radius $\epsilon > 0$ centered at x is

$$B_\epsilon(x) = \begin{cases} \{x\} & \text{if } \epsilon \leq 1, \\ X & \text{if } \epsilon > 1. \end{cases}$$

Therefore, the base for the topology generated by d contains all of the singletons (single-point sets), and so $\mathscr{T}_d$ is the discrete topology. (For this reason, d is called the *discrete metric* on X.)

Example 2.7.10. Let $X = \mathbb{R} \times \mathbb{R}$ and let d be the metric of Example 2.7.4. What is the metric topology $\mathscr{T}_d$?

Solution. For any $(a, b) \in \mathbb{R} \times \mathbb{R}$ and $\epsilon > 0$, the open ball $B_\epsilon((a, b))$ is the interior of a square centered at (a, b), with sides parallel to the x- and y-axes, and each side having length 2ϵ. The topology, therefore, is generated by the collection of all interiors of squares with sides parallel to the coordinate axes. (See Exercise 5 in this section and Exercise 7 in Section 2.3.)

One of the significant properties of a metric space is made explicit in the next theorem.

Theorem 2.7.11. *A metric space is a Hausdorff space.*

Proof. Let $(X, \mathscr{T}_d)$ be a metric space with metric d. Suppose x and y are two distinct points of X. Since d is a metric, $d(x, y) > 0$. Let $\epsilon = \frac{1}{2}d(x, y)$. Then $B_\epsilon(x)$ and $B_\epsilon(y)$ are disjoint neighborhoods of x and y, respectively. The open balls are neighborhoods of x and y, by the definition of the metric topology. To show that they are disjoint, assume there exists some $z \in B_\epsilon(x) \cap B_\epsilon(y)$. Note that $z \in B_\epsilon(x)$ implies $d(x, z) < \epsilon$. Similarly, $z \in B_\epsilon(y)$ implies that $d(y, z) < \epsilon$. Therefore, by an application of the triangle inequality, we have

$$d(x, y) \leq d(x, z) + d(z, y) < \epsilon + \epsilon = 2\epsilon = 2\left(\frac{1}{2}d(x, y)\right) = d(x, y).$$

We have arrived at the contradiction $d(x, y) < d(x, y)$. We are forced to reject the assumption that $B_\epsilon(x) \cap B_\epsilon(y) \neq \emptyset$, and conclude that $B_\epsilon(x)$ and $B_\epsilon(y)$ are disjoint. Therefore, $(X, \mathscr{T}_d)$ is a Hausdorff space.   Q.E.D.

It is useful to now introduce some definitions that will be needed later.

Definitions 2.7.12. Suppose E is a set of real numbers. If M is the least number such that $x \leq M$ for all $x \in E$, then we call M the *supremum* (or *least upper bound*) of E and we write $M = \sup(E)$ (or $M = \text{lub}(E)$). Similarly, if m is the greatest number such that $m \leq x$ for all $x \in E$, then we call m the *infimum* (or *greatest lower bound*) of E and we write $m = \inf(E)$ (or $m = \text{glb}(E)$).

Definitions 2.7.13. Let $(X, \mathcal{T}_d)$ be a metric space and A a nonempty subset of X. If the set $\{d(x, y) \mid \{x, y\} \subseteq A\}$ of real numbers has an upper bound, we say that A is a *bounded set*, and we call the least upper bound the *diameter* of A. We denote the diameter of A by $d(A)$, so that $d(A) = \sup\{d(x, y) \mid \{x, y\} \subseteq A\}$. For convenience, we define $d(\emptyset) = 0$.

See Figure 2.7.b for an illustration of Definitions 2.7.13.

Example 2.7.14. Let $\mathbb{R}$ be the set of real numbers and let $A = [0, 2]$. Find $d(A)$ in the following two cases:

(a) d is the metric of Example 2.7.2. (The standard metric.)

(b) d is the metric of Example 2.7.3. (The discrete metric.)

Solution. (a) Suppose that $\{x, y\} \subseteq A$. Then $\{x, y\} \subseteq [0, 2]$, and so $d(x, y) = |x - y| \leq 2$. It follows that $d(A) \leq 2$. Since $d(0, 2) = 2$, we conclude that $d(A) = 2$.

(b) With the discrete metric, $\{d(x, y) \mid \{x, y\} \subseteq A\} = \{0, 1\}$. Thus $d(A) = 1$, because $\sup\{0, 1\} = 1$.

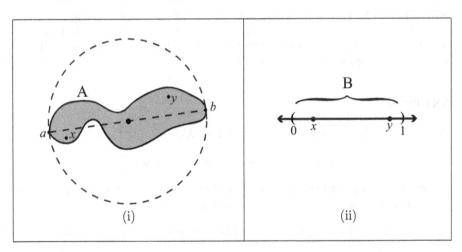

(i)   (ii)

Figure 2.7.b: (i) The set A is the shaded region that includes the points inside the closed curve together with its boundary. If x and y are points in the set A, then $d(x, y) \leq d(a, b) = d(A)$. The length of the dashed line segment between a and b is the diameter of the set A. This line segment is the diameter of the dashed circle which encloses A. (ii) The set B is the open interval $(0, 1)$. In this case, $d(B) = 1$, even though $d(x, y) < 1$ for any x and y in B.

Definition 2.7.15. Let $(X, \mathscr{T}_d)$ be a metric space and let A be a nonempty subset of X. If $c \in X$, then the *distance* between c and A is denoted $d(c, A)$ and is defined by

$$d(c, A) = \inf\{d(c, a) \mid a \in A\}.$$

Observe that the set $\{d(c, a) \mid a \in A\}$ is always bounded below by 0. If $c \in A$, then $d(c, A) = 0$, but the converse of this statement need not be true. To see this, consider the real numbers with the standard metric of Example 2.7.2. In this metric, if A is the open interval $(0, 2)$, then $d(2, A) = 0$, even though $2 \notin A$.

Theorem 2.7.16. *Let $(X, \mathscr{T}_d)$ be a metric space and let V be a nonempty subset of X. The set V is open if and only if for each $x \in V$, the distance $d(x, V') > 0$.*

Proof. Suppose first that V is open and let $x \in V$. Since V is the union of open balls (by Theorem 2.7.6), there exists some $y \in V$ and $\epsilon > 0$ such that $x \in B_\epsilon(y)$ and $B_\epsilon(y) \subseteq V$. Thus $d(x, y) < \epsilon$. Let $\delta = \epsilon - d(x, y)$, so that $\delta > 0$. If $z \in V'$, then $z \notin B_\epsilon(y)$, and so $d(z, y) \geq \epsilon$. By the triangle inequality, $d(y, z) \leq d(y, x) + d(x, z)$, and hence, for all $z \in V'$,

$$d(x, z) \geq d(y, z) - d(y, x) \geq \epsilon - d(y, x) = \delta.$$

We have shown that δ is a lower bound of the set $\{d(x, z) \mid z \in V'\}$. Therefore, $d(x, V') = \inf\{d(x, z) \mid z \in V'\} \geq \delta$, and so $d(x, V') > 0$.

Conversely, suppose for each $x \in V$, we have $d(x, V') > 0$. We will show that V is an open set. Let $x_0 \in V$ and let $\epsilon = d(x_0, V')$, which is nonzero, by assumption. Because $\inf\{d(x_0, z) \mid z \in V'\} = \epsilon$, it follows that $d(x_0, z) \geq \epsilon$ for each $z \in V'$. Consequently, if $d(x_0, z) < \epsilon$, it must be the case that $z \in V$. Therefore, $B_\epsilon(x_0) \subseteq V$. We conclude that V is the union of open balls (by Lemma 2.3.4), and so V is an open set.   Q.E.D.

Exercises

1. Let $\mathbb{R}$ be the set of real numbers and let $X = \mathbb{R} \times \mathbb{R}$. Define

   $$d\big((x_1, x_2), (y_1, y_2)\big) = |x_1 - y_1| + |x_2 - y_2|,$$

   where (x_1, x_2) and (y_1, y_2) are in X. Prove that d is a metric on X. (This metric is sometimes called the *taxicab metric* on $\mathbb{R}$.)

2. Let X and d be as in Exercise 1. Sketch the open ball $B_2\big((0, 0)\big)$.

3. Let $C[a, b]$ denote the set of all continuous real valued functions defined on the closed bounded interval $[a, b]$, where $a < b$. For f and g in $C[a, b]$, define

   $$d(f, g) = \int_a^b |f(x) - g(x)| \, dx.$$

   Prove that d is a metric on $C[a, b]$.

4. Let $C[a, b]$ denote the set of all continuous real valued functions defined on the closed bounded interval $[a, b]$, where $a < b$. For f and g in $C[a, b]$, define

$$d(f, g) = \left| \int_a^b \left[f(x) - g(x) \right] dx \right|.$$

Is d a metric on $C[a, b]$? If so, prove that it is. If not, explain why it is not. (Compare to Exercise 3.)

5. Let X and d be as in Example 2.7.4. That is, let $X = \mathbb{R} \times \mathbb{R}$ and

$$d((x_1, x_2), (y_1, y_2)) = \max\{|x_1 - y_1|, |x_2 - y_2|\},$$

for (x_1, x_2) and (y_1, y_2) in X. Sketch the "circle" with center $(1, 2)$ and radius 2. That is, sketch the set $\{(x, y) \mid d((x, y), (1, 2)) = 2\}$.

6. Let $C[a, b]$ be the set of all continuous real valued functions defined on the closed bounded interval $[a, b]$, where $a < b$. For f and g in $C[a, b]$, define

$$d(f, g) = \sup \{|f(x) - g(x)| \mid a \le x \le b\}.$$

Prove that d is a metric on $C[a, b]$. This metric is known as the *sup metric* on $C[a, b]$. (See Figure 2.7.c for an illustration of an open ball of radius ϵ in this space.)

7. The standard metric defined on $\mathbb{R} \times \mathbb{R}$ is the *Euclidean metric*, which is the distance function defined in elementary mathematics. That is, for (x_1, x_2) and (y_1, y_2) in $\mathbb{R} \times \mathbb{R}$,

$$d((x_1, x_2), (y_1, y_2)) = \sqrt{(x_1 - y_1)^2 + (x_2 - y_2)^2}.$$

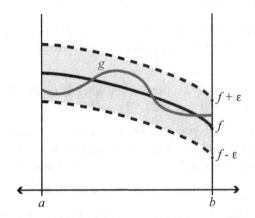

Figure 2.7.c: An open ball in $C[a, b]$ with the sup metric. The center is f and the radius is ϵ. Since $|f(x) - g(x)| < \epsilon$ for all $x \in [a, b]$, the function g is in this open ball.

We will show in the next section that d is a metric on $\mathbb{R} \times \mathbb{R}$. Find $d\big((-1,3),(2,7)\big)$ if

   (a) d is the Euclidean metric defined above.
   (b) d is the discrete metric of Example 2.7.3.
   (c) d is the metric of Example 2.7.4.
   (d) d is the metric of Exercise 1.

8. In a Cartesian coordinate plane (i.e., in the xy-plane), sketch the graph of the set $\{(x,y) \mid d\big((x,y),(0,0)\big) \leq 2\}$, which is the set of points no farther than 2 units from the origin, for each of the metrics d in Exercise 7.

9. Let A be the interior of the ellipse $\dfrac{x^2}{25} + \dfrac{y^2}{16} = 1$ and let c be the point $(7,0)$. Find $d(c,A)$ for each of the metrics in Exercise 7.

10. Let X be a set with at least two elements and let d be the discrete metric on X. (See Example 2.7.3.) Show that if A is a subset of X having at least two members, then the diameter of A is 1.

11. Let $(X, \mathscr{T}_d)$ be a metric space and let $A \subseteq X$. Prove that the diameter of A is positive if and only if A has at least two elements.

12. Let $(X, \mathscr{T}_d)$ be a metric space and let $A \subseteq X$. Prove that a point $c \in X$ is an *accumulation point* of A if and only if $d(c, A \backslash \{c\}) = 0$. (See Exercise 11 of Section 2.5.)

13. Let $(X, \mathscr{T}_d)$ be a metric space and suppose A is a finite subset of X. Prove that A is closed.

14. Let $(X, \mathscr{T}_d)$ be a metric space and suppose A is a nonempty subset of X. Prove that A is closed if and only if $d(x, A) > 0$ for each $x \in A'$.

15. Let $(X, \mathscr{T}_d)$ be a metric space. Suppose that U is a nonempty open subset of X and let $x \in U$. Show that there exists a real number $\epsilon > 0$ such that $B_\epsilon(x) \subseteq U$.

16. If A is a finite set of real numbers, let $\max(A)$ denote the member c of A with the property that $a \leq c$ for all $a \in A$.

   (a) Verify that if x is any real number, then $|x| = \max\{x, -x\}$.
   (b) Let S and T be finite sets of real numbers and define $S + T$ to be the set of real numbers $S + T = \{s + t \mid s \in S \text{ and } t \in T\}$. Show that
   $$\max(S + T) = \max(S) + \max(T).$$
   (c) Suppose that A is a finite set of real numbers and suppose $B \subseteq A$ is nonempty. Show that $\max(B) \leq \max(A)$.
   (d) Using the results of (a), (b), and (c), give a different proof of the triangle inequality for the metric in Example 2.7.4.

2.8 Euclidean Spaces

In this section, we introduce the most important class of metric spaces, the *Euclidean spaces*. We briefly review some concepts that the student likely encountered in calculus and linear algebra courses.

In elementary mathematics, the distance formula in the Cartesian plane is obtained by an application of the Pythagorean Theorem. The distance between the two points (x_1, x_2) and (y_1, y_2) is found to be given by the formula

$$d\big((x_1, x_2), (y_1, y_2)\big) = \sqrt{(x_1 - y_1)^2 + (x_2 - y_2)^2}.$$

We now extend this to n dimensions. Let $\mathbb{R}^n$ denote the set of all ordered n-tuples, or n-dimensional *vectors*, of the form $x = (x_1, x_2, \ldots, x_n)$. For the n-tuples $x = (x_1, x_2, \ldots, x_n)$ and $y = (y_1, y_2, \ldots, y_n)$, define $d : \mathbb{R}^n \times \mathbb{R}^n \to [0, \infty)$ by

$$d(x, y) = \sqrt{(x_1 - y_1)^2 + (x_2 - y_2)^2 + \cdots + (x_n - y_n)^2}.$$

To prove that d is a metric on $\mathbb{R}^n$, it will be useful to review some of the operations defined for the space $\mathbb{R}^n$.

We recall that addition and subtraction in $\mathbb{R}^n$ are done coordinate-wise.

Definition 2.8.1. Let $x = (x_1, \ldots, x_n)$ and $y = (y_1, \ldots, y_n)$ be two elements of $\mathbb{R}^n$. We define the *dot product* (or *scalar product*) to be the real number $x \cdot y$ given by the formula

$$x \cdot y = \sum_{i=1}^{n} x_i y_i = x_1 y_1 + \cdots + x_n y_n.$$

Example 2.8.2. Let $x = (-2, 3, 1, 0)$ and $y = (-3, 2, 5, -2)$ be two elements of $\mathbb{R}^4$. Compute $x \cdot y$.

Solution. By definition, $x \cdot y = (-2)(-3) + (3)(2) + (1)(5) + (0)(-2) = 17$.

Definition 2.8.3. If $x \in \mathbb{R}^n$, then the *length* or *norm* of x is denoted $\|x\|$ and is defined by the formula $\|x\| = \sqrt{x \cdot x}$.

Observe that the norm of a vector is always a nonnegative real number.

Example 2.8.4. Let $x = (-3, 2, 0, 5, -1)$ and compute $\|x\|$.

Solution. From the definition, we compute:

$$\|x\| = \sqrt{(-3)^2 + (2)^2 + (0)^2 + (5)^2 + (-1)^2} = \sqrt{39}.$$

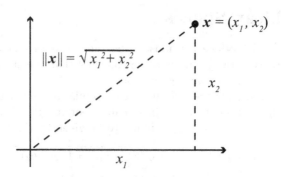

Figure 2.8.a: The norm of a vector $x = (x_1, x_2)$ in $\mathbb{R}^2$ can be realized as the length of the hypotenuse of a right triangle having sides of length $|x_1|$ and $|x_2|$.

Subtraction in $\mathbb{R}^n$ is coordinate-wise: if $x = (x_1, \ldots, x_n)$ and $y = (y_1, \ldots, y_n)$ are vectors in $\mathbb{R}^n$, then

$$\|x - y\| = \sqrt{(x_1 - y_1)^2 + (x_2 - y_2)^2 + \cdots + (x_n - y_n)^2}.$$

Thus, we formally define a metric on $\mathbb{R}^n$ as follows:

Definition 2.8.5. The *(Euclidean) distance* between x and y in $\mathbb{R}^n$ is given by $d(x, y) = \|x - y\|$. The metric d is called the *Euclidean metric* on $\mathbb{R}^n$.

We must show that d is a metric. It is easy to verify that $d(x, y) = 0$ if and only if $x = y$. It is also straight forward to show that $d(x, y) = d(y, x)$ for all x and y in $\mathbb{R}^n$. It is more difficult to verify the triangle inequality. The next theorem, a significant result in mathematical analysis, will simplify matters considerably.

Let us start in two-dimensional space. Suppose that $x = (x_1, x_2)$ and $y = (y_1, y_2)$ are two vectors in $\mathbb{R}^2$. Consider the triangle $\triangle ABC$, where the coordinates of A, B, and C are $(0,0)$, (x_1, x_2), and (y_1, y_2), respectively. Let θ be the angle at the origin. (See Figure 2.8.b.) Using the Law of Cosines, it is easy to show that

$$\cos \theta = \frac{x \cdot y}{\|x\| \|y\|}.$$

Because $|\cos \theta| \leq 1$, we conclude that $|x \cdot y| \leq \|x\| \|y\|$. (See Exercise 4 at the end of this section.) Remarkably, even though this inequality was proved by means of planar geometry, it generalizes to the n-dimensional case.

Theorem 2.8.6 (Cauchy-Schwarz Inequality). *If x and y are vectors in $\mathbb{R}^n$, then $|x \cdot y| \leq \|x\| \|y\|$.*

Proof. Begin by choosing vectors $x = (x_1, \ldots, x_n)$ and $y = (y_1, \ldots, y_n)$ in $\mathbb{R}^n$. Now, let t be a real variable and define a polynomial function s in t by

$$s(t) = \sum_{i=1}^{n} (x_i t - y_i)^2.$$

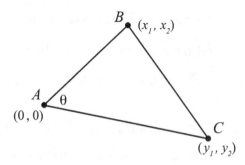

Figure 2.8.b: Illustration for the geometric proof of the Cauchy-Schwarz Inequality when $n = 2$. The Law of Cosines can be used to write $\cos\theta$ in terms of the vectors (x_1, x_2) and (y_1, y_2).

Certainly, for each $i \in \{1, \ldots, n\}$ and real number t, we have that $(x_i t - y_i)^2 \geq 0$, and so $s(t) \geq 0$ for each $t \in \mathbb{R}$. Expanding the polynomial, we see

$$s(t) = \left(\sum_{i=1}^{n} x_i^2 \right) t^2 - 2 \left(\sum_{i=1}^{n} x_i y_i \right) t + \sum_{i=1}^{n} y_i^2.$$

We next recall the definitions of the norm and the dot product and rewrite the polynomial as

$$s(t) = \|\boldsymbol{x}\|^2 t^2 - 2(\boldsymbol{x} \cdot \boldsymbol{y}) t + \|\boldsymbol{y}\|^2.$$

Because $s(t) \geq 0$ for each $t \in \mathbb{R}$, we conclude that the quadratic polynomial

$$\|\boldsymbol{x}\|^2 t^2 - 2(\boldsymbol{x} \cdot \boldsymbol{y}) t + \|\boldsymbol{y}\|^2$$

can have at most one real zero (in t). Therefore, the discriminant of the quadratic polynomial cannot be positive, and thence

$$[-2(\boldsymbol{x} \cdot \boldsymbol{y})]^2 - 4\|\boldsymbol{x}\|^2\|\boldsymbol{y}\|^2 \leq 0.$$

Simplifying, we get $(\boldsymbol{x} \cdot \boldsymbol{y})^2 \leq \|\boldsymbol{x}\|^2\|\boldsymbol{y}\|^2$. Taking square roots, we conclude that $|\boldsymbol{x} \cdot \boldsymbol{y}| \leq \|\boldsymbol{x}\|\|\boldsymbol{y}\|$.   Q.E.D.

The Cauchy-Schwarz Inequality leads to the following valuable result, which will allow us to prove the triangle inequality for the Euclidean metric.

Theorem 2.8.7 (Minkowski Inequality). *If $\boldsymbol{x}$ and $\boldsymbol{y}$ are vectors in $\mathbb{R}^n$, then*

$$\|\boldsymbol{x} + \boldsymbol{y}\| \leq \|\boldsymbol{x}\| + \|\boldsymbol{y}\|.$$

Proof. Let $\boldsymbol{x} = (x_1, \ldots, x_n)$ and $\boldsymbol{y} = (y_1, \ldots, y_n)$. We then have

$$\|\boldsymbol{x} + \boldsymbol{y}\|^2 = \sum_{i=1}^{n}(x_i + y_i)^2 = \sum_{i=1}^{n}(x_i^2 + 2x_i y_i + y_i^2)$$

$$= \sum_{i=1}^{n} x_i^2 + 2\left(\sum_{i=1}^{n} x_i y_i \right) + \sum_{i=1}^{n} y_i^2.$$

Applying the definitions of the norm and the dot product, we see

$$\|\boldsymbol{x} + \boldsymbol{y}\|^2 = \|\boldsymbol{x}\|^2 + 2(\boldsymbol{x} \cdot \boldsymbol{y}) + \|\boldsymbol{y}\|^2.$$

Because $\boldsymbol{x} \cdot \boldsymbol{y} \le |\boldsymbol{x} \cdot \boldsymbol{y}|$, we may now apply the Cauchy-Schwarz Inequality to conclude

$$\|\boldsymbol{x} + \boldsymbol{y}\|^2 \le \|\boldsymbol{x}\|^2 + 2\|\boldsymbol{x}\|\|\boldsymbol{y}\| + \|\boldsymbol{y}\|^2 = (\|\boldsymbol{x}\| + \|\boldsymbol{y}\|)^2.$$

Taking the square root of both sides, we obtain the desired inequality. Q.E.D.

We are now ready to prove the triangle inequality for the Euclidian metric. Let $\boldsymbol{x}$, $\boldsymbol{y}$, and $\boldsymbol{z}$ be in $\mathbb{R}^n$. Then,

$$\begin{aligned} d(\boldsymbol{x}, \boldsymbol{y}) = \|\boldsymbol{x} - \boldsymbol{y}\| &= \|(\boldsymbol{x} - \boldsymbol{z}) + (\boldsymbol{z} - \boldsymbol{y})\| \\ &\le \|\boldsymbol{x} - \boldsymbol{z}\| + \|\boldsymbol{z} - \boldsymbol{y}\| = d(\boldsymbol{x}, \boldsymbol{z}) + d(\boldsymbol{z}, \boldsymbol{y}). \end{aligned}$$

The *standard topology* on $\mathbb{R}^n$ is the topology on $\mathbb{R}^n$ generated by the Euclidean metric d. As usual, we consider the collection of all open balls $B_\epsilon(\boldsymbol{x})$, where $\boldsymbol{x} \in \mathbb{R}^n$ and $\epsilon > 0$, as a base for a topology $\mathscr{T}_d$. Thus, a subset of $\mathbb{R}^n$ is open if and only if it is the union of a collection of d-open balls. The resulting metric space $(X, \mathscr{T}_d)$ is known as (n-dimensional) *Euclidean space*.

Given a topological space $(X, \mathscr{T})$, it is natural to ask whether or not a metric d can be defined on X such that the induced topology $\mathscr{T}_d$ is in fact the original topology $\mathscr{T}$.

Definition 2.8.8. A topological space $(X, \mathscr{T})$ is called *metrizable* if there is a metric d on X which generates the topology $\mathscr{T}$.

In more advanced topology texts, a number of theorems give necessary and sufficient conditions for a topological space to be metrizable. Here we will only consider a few basic properties. (See Exercises 9, 10, and 11 at the end of this section.)

Exercises

1. Let $\boldsymbol{x} = (-1, 3, 4, 0, 2)$ and $\boldsymbol{y} = (2, -1, 3, 7, -3)$.

    (a) Find $\|\boldsymbol{x}\|$, $\|\boldsymbol{y}\|$, and $\|\boldsymbol{x} + \boldsymbol{y}\|$ and verify that $\|\boldsymbol{x} + \boldsymbol{y}\| \le \|\boldsymbol{x}\| + \|\boldsymbol{y}\|$.

    (b) Find the dot product $\boldsymbol{x} \cdot \boldsymbol{y}$.

    (c) Verify the Cauchy-Schwarz Inequality.

    (d) Find $d(\boldsymbol{x}, \boldsymbol{y})$.

2. Repeat Exercise 1 for $\boldsymbol{x} = (2, -1, 0, 3, 1, 5)$ and $\boldsymbol{y} = (1, 3, 2, -5, 9, -1)$.

3. Repeat Exercise 1 for $\boldsymbol{x} = (-3, 0, 2, 5, 1, 9, 3)$ and $\boldsymbol{y} = (2, 5, 0, 3, -2, 1, 4)$.

4. Suppose that $x = (x_1, x_2)$ and $y = (y_1, y_2)$ are two nonzero vectors in $\mathbb{R}^2$. Consider the triangle $\triangle ABC$, where the coordinates of A, B, and C are $(0,0)$, (x_1, x_2), and (y_1, y_2), respectively. Let θ be the angle at the origin. (See Figure 2.8.b.) Use the Law of Cosines to show that

$$\cos \theta = \frac{x \cdot y}{\|x\| \|y\|}.$$

5. Consider a triangle with vertices at the coordinates $(0,0)$, $(2, 2\sqrt{3})$, and $(3\sqrt{3}, 3)$. Use Exercise 4 to show that the measure of the angle at the origin is $30°$.

6. Suppose x and y are in $\mathbb{R}^n$. Prove that $\|x + y\|^2 = \|x\|^2 + \|y\|^2$ if and only if $x \cdot y = 0$.

7. Suppose x and y are in $\mathbb{R}^n$. Prove the *Parallelogram Law:*

$$\|x + y\|^2 + \|x - y\|^2 = 2\|x\|^2 + 2\|y\|^2.$$

8. Let $\mathscr{H}$ denote the collection of all infinite sequences $(x_1, x_2, x_3, \ldots)$ for which $x_i \in \mathbb{R}$ for all $i \in \mathbb{N}$, and such that the infinite series $\sum_{i=1}^{\infty} x_i^2$ converges to a finite limit. For each $x \in \mathscr{H}$, define

$$\|x\| = \left(\sum_{i=1}^{\infty} x_i^2 \right)^{1/2},$$

where $x = (x_1, x_2, x_3, \ldots)$. Further, define a function $d : \mathscr{H} \times \mathscr{H} \to [0, \infty)$ by $d(x, y) = \|x - y\|$ for x and y in $\mathscr{H}$. Prove d is a well-defined function and a metric on $\mathscr{H}$. The metric space $(\mathscr{H}, d)$ is known as *Hilbert space.*

9. Show that every discrete space is metrizable.

10. Show that $(\mathbb{R}, \mathscr{T})$, where $\mathbb{R}$ is the set of real numbers and $\mathscr{T}$ is the standard topology on $\mathbb{R}$, is metrizable.

11. Suppose X has more that one point and $\mathscr{T}$ is the indiscrete topology on X. Show that $(X, \mathscr{T})$ is not metrizable.

Chapter 3

Continuous Functions

3.1 Review of the Function Concept

We have assumed so far that the student was familiar with the terms "function" and "sequence." These terms, of course, were used in basic calculus and in the first two chapters we have used them freely. We now will use these concepts extensively, however, and so we formally define these terms, and others. A *function* is a rule that assigns to each element of a set D a unique element of a set C. We call D the *domain* of the function and we call C the *codomain* of the function. The sets D and C need not be distinct.

If the function is called f, and if $y \in C$ is the unique member of C that is assigned to $x \in D$, then we write $y = f(x)$. In this case, we say $f(x)$ is the *image* of x, and we call x and y the *independent variable* and *dependent variable*, respectively. The *range* of f is the set $R = \{f(x) \mid x \in D\}$. Because it is the collection of all images, the set R is also called the *image of f*. Because $R \subseteq C$, we say that f maps D *into* C. If $R = C$, the function is said to be *onto*.

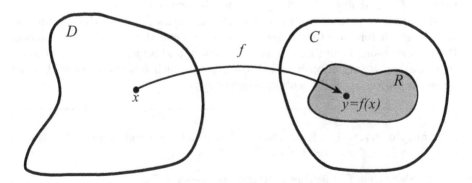

Figure 3.1.a: The function f assigns to every x in the domain D one and only one y in the codomain C. The set R is the range (or image) of the function f.

We often write $f : D \to C$ to indicate that f is a function with domain D and codomain C.

Suppose that X, Y, Z, and W are sets and let $f : X \to Y$ and $g : Z \to W$ be functions. Define a new set S as follows:

$$S = \{x \mid x \in X \text{ and } f(x) \in Y \cap Z\}.$$

The *composition* of g with f is the function $g \circ f : S \to W$ defined by the formula

$$(g \circ f)(x) = g(f(x)),$$

for all $x \in S$.

A function $f : X \to Y$ is called a *one-to-one* function if

$$f(x_1) = f(x_2) \implies x_1 = x_2.$$

If f is both one-to-one and onto, then f has an *inverse function*, denoted f^{-1}, that has domain Y and range X and is defined so that $f^{-1}(f(x)) = x$ for all $x \in X$ and $f(f^{-1}(y)) = y$ for all $y \in Y$.

Let $f : X \to Y$ be a function. If $A \subseteq X$ is a subset of the domain of f, then the *image of A* is denoted $f(A)$ and is defined to be the set

$$f(A) = \{f(x) \mid x \in A\}.$$

If $B \subseteq Y$ is a subset of the codomain of f, then the *inverse image* (or *preimage*) of B is denoted $f^{-1}(B)$ and is defined to be the set

$$f^{-1}(B) = \{x \mid f(x) \in B\}.$$

It is important to note that the notation $f^{-1}(B)$ does not imply that the inverse function f^{-1} is defined, because f may not be one-to-one.

It is common in the literature to see other terms for functions that are onto or one-to-one. A function that is onto is sometimes called a *surjection*, a function that is one-to-one is also known as an *injection*, and a function that is both an injection and a surjection is called a *bijection*. Such functions are said to be *surjective*, *injective*, and *bijective*, respectively.

Example 3.1.1. Let $\mathbb{R}$ be the set of real numbers and define a function $f : \mathbb{R} \to \mathbb{R}$ by $f(x) = x^2$ for $x \in \mathbb{R}$.

(a) Show that f is neither onto nor one-to-one.

(b) Find $f(\{-3, -1, 2, 4\})$, the image of the set $\{-3, -1, 2, 4\}$.

(c) Find the preimages: $f^{-1}([4, 9])$, $f^{-1}([-1, 9])$, and $f^{-1}([-4, -1])$.

Solution. (a) The range of f is contained in the interval $[0, \infty)$ because $x^2 \geq 0$ for all real numbers x. Since $[0, \infty)$ is a proper subset of $\mathbb{R}$, the function f is not onto. To see that the range of f is $[0, \infty)$, simply observe that if $x \geq 0$, then $\sqrt{x} \in \mathbb{R}$ and $f(\sqrt{x}) = x$. Finally, note that f is not one-to-one because $f(-2) = 4 = f(2)$, but $-2 \neq 2$.

(b) By definition,

$$f(\{-3, -1, 2, 4\}) = \{f(-3), f(-1), f(2), f(4)\} = \{9, 1, 4, 16\}.$$

(c) We use the definition of the preimage:

$$f^{-1}([4, 9]) = \{x \mid f(x) \in [4, 9]\} = \{x \mid 4 \leq x^2 \leq 9\} = [-3, -2] \cup [2, 3],$$
$$f^{-1}([-1, 9]) = \{x \mid f(x) \in [-1, 9]\} = \{x \mid -1 \leq x^2 \leq 9\}$$
$$= \{x \mid 0 \leq x^2 \leq 9\} = [-3, 3],$$
$$f^{-1}([-4, -1]) = \{x \mid f(x) \in [-4, -1]\} = \{x \mid -4 \leq x^2 \leq -1\} = \emptyset.$$

Notice that the last preimage is the empty set because $x^2 \geq 0$ for all $x \in \mathbb{R}$.

Example 3.1.2. Let $\mathbb{R}$ be the set of real numbers and define a function $g : \mathbb{R} \to \mathbb{R}$ by $g(x) = 3x + 2$ for $x \in \mathbb{R}$.

(a) Show that g is onto and one-to-one.

(b) Find a formula for $g^{-1}(x)$.

Solution. (a) Let $c \in \mathbb{R}$. Solve the equation $3x + 2 = c$ for x and get $x = \dfrac{c - 2}{3}$. Because

$$g\left(\frac{c-2}{3}\right) = 3\left(\frac{c-2}{3}\right) + 2 = c,$$

it follows that g is onto. To see that g is one-to-one, observe that

$$[g(x_1) = g(x_2)] \implies [3x_1 + 2 = 3x_2 + 2] \implies [3x_1 = 3x_2] \implies [x_1 = x_2].$$

(b) In (a), we showed g was one-to-one and onto. Consequently, the inverse function g^{-1} exists and for each $x \in \mathbb{R}$, we have $g(g^{-1}(x)) = x$. Thus,

$$x = g(g^{-1}(x)) = 3(g^{-1}(x)) + 2.$$

Therefore, $3(g^{-1}(x)) = x - 2$, and so

$$g^{-1}(x) = \frac{x - 2}{3}.$$

Example 3.1.3. Let $\mathbb{R}$ be the set of real numbers and define the two functions $f : \mathbb{R} \to \mathbb{R}$ and $g : [3, \infty) \to \mathbb{R}$ by

$$f(x) = 2x + 1 \text{ and } g(x) = \sqrt{x - 3}.$$

Find formulas for $(f \circ g)(x)$ and $(g \circ f)(x)$. Identify the domain of each composition.

Solution. Note that $(f \circ g)(x) = f(g(x))$, and hence $g(x)$ must be in the domain of f. The domain of f is the set of all real numbers, and so $(f \circ g)(x)$ is defined so long as $g(x)$ is a real number. Therefore, since $g(x)$ is a real number when $x \geq 3$, the domain of $f \circ g$ is the set $[3, \infty)$. To find a formula for $(f \circ g)(x)$, we compute directly:

$$(f \circ g)(x) = f\big(g(x)\big) = 2g(x) + 1 = 2\sqrt{x - 3} + 1.$$

We now find a formula for $(g \circ f)(x)$. Observe that $(g \circ f)(x) = g(f(x))$, and so the number $f(x)$ must be in the domain of g. Because the domain of g is the set $[3, \infty)$, we must have $f(x) \in [3, \infty)$, or equivalently $f(x) \geq 3$. That is to say, $2x + 1 \geq 3$. Consequently, in order for $(g \circ f)(x)$ to be defined, it must be that $x \geq 1$. Therefore, $g \circ f : [1, \infty) \to \mathbb{R}$ and

$$(g \circ f)(x) = g\big(f(x)\big) = \sqrt{f(x) - 3} = \sqrt{(2x + 1) - 3} = \sqrt{2x - 2}.$$

Example 3.1.4. Let $\mathbb{R}$ be the set of real numbers and let f and g be as in Example 3.1.3. Define a function $h : \mathbb{R} \to \mathbb{R}$ by $h(x) = 2x^2 + 1$ for all $x \in \mathbb{R}$. (Compare to f.) Find a formula for $(h \circ g)(x)$ and identify the domain of the composition.

Solution. We argue as we did in Example 3.1.3. By the definition of composition, $(h \circ g)(x) = h(g(x))$. Thus, the composition is defined only when $g(x)$ is in the domain of h. The domain of h is the set $\mathbb{R}$, and so $(h \circ g)(x)$ is defined so long as $g(x)$ is a real number. As before, $g(x)$ is a real number when $x \geq 3$, and so the domain of $h \circ g$ is the set $[3, \infty)$. Therefore, $h \circ g : [3, \infty) \to \mathbb{R}$ and

$$(h \circ g)(x) = h\big(g(x)\big) = 2\big(g(x)\big)^2 + 1 = 2(x - 3) + 1 = 2x - 5.$$

It is important to notice that the domain of $h \circ g$ is only $[3, \infty)$, despite the fact that $2x - 5$ is defined for every real number x. This is a consequence of the definition of the composition.

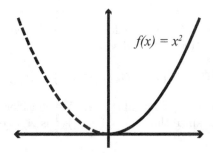

$$f(x) = x^2$$

Figure 3.1.b: The function f is defined by the formula $f(x) = x^2$ for all x in $(-\infty, \infty)$. The restriction of f to $[0, \infty)$ is represented here graphically. To indicate that f is restricted to the set of nonnegative real numbers, the portion of the graph over the negative numbers is drawn as a dashed curve.

It is often the case that a function lacks certain desirable properties. For example, a function may fail to be one-to-one on its domain. The function may, however, have the desired property on some subset of its domain. In such a case, we consider a new function which is closely related to the original function.

Definition 3.1.5. Let X and Y be sets and let $f : X \to Y$ be a function. If $Z \subseteq X$, then the *restriction of f to Z* is denoted $f|_Z$ and is the function $f|_Z : Z \to Y$ defined by the rule $f|_Z(x) = f(x)$ for all $x \in Z$.

Example 3.1.6. Let $\mathbb{R}$ be the set of real numbers and let $f(x) = x^2$ for all $x \in \mathbb{R}$. If $A = [0, \infty)$, then show that $f|_A$ is one-to-one and find $(f|_A)^{-1}$.

Solution. In Example 3.1.1, we showed that f is not one-to-one on $\mathbb{R}$. We also showed that A is the range of f. Suppose that x_1 and x_2 are in A and suppose that $f(x_1) = f(x_2)$. Then both x_1 and x_2 are nonnegative and $x_1^2 = x_2^2$. It follows that

$$0 = x_1^2 - x_2^2 = (x_1 - x_2)(x_1 + x_2).$$

Thus, either $x_1 = x_2$ or $x_1 = -x_2$. Since both numbers are in A (and so nonnegative), we conclude that $x_1 = x_2$. We have demonstrated that

$$f|_A(x_1) = f|_A(x_2) \implies x_1 = x_2,$$

and hence $f|_A$ is one-to-one.

It is straightforward to show that the range of $f|_A$ is all of A (see Example 3.1.1), and so the inverse function $(f|_A)^{-1}$ exists. We now find a formula for $(f|_A)^{-1}$. For each $x \in A$, we have

$$x = f|_A\left((f|_A)^{-1}(x)\right).$$

Because the domain of $f|_A$ is A, it follows that the range of $(f|_A)^{-1}$ is A. Consequently, $(f|_A)^{-1}(x) \in A$, and so

$$x = f|_A((f|_A)^{-1}(x)) = f((f|_A)^{-1}(x)) = ((f|_A)^{-1}(x))^2.$$

Therefore, $(f|_A)^{-1}(x) = \sqrt{x}$ for all $x \in A$. (We consider only the nonnegative square root of x since the range of $(f|_A)^{-1}$ is $A = [0, \infty)$.)

We conclude this section with a discussion of a special type of function called a *sequence*. (Compare to Definitions 2.6.6.)

Definition 3.1.7. A *sequence* is any function whose domain is a set of consecutive integers.

Remark 3.1.8. A sequence may have a domain of any set of consecutive integers, but it is common for the domain to be $\mathbb{N}$, the set of natural numbers.

We use several conventions that distinguish sequences from general functions. Instead of the usual functional notation, if the sequence is x, it is customary to write x_n instead of $x(n)$. We call x_n the nth term of the sequence. The sequence itself is denoted $\{x_n\}_{n=1}^{\infty}$. If $\{x_n\}_{n=1}^{\infty}$ is a sequence and $\{n_j\}_{j=1}^{\infty}$ is an increasing sequence of natural numbers (that is, $1 \leq n_1 < n_2 < n_3 < \cdots$), then the composition $\{x_{n_j}\}_{j=1}^{\infty}$ is called a *subsequence* of $\{x_n\}_{n=1}^{\infty}$.

Definitions 3.1.9. Let $(X, \mathcal{T}_d)$ be a metric space and let $\{x_n\}_{n=1}^{\infty}$ be a sequence with range a subset of X. (That is, a sequence with terms in X.) We say the sequence *converges* to $x_0 \in X$ provided that for each positive real number ϵ, there exists a positive integer N such that $d(x_n, x_0) < \epsilon$ whenever $n \geq N$. We say that x_0 is the *limit* of the sequence and we write $\lim_{n \to \infty} x_n = x_0$.

The concepts of convergence and limit were introduced in Definitions 2.6.6 in a more general setting. The definitions agree when $(X, \mathcal{T}_d)$ is a metric space because the collection $\{B_\epsilon(x) : x \in X \text{ and } \epsilon > 0\}$ of all open balls in X is a base for the metric topology $\mathcal{T}_d$. (See Theorem 2.7.6.)

By Theorem 2.7.11, a metric space is a Hausdorff space. Consequently, if $(X, \mathcal{T}_d)$ is a metric space and $\{x_n\}_{n=1}^{\infty}$ is a sequence whose terms form a subset of X, then if this sequence converges, its limit is unique. (See Theorem 2.6.8.)

Suppose now that $(Y, \mathcal{T}_{d'})$ is another metric space. If $f : X \to Y$ is a function and $\{x_n\}_{n=1}^{\infty}$ is a sequence whose terms form a subset of X, then is it necessarily true that

$$\lim_{n \to \infty} x_n = x_0 \quad \Longrightarrow \quad \lim_{n \to \infty} f(x_n) = f(x_0)?$$

(See Exercise 8, below.)

Exercises

1. Let $f : \mathbb{R} \to \mathbb{R}$ be defined by the formula $f(x) = x^4$ for all $x \in \mathbb{R}$.

   (a) Show that f is neither one-to-one nor onto.
   (b) Find $f(\{-1, 0, 2, 3\})$.
   (c) Find $f^{-1}(\{1, 16, 625\})$, $f^{-1}([-2, 16])$, and $f^{-1}([-5, -2])$.

2. Let $f : \mathbb{R} \to \mathbb{R}$ be defined by the formula $f(x) = 3x - 1$ for all $x \in \mathbb{R}$.

   (a) Show that f is both one-to-one and onto.
   (b) Find $f(\{-2, 0, 2, 3, 6\})$.
   (c) Find a formula for $f^{-1}(x)$.
   (d) Find $f^{-1}(\{-4, 5, 14\})$ and $f^{-1}([-5, 8])$.

3. Let $f : \mathbb{R} \to \mathbb{R}$ be defined by the formula $f(x) = x^3 + 2$ for all $x \in \mathbb{R}$.

   (a) Show that f is both one-to-one and onto.
   (b) Find $f(\{-3, 1, 2, 4, 6\})$.
   (c) Find a formula for $f^{-1}(x)$.

4. Let $f : \mathbb{R} \to \mathbb{R}$ be defined by the formula $f(x) = x^2 - 2x + 1$ for all $x \in \mathbb{R}$.

   (a) Sketch the graph of the equation $y = f(x)$.
   (b) It is clear from the graph that f is neither one-to-one nor onto. Prove these facts.
   (c) Let $B = [1, \infty)$. Show that $f|_B$ is one-to-one.
   (d) Find the image of $(f|_B)^{-1}$ and derive a formula for $(f|_B)^{-1}(x)$. (*Note that a function is always onto its image.*)

5. Consider the sine function $\sin : \mathbb{R} \to \mathbb{R}$.

   (a) Show that $\sin$ is neither one-to-one nor onto.
   (b) Find $\sin\left(\left\{-\dfrac{5\pi}{6}, \dfrac{2\pi}{3}, \dfrac{3\pi}{2}\right\}\right)$.
   (c) Let $C = [-\pi/2, \pi/2]$ and show that $\sin|_C$ is one-to-one.
   (d) Find $(\sin|_C)^{-1}(1/2)$ and $(\sin|_C)^{-1}(-\sqrt{3}/2)$

6. Let $A = \{1, 2, 3, 5, 7\}$ and $B = \{a, b, c, d, e\}$. Define a function $f : A \to B$ by means of the following table:

x	1	2	3	5	7
$f(x)$	b	e	a	e	d

(a) Show that f is neither one-to-one nor onto.

(b) Find two distinct subsets C_1 and C_2 of A, each containing four elements, such that $f|_{C_1}$ and $f|_{C_2}$ are both one-to-one.

7. In many textbooks, a function is defined as a set of ordered pairs, no two distinct members of which have the same first element. For each of the following sets of ordered pairs, determine whether the set defines a function. For each affirmative answer, describe the domain and range of the function.

   (a) $\{(1,5),(3,7),(8,2),(11,-1),(14,2)\}$

   (b) $\{(-1,0),(3,2),(-1,3),(5,7),(9,-4)\}$

   (c) $\{(x,y) \mid x$ and y are real numbers and $y = 2x - 4\}$

   (d) $\{(x,y) \mid x$ and y are real numbers and $x = y^2\}$

   (e) $\{(x,y) \mid x$ and y are real numbers and $y = x^2 - 4\}$

8. Give an example of a metric space $(X, \mathscr{T}_d)$, a function $f : X \to X$, and a sequence $\{x_n\}_{n=1}^{\infty}$ with terms in X such that $\lim_{n \to \infty} x_n = x_0$ but $\lim_{n \to \infty} f(x_n) \neq f(x_0)$.

3.2   More on Image and Inverse Image

Suppose that $f : X \to Y$ is a function between two sets X and Y, and let $B \subseteq Y$. In the preceding section, we defined the inverse image $f^{-1}(B)$ to be the set $\{x \mid f(x) \in B\}$. *We must stress again that, in this context, f^{-1} does not denote the inverse of the function f, because f may not be a one-to-one function.*

In this section, we consider several questions about inverse images. For example, suppose B and C are two subsets of Y. We may first form the union $B \cup C$ and then find $f^{-1}(B \cup C)$, or we may first find $f^{-1}(B)$ and $f^{-1}(C)$ and then form their union $f^{-1}(B) \cup f^{-1}(C)$. What is the relationship (if any) between the sets $f^{-1}(B \cup C)$ and $f^{-1}(B) \cup f^{-1}(C)$? In this section, we state several theorems answering this and similar questions.

Let us begin, however, by considering several examples. These examples illustrate the properties that we will later prove are true in general.

Example 3.2.1. Let $f : \mathbb{R} \to \mathbb{R}$ be defined by the formula $f(x) = x^2$ for all $x \in \mathbb{R}$. Let $A = \{-5, -3, 0, 1, 4, 16\}$, $B = \{1, 16, 25\}$, and $C = \{1, 3\}$.

(a) Compute $f^{-1}(\{16\})$ and $f^{-1}(\{-5\})$.

(b) Compute $f^{-1}(A)$ and $f^{-1}(B)$.

(c) Compute $f^{-1}(A \cup B)$ and $f^{-1}(A) \cup f^{-1}(B)$.

(d) Compute $f^{-1}(A \cap B)$ and $f^{-1}(A) \cap f^{-1}(B)$.

(e) Compute $f(f^{-1}(A))$ and $f(f^{-1}(B))$

(f) Compute $f^{-1}(f(C))$.

Solution. (a) By definition,

$$f^{-1}(\{16\}) = \{x \mid f(x) \in \{16\}\} = \{x \mid x^2 = 16\} = \{-4, 4\}.$$

Next, observe that

$$f^{-1}(\{-5\}) = \{x \mid f(x) \in \{-5\}\} = \{x \mid x^2 = -5\} = \emptyset,$$

because the domain of f is $\mathbb{R}$ and there are no real numbers x such that $x^2 = -5$.

(b) Computing directly, we obtain:

$$f^{-1}(A) = \{x \mid x^2 \in \{-5, -3, 0, 1, 4, 16\}\} = \{-4, -2, -1, 0, 1, 2, 4\},$$

$$f^{-1}(B) = \{x \mid x^2 \in \{1, 16, 25\}\} = \{-5, -4, -1, 1, 4, 5\}.$$

(c) Observe that $A \cup B = \{-5, -3, 0, 1, 4, 16, 25\}$. Thus,

$$f^{-1}(A \cup B) = \{x \mid x^2 \in \{-5, -3, 0, 1, 4, 16, 25\}\}$$

$$= \{-5, -4, -2, -1, 0, 1, 2, 4, 5\},$$

$$f^{-1}(A) \cup f^{-1}(B) = \{-4, -2, -1, 0, 1, 2, 4\} \cup \{-5, -4, -1, 1, 4, 5\}$$

$$= \{-5, -4, -2, -1, 0, 1, 2, 4, 5\}.$$

Notice that, at least in this example, $f^{-1}(A \cup B) = f^{-1}(A) \cup f^{-1}(B)$.

(d) We have that $A \cap B = \{1, 16\}$, and so

$$f^{-1}(A \cap B) = \{x \mid x^2 \in \{1, 16\}\} = \{-4, -1, 1, 4\},$$

$$f^{-1}(A) \cap f^{-1}(B) = \{-4, -2, -1, 0, 1, 2, 4\} \cap \{-5, -4, -1, 1, 4, 5\}$$

$$= \{-4, -1, 1, 4\}.$$

Thus, we have also for this example that $f^{-1}(A \cap B) = f^{-1}(A) \cap f^{-1}(B)$.

(e) Once again, we compute directly:

$$f(f^{-1}(A)) = f(\{-4, -2, -1, 0, 1, 2, 4\})\}$$

$$= \{f(-4), f(-2), f(-1), f(0), f(1), f(2), f(4)\}$$

$$= \{0, 1, 4, 16\},$$

$$f\left(f^{-1}(B)\right) = f\left(\{-5, -4, -1, 1, 4, 5\}\right)\}$$

$$= \{f(-5),\, f(-4),\, f(-1),\, f(1),\, f(4),\, f(5)\}$$

$$= \{1, 16, 25\}.$$

Notice that $f\left(f^{-1}(A)\right)$ is a proper subset of A, but $f\left(f^{-1}(B)\right) = B$.
(f) Recall that $C = \{1, 3\}$. Hence,

$$f(C) = f\left(\{1, 3\}\right) = \{f(1),\, f(3)\} = \{1, 9\}.$$

Thus,
$$f^{-1}\left(f(C)\right) = \{x \mid x^2 \in \{1, 9\}\} = \{-3, -1, 1, 3\}.$$

In this case, we have that C is a proper subset of $f^{-1}\left(f(C)\right)$.

Example 3.2.2. Let $X = \{1, 2, 3, 4, 5, 6, 7\}$ and $Y = \{a, b, c, d, e, f, g\}$.
Define a function $F : X \to Y$ by means of the following table:

x	1	2	3	4	5	6	7
$F(x)$	c	e	a	c	g	a	c

Compute $F^{-1}(\{y\})$ for each $y \in Y$. Does the function F have an inverse?

Solution. By definition, $F^{-1}(\{y\}) = \{x \mid F(x) = y\}$. Thus:

$$F^{-1}(\{a\}) = \{3, 6\},\ F^{-1}(\{b\}) = \emptyset,\ F^{-1}(\{c\}) = \{1, 4, 7\},$$

$$F^{-1}(\{d\}) = \emptyset,\ F^{-1}(\{e\}) = \{2\},\ F^{-1}(\{f\}) = \emptyset,\ F^{-1}(\{g\}) = \{5\}.$$

The function F is not one-to-one; hence, F does not have an inverse function.

Example 3.2.3. Let $F : X \to Y$ be the function from Example 3.2.2. Let
A be the set $\{a, b, g\}$ and show that $F^{-1}(A') = \left(F^{-1}(A)\right)'$.

Solution. Note first that $A \subseteq Y$, and so $A' = Y \setminus A = \{c, d, e, f\}$.
Consequently,

$$F^{-1}(A') = \{x \mid F(x) \in \{c, d, e, f\}\} = \{1, 2, 4, 7\}.$$

Next, observe that

$$F^{-1}(A) = \{x \mid F(x) \in \{a, b, g\}\} = \{3, 5, 6\}.$$

Thus, because $F^{-1}(A) \subseteq X$, we have

$$\left(F^{-1}(A)\right)' = X \setminus \{3, 5, 6\} = \{1, 2, 4, 7\}.$$

Therefore, the two sets are the same.

Example 3.2.4. Once again, let $F : X \to Y$ be the function from Example 3.2.2. Let $A = \{1, 2, 3, 7\}$ and $B = \{2, 3, 4\}$. Show that $F(A \cup B) = F(A) \cup F(B)$ and that $F(A \cap B) \subseteq F(A) \cap F(B)$.

Solution. First, observe that $A \cup B = \{1, 2, 3, 4, 7\}$. Because we have $F(A) = \{a, c, e\}$, $F(B) = \{a, c, e\}$, and $F(A \cup B) = \{a, c, e\}$, it follows that $F(A \cup B) = F(A) \cup F(B)$, which is the first fact we wished to verify. Next, observe that $A \cap B = \{2, 3\}$. Consequently,

$$F(A \cap B) = F(\{2, 3\}) = \{a, e\}.$$

Since $F(A) \cap F(B) = \{a, c, e\}$, we conclude that $F(A \cap B) \subseteq F(A) \cap F(B)$.

The preceding examples illustrate several properties of images and preimages. We now state and prove these properties in a general context.

Theorem 3.2.5. *Let X and Y be sets and suppose $f : X \to Y$ is a function. If A and B are subsets of X, and C and D are subsets of Y, then:*

(a) $f^{-1}(C \cup D) = f^{-1}(C) \cup f^{-1}(D)$,

(b) $f^{-1}(C \cap D) = f^{-1}(C) \cap f^{-1}(D)$,

(c) $f(A \cup B) = f(A) \cup f(B)$,

(d) $f(A \cap B) \subseteq f(A) \cap f(B)$,

(e) $f^{-1}(C') = \left(f^{-1}(C)\right)'$,

(f) $f\left(f^{-1}(C)\right) \subseteq C$,

(g) $A \subseteq f^{-1}\left(f(A)\right)$.

Proof. In this proof, we make extensive use of our set theory notation, especially the double implication ($\Leftrightarrow$). Recall that two sets U and V are identical if and only if $x \in U \Leftrightarrow x \in V$.

(a) We will show the sets are the same by showing mutual inclusion:

$$
\begin{aligned}
x \in f^{-1}(C \cup D) &\Longleftrightarrow f(x) \in C \cup D \\
&\Longleftrightarrow [f(x) \in C] \vee [f(x) \in D] \\
&\Longleftrightarrow [x \in f^{-1}(C)] \vee [x \in f^{-1}(D)] \\
&\Longleftrightarrow x \in f^{-1}(C) \cup f^{-1}(D).
\end{aligned}
$$

(b) Again, we will show mutual inclusion:

$$
\begin{aligned}
x \in f^{-1}(C \cap D) &\Longleftrightarrow f(x) \in C \cap D \\
&\Longleftrightarrow [f(x) \in C] \wedge [f(x) \in D] \\
&\Longleftrightarrow [x \in f^{-1}(C)] \wedge [x \in f^{-1}(D)] \\
&\Longleftrightarrow x \in f^{-1}(C) \cap f^{-1}(D).
\end{aligned}
$$

(c) We first show that $f(A \cup B) \subseteq f(A) \cup f(B)$:

$$
\begin{aligned}
y \in f(A \cup B) &\Longrightarrow y = f(x) \text{ for some } x \in A \cup B \\
&\Longrightarrow [y = f(x) \text{ for } x \in A] \vee [y = f(x) \text{ for } x \in B] \\
&\Longrightarrow [y \in f(A)] \vee [y \in f(B)] \\
&\Longrightarrow y \in f(A) \cup f(B).
\end{aligned}
$$

Conversely:

$$
y \in f(A) \cup f(B)
$$

$$
\begin{aligned}
&\Longrightarrow [y \in f(A)] \vee [y \in f(B)] \\
&\Longrightarrow [y = f(x_1) \text{ for } x_1 \in A] \vee [y = f(x_2) \text{ for } x_2 \in B].
\end{aligned}
$$

Thus, there is some $x \in \{x_1, x_2\} \subseteq A \cup B$ so that $y = f(x)$. Therefore,

$$
\begin{aligned}
y \in f(A) \cup f(B) &\Longrightarrow y = f(x) \text{ for some } x \in A \cup B \\
&\Longrightarrow y \in f(A \cup B).
\end{aligned}
$$

(d) We show $f(A \cap B) \subseteq f(A) \cap f(B)$:

$$
\begin{aligned}
y \in f(A \cap B) &\Longrightarrow y = f(x) \text{ for some } x \in A \cap B \\
&\Longrightarrow [y = f(x) \text{ for } x \in A] \wedge [y = f(x) \text{ for } x \in B] \\
&\Longrightarrow [y \in f(A)] \wedge [y \in f(B)] \\
&\Longrightarrow y \in f(A) \cap f(B).
\end{aligned}
$$

(e) Once again, we show the sets are equal by showing mutual inclusion:

$$x \in f^{-1}(C') \iff f(x) \in C'$$
$$\iff f(x) \notin C$$
$$\iff x \notin f^{-1}(C)$$
$$\iff x \in \left(f^{-1}(C)\right)'.$$

(f) Suppose $y \in f\left(f^{-1}(C)\right)$. Then $y = f(x)$ for some $x \in f^{-1}(C)$. This last containment implies that $f(x) \in C$, and hence $y \in C$. Therefore, $f\left(f^{-1}(C)\right) \subseteq C$.

(g) Suppose $x \in A$. We have,

$$x \in A \implies f(x) \in f(A) \implies x \in f^{-1}(f(A)).$$

Thus, $A \subseteq f^{-1}(f(A))$.

Q.E.D.

We comment that the examples of this section show that the containments in (d), (f), and (g) of Theorem 3.2.5 cannot be replaced with equalities. In each case, we have seen examples where the containments are proper. (That is to say, the given sets were proper subsets).

We now generalize (a), (b), (c), and (d) of Theorem 3.2.5.

Theorem 3.2.6. *Let X and Y be sets and suppose $f : X \to Y$ is a function. If $\mathscr{T}$ is a collection of subsets of X, and $\mathscr{R}$ is a collection of subsets of Y, then:*

(a) $f^{-1}\left(\bigcup_{C \in \mathscr{R}} C \right) = \bigcup_{C \in \mathscr{R}} f^{-1}(C)$,

(b) $f^{-1}\left(\bigcap_{C \in \mathscr{R}} C \right) = \bigcap_{C \in \mathscr{R}} f^{-1}(C)$,

(c) $f\left(\bigcup_{A \in \mathscr{T}} A \right) = \bigcup_{A \in \mathscr{T}} f(A)$,

(d) $f\left(\bigcap_{A \in \mathscr{T}} A \right) \subseteq \bigcap_{A \in \mathscr{T}} f(A)$.

Proof. The proof of (a) is similar to the proof of Theorem 3.2.5(a):

$$x \in f^{-1}\left(\bigcup_{C \in \mathscr{R}} C \right) \iff f(x) \in \bigcup_{C \in \mathscr{R}} C$$
$$\iff f(x) \in C \text{ for some } C \in \mathscr{R}$$
$$\iff x \in f^{-1}(C) \text{ for some } C \in \mathscr{R}$$
$$\iff x \in \bigcup_{C \in \mathscr{R}} f^{-1}(C).$$

We leave the proofs of (b), (c), and (d) as an exercise. (See Exercise 5 at the end of this section.)

Q.E.D.

Exercises

1. Let $\mathbb{R}$ be the set of real numbers and define $f : \mathbb{R} \to \mathbb{R}$ by $f(x) = x^2$ for $x \in \mathbb{R}$. Verify the conclusions of Theorem 3.2.5 for the function f if $A = [-3, -1]$, $B = [1, 4]$, $C = [-1, 9]$, and $D = [4, 25]$.

2. Let $X = \{a, b, c, d, e, f, g, h, i\}$ and $Y = \{1, 2, 3, 4, 5, 6, 7, 8, 9\}$. Define a function $F : X \to Y$ by the following table:

x	a	b	c	d	e	f	g	h	i
$F(x)$	3	1	4	9	5	1	3	1	7

   Verify the conclusions of Theorem 3.2.5 are satisfied for the function F if $A = \{a, c, g, h\}$, $B = \{b, d, e, f\}$, $C = \{1, 3, 5, 7, 9\}$, and $D = \{2, 4, 5, 8, 9\}$.

3. Let $\mathbb{R}$ be the set of real numbers and define $f : \mathbb{R} \to \mathbb{R}$ by $f(x) = \sin x$ for $x \in \mathbb{R}$. Verify the conclusions of Theorem 3.2.5 for the function f if $A = [-\frac{\pi}{3}, \frac{2\pi}{3}]$, $B = [-\frac{\pi}{6}, \frac{3\pi}{4}]$, $C = [\frac{1}{2}, 5]$, and $D = [-\frac{1}{2}, \frac{\sqrt{3}}{2}]$.

4. Let $X = \mathbb{R} \setminus \left\{ \frac{(2n+1)\pi}{2} \mid n \in \mathbb{Z} \right\}$, where $\mathbb{Z}$ is the set of integers, and let $Y = \mathbb{R}$. Define $f : X \to Y$ by $f(x) = \tan x$ for $x \in X$. Verify that the conclusions of Theorem 3.2.5 hold for the function f if

$$A = \left[-\tfrac{\pi}{4}, \tfrac{\pi}{3} \right] \cup \left[\tfrac{5\pi}{6}, \tfrac{4\pi}{3} \right], \qquad B = \left[-\tfrac{5\pi}{6}, -\tfrac{3\pi}{4} \right] \cup \left[-\tfrac{\pi}{6}, 0 \right],$$

$$C = \left[0, \sqrt{3} \right], \qquad\qquad D = \left[-1, \tfrac{\sqrt{3}}{3} \right].$$

5. Prove (b), (c), and (d) of Theorem 3.2.6.

6. Let X and Y be sets and let $f : X \to Y$ be a function. Under what condition is it true that $A = f^{-1}(f(A))$ for each subset A of X? Justify your answer.

7. Let X and Y be sets and let $f : X \to Y$ be a function. Under what condition is it true that $B = f(f^{-1}(B))$ for each subset B of Y? Justify your answer.

3.3   Continuous Functions

We begin this section by briefly reviewing the concept of continuity of a function as defined in calculus. Let $\mathbb{R}$ be the set of real numbers and let $f : \mathbb{R} \to \mathbb{R}$ be a function. We say that f is *continuous at* x_0 if for each $\epsilon > 0$ there exists a $\delta > 0$ such that

$$|x - x_0| < \delta \implies |f(x) - f(x_0)| < \epsilon.$$

If f is continuous at each $x \in \mathbb{R}$, then we say f is *continuous on* $\mathbb{R}$. Recall also that a set U of real numbers is said to be *open* if for each $x \in U$ there is a number $\delta > 0$ such that $(x - \delta, x + \delta) \subseteq U$. Using these definitions, we prove the next theorem.

Theorem 3.3.1. *Let $\mathbb{R}$ be the set of real numbers. A function $f : \mathbb{R} \to \mathbb{R}$ is continuous on $\mathbb{R}$ if and only if $f^{-1}(U)$ is an open set whenever U is an open set.*

Proof. First, assume f is continuous and let U be an open set. If $f^{-1}(U) = \emptyset$, then it is open, so assume instead that $f^{-1}(U) \neq \emptyset$. Let $x_0 \in f^{-1}(U)$. Then $f(x_0) \in U$, which is an open set (by assumption), and so there exists a number $\epsilon > 0$ such that $(f(x_0) - \epsilon, f(x_0) + \epsilon) \subseteq U$. By the continuity of f, there exists a $\delta > 0$ such that

$$|x - x_0| < \delta \implies |f(x) - f(x_0)| < \epsilon.$$

We claim that $(x_0 - \delta, x_0 + \delta) \subseteq f^{-1}(U)$. To see this is the case, suppose that x is chosen in the interval $(x_0 - \delta, x_0 + \delta)$. It follows that $|x - x_0| < \delta$. Thus, by the continuity of f, we conclude that $|f(x) - f(x_0)| < \epsilon$. This last inequality is equivalent to the statement $f(x) \in (f(x_0) - \epsilon, f(x_0) + \epsilon)$. By the choice of ϵ, then, we have that $f(x) \in U$, or in other words, that $x \in f^{-1}(U)$. Since the choice of $x \in (x_0 - \delta, x_0 + \delta)$ was arbitrary, we conclude that

$$(x_0 - \delta, x_0 + \delta) \subseteq f^{-1}(U).$$

We have now shown that for each $x_0 \in f^{-1}(U)$, there can be found a $\delta > 0$ such that $(x_0 - \delta, x_0 + \delta) \subseteq f^{-1}(U)$. Therefore, the set $f^{-1}(U)$ is open, as required.

Conversely, suppose that for each open set U, the set $f^{-1}(U)$ is open. We wish to show that f is continuous. Let $x_0 \in \mathbb{R}$ and suppose $\epsilon > 0$ is given. The interval $(f(x_0) - \epsilon, f(x_0) + \epsilon)$ is an open set, and consequently the preimage

$$f^{-1}\big[(f(x_0) - \epsilon, f(x_0) + \epsilon)\big]$$

is an open set. Because x_0 is in this set, and because it is an open set, there exists some $\delta > 0$ such that

$$(x_0 - \delta, x_0 + \delta) \subseteq f^{-1}\big[(f(x_0) - \epsilon, f(x_0) + \epsilon)\big].$$

We now observe the following implications:

$$|x - x_0| < \delta \implies x \in (x_0 - \delta, x_0 + \delta)$$

$$\implies x \in f^{-1}\big[(f(x_0) - \epsilon, f(x_0) + \epsilon)\big]$$

$$\implies f(x) \in (f(x_0) - \epsilon, f(x_0) + \epsilon)$$

$$\implies |f(x) - f(x_0)| < \epsilon.$$

It follows that f is continuous at x_0. Since this is true for every $x_0 \in \mathbb{R}$, we conclude that f is continuous on $\mathbb{R}$.   Q.E.D.

Our objective is to define continuity of a function on an arbitrary topological space. We wish to make the definition in such a way that it agrees with the usual definition when applied to the set $\mathbb{R}$ of real numbers with the standard topology. In general, we do not have a notion of distance in arbitrary topological spaces, but we always have a notion of open sets. (See Section 2.2.) Guided by the previous theorem, we define continuity of a function in the following way.

Definition 3.3.2. Let $(X, \mathscr{T})$ and $(Y, \mathscr{R})$ be two topological spaces and suppose that $f : X \to Y$ is a function. We say that f is $\mathscr{T}$-to-$\mathscr{R}$ *continuous* if $f^{-1}(U)$ is open in X whenever U is an open set in Y. That is, f is $\mathscr{T}$-to-$\mathscr{R}$ continuous if

$$U \in \mathscr{R} \implies f^{-1}(U) \in \mathscr{T}.$$

If no confusion arises, we say f is *continuous* rather than $\mathscr{T}$-to-$\mathscr{R}$ continuous.

Example 3.3.3. Let $X = \{1, 2, 3\}$ and $Y = \{a, b, c\}$. Define a topology $\mathscr{T}$ on X and a topology $\mathscr{R}$ on Y by

$$\mathscr{T} = \{\emptyset, X, \{1\}, \{2\}, \{1, 2\}\} \quad \text{and} \quad \mathscr{R} = \{\emptyset, Y, \{a\}, \{b, c\}\}.$$

Define functions $f : X \to Y$ and $g : X \to Y$ by means of the following table:

x	1	2	3
$f(x)$	a	c	a
$g(x)$	a	a	a

Is f $\mathscr{T}$-to-$\mathscr{R}$ continuous? Is g $\mathscr{T}$-to-$\mathscr{R}$ continuous?

Solution. Note that $\{a\} \in \mathscr{R}$, but $f^{-1}(\{a\}) = \{1, 3\} \notin \mathscr{T}$. Therefore, f is not $\mathscr{T}$-to-$\mathscr{R}$ continuous.

We claim that g is $\mathscr{T}$-to-$\mathscr{R}$ continuous. To prove this, we must show that $g^{-1}(U) \in \mathscr{T}$ for every $U \in \mathscr{R}$. We compute each preimage directly:

$$g^{-1}(\emptyset) = \emptyset, \quad g^{-1}(Y) = X, \quad g^{-1}(\{a\}) = X, \quad g^{-1}(\{b, c\}) = \emptyset.$$

We have shown that $g^{-1}(U) \in \mathscr{T}$ for each $U \in \mathscr{R}$, and so g is a $\mathscr{T}$-to-$\mathscr{R}$ continuous function.

Remark 3.3.4. In order to show that f is not continuous, it was sufficient to find one $U \in \mathscr{R}$ such that $f^{-1}(U) \notin \mathscr{T}$. On the other hand, to show that g is continuous, we had to verify that $g^{-1}(U) \in \mathscr{T}$ for *every* set $U \in \mathscr{R}$.

It is important to note that continuity depends not only on the definition of the function, but also on the topologies assigned to the domain and codomain of the function. We illustrate this point by means of an example.

Example 3.3.5 (The Dirichlet function). Let $\mathbb{R}$ be the set of real numbers and let $f : \mathbb{R} \to \mathbb{R}$ be defined as follows:

$$f(x) = \begin{cases} 1 & \text{if } x \text{ is rational,} \\ -1 & \text{if } x \text{ is irrational.} \end{cases}$$

This function is clearly not continuous if the domain and codomain are given the standard topology on $\mathbb{R}$. Let $\mathscr{T}$ be the indiscrete topology on $\mathbb{R}$ (that is, let $\mathscr{T} = \{\emptyset, \mathbb{R}\}$), let $\mathscr{S}$ be the discrete topology on $\mathbb{R}$, and let $\mathscr{R}$ denote an arbitrary topology on $\mathbb{R}$. Show:

(a) f is $\mathscr{R}$-to-$\mathscr{T}$ continuous, and

(b) f is $\mathscr{S}$-to-$\mathscr{R}$ continuous.

Solution. (a) Since $\mathscr{T}$ has only two open sets, we can verify directly that the preimages under f are open:

$$f^{-1}(\emptyset) = \emptyset \text{ and } f^{-1}(\mathbb{R}) = \mathbb{R}.$$

Whichever topology is chosen to be $\mathscr{R}$, it must contain both $\emptyset$ and $\mathbb{R}$. Thus, f is $\mathscr{R}$-to-$\mathscr{T}$ continuous.

(b) Let $U \in \mathscr{R}$. Since $\mathscr{S}$ is the discrete topology on $\mathbb{R}$, it contains every subset of $\mathbb{R}$. Certainly, the set $f^{-1}(U)$ is a subset of $\mathbb{R}$, and consequently $f^{-1}(U) \in \mathscr{S}$. It follows that f is $\mathscr{S}$-to-$\mathscr{R}$ continuous.

In elementary calculus, one learns that the composition of continuous functions is continuous. We wish to prove this statement for general topological spaces. To that end, we provide the following theorem.

Theorem 3.3.6. *Let X, Y, and Z be sets and suppose that $f : X \to Y$ and $g : Y \to Z$ are functions. If $A \subseteq Z$, then*

$$(g \circ f)^{-1}(A) = f^{-1}\big(g^{-1}(A)\big).$$

We pause once again to remind the reader that f and g are not assumed to be one-to-one functions, and so use of the symbols f^{-1} and g^{-1} does not imply the existence of inverse functions.

Proof. We show that the sets $(g \circ f)^{-1}(A)$ and $f^{-1}(g^{-1}(A))$ coincide by showing mutual inclusion:

$$x \in (g \circ f)^{-1}(A) \iff (g \circ f)(x) \in A$$

$$\iff g\big(f(x)\big) \in A$$

$$\iff f(x) \in g^{-1}(A) \iff x \in f^{-1}\big(g^{-1}(A)\big).$$

Since inclusion in one set implies inclusion in the other, the two sets must be the same.   Q.E.D.

We can now prove the following theorem.

Theorem 3.3.7. *The composition of two continuous functions is a continuous function.*

Proof. Let $(X, \mathscr{R})$, $(Y, \mathscr{S})$, and $(Z, \mathscr{T})$ be three topological spaces. Suppose that $f : X \to Y$ is a $\mathscr{R}$-to-$\mathscr{S}$ continuous function and $g : Y \to Z$ is a $\mathscr{S}$-to-$\mathscr{T}$ continuous function. We will show that $g \circ f : X \to Z$ is a $\mathscr{R}$-to-$\mathscr{T}$ continuous function.

Let $U \in \mathscr{T}$. By assumption, the function g is continuous from $\mathscr{S}$ to $\mathscr{T}$, and so $g^{-1}(U) \in \mathscr{S}$. Similarly, the function f is continuous from $\mathscr{R}$ to $\mathscr{S}$, and so we conclude that $f^{-1}(g^{-1}(U)) \in \mathscr{R}$. By Theorem 3.3.6, $f^{-1}(g^{-1}(U)) = (g \circ f)^{-1}(U)$, and consequently $(g \circ f)^{-1}(U) \in \mathscr{R}$ whenever $U \in \mathscr{T}$. Therefore, the composition function $g \circ f$ is $\mathscr{R}$-to-$\mathscr{T}$ continuous.   Q.E.D.

When a student first encounters the notion of continuity in a calculus course, it is usually "continuity at a point" in $\mathbb{R}$. Only then is a "continuous function on $\mathbb{R}$" defined as a function which is "continuous at all points." (Recall the first paragraph of this section, for example.) In Definition 3.3.2, we defined continuity of a function as a global property of the function, as it relates to all open sets in the domain and codomain. But here, too, we have a notion of "continuity at a point."

Definition 3.3.8. Let $(X, \mathscr{T})$ and $(Y, \mathscr{R})$ be two topological spaces and suppose that $f : X \to Y$ is a function. We say that f is *continuous at the point* $x_0 \in X$ if for every open set U containing $f(x_0)$, there exists an open neighborhood N of x_0 such that $f(N) \subseteq U$.

In the next section, we will see that a function $f : X \to Y$ is continuous if and only if it is continuous at each point. (See Theorem 3.4.1.) For now, we will content ourselves with the following theorem, which shows that, in certain cases, the answer to the question raised at the end of Section 3.1 is affirmative. That is, there are some occasions when $\lim_{n \to \infty} x_n = x_0$ implies that $\lim_{n \to \infty} f(x_n) = f(x_0)$. (See also Exercise 8 in Section 3.1.)

Theorem 3.3.9. *Suppose $(X, \mathscr{T}_{d_X})$ and $(Y, \mathscr{R}_{d_Y})$ are metric spaces and assume $x_0 \in X$ is an accumulation point of X. A function $f : X \to Y$ is continuous at x_0 if and only if $\lim_{n \to \infty} f(x_n) = f(x_0)$ for any sequence $\{x_n\}_{n=1}^{\infty}$ with terms in X such that $\lim_{n \to \infty} x_n = x_0$.*

Proof. First, suppose that f is continuous at x_0 and let $\{x_n\}_{n=1}^{\infty}$ be a sequence in X that converges to x_0. We will show that $\lim_{n \to \infty} f(x_n) = f(x_0)$. Let $\epsilon > 0$ be given. By definition, the open ball

$$B_\epsilon(f(x_0)) = \{y : y \in Y \text{ and } d_Y(f(x_0), y) < \epsilon\}$$

3.3. *CONTINUOUS FUNCTIONS*

is an open set containing $f(x_0)$. Thus, since we assumed that f is continuous at x_0, there exists a neighborhood N of x_0 such that $f(N) \subseteq B_\epsilon(f(x_0))$. We also assumed that $\lim_{n\to\infty} x_n = x_0$, and so there exists a positive integer M such that $x_n \in N$ whenever $n \geq M$. It follows that $f(x_n) \in f(N)$, and consequently $f(x_n) \in B_\epsilon(f(x_0))$, whenever $n \geq M$. Therefore, we have that $d_Y(f(x_n), f(x_0)) < \epsilon$ whenever $n \geq M$, and so $\lim_{n\to\infty} f(x_n) = f(x_0)$, as required.

Conversely, suppose that $\lim_{n\to\infty} f(x_n) = f(x_0)$ whenever $\{x_n\}_{n=1}^\infty$ is a sequence in X such that $\lim_{n\to\infty} x_n = x_0$. Our goal is to prove that f is continuous at x_0. Suppose to the contrary that f is not continuous at x_0. Then there is some $\epsilon > 0$ such that there exists no neighborhood N of x for which $f(N) \subseteq B_\epsilon(f(x_0))$. For each $n \in \mathbb{N}$, the set $B_{1/n}(x_0)$ is a neighborhood of x_0. Consequently, for each $n \in \mathbb{N}$, we have that

$$f(B_{1/n}(x_0)) \setminus B_\epsilon(f(x_0)) \neq \emptyset.$$

For each $n \in \mathbb{N}$, choose some $x_n \in B_{1/n}(x_0)$ such that $f(x_n) \notin B_\epsilon(f(x_0))$. It follows that $\lim_{n\to\infty} x_n = x_0$, but $\lim_{n\to\infty} f(x_n) \neq f(x_0)$. This contradicts our assumption, and hence we conclude that f is continuous at x_0. Q.E.D.

The reader will likely recall that Euclidean geometry studies properties of figures that are left unchanged (or *invariant*) under rigid motions. In topology, we study properties that are preserved under what are called "topological transformations." In the definition below, we will introduce the concept of *homeomorphism*. We will see that these functions preserve the structure of a topological space. Much of the remainder of our work will consist of studying "topological properties," or properties that are left invariant under these homeomorphisms.

Definitions 3.3.10. Two topological spaces $(X, \mathscr{T})$ and $(Y, \mathscr{R})$ are said to be *homeomorphic* (or *topologically equivalent*) if there exists a function $h : X \to Y$ that is one-to-one and onto such that both h and h^{-1} are continuous. The function h is called a *homeomorphism*.

Example 3.3.11. Let $X = \{1, 2, 3\}$ and $Y = \{a, b, c\}$. Define a topology $\mathscr{T}$ on X and a topology $\mathscr{R}$ on Y by

$$\mathscr{T} = \{\emptyset, X, \{1\}, \{2\}, \{1, 2\}\} \quad \text{and} \quad \mathscr{R} = \{\emptyset, Y, \{a\}, \{c\}, \{a, c\}\}.$$

Define a function $h : X \to Y$ by means of the following table:

x	1	2	3
$h(x)$	b	a	c

(a) Is h a homeomorphism?

(b) Are $(X, \mathscr{T})$ and $(Y, \mathscr{R})$ homeomorphic?

Solution. (a) It is easy to see that h is one-to-one and onto. However, the set $\{c\}$ is in $\mathscr{R}$, but $h^{-1}(\{c\}) = \{3\}$ and $\{3\} \notin \mathscr{T}$. Thus, h is not $\mathscr{T}$-to-$\mathscr{R}$ continuous, and hence is not a homeomorphism.

(b) Although h is not a homeomorphism, the two spaces $(X, \mathscr{T})$ and $(Y, \mathscr{R})$ are homeomorphic. To see this, define $k : X \to Y$ by the table:

x	1	2	3
$k(x)$	a	c	b

It is easy to see that k is both one-to-one and onto, and it is routine to check that both k and k^{-1} are continuous.

Observe also that, in addition to being a one-to-one function from X to Y, the function k induces a one-to-one correspondence between the members of $\mathscr{T}$ and $\mathscr{R}$.

Informally, the idea behind Definitions 3.3.10 is that continuous functions in some way preserve open sets. Thus, a homeomorphism between two spaces $(X, \mathscr{T})$ and $(Y, \mathscr{R})$ ensures that the topologies on X and Y are, in some sense, "the same." We are interested, therefore, in studying properties of a space that depend only on the topology of that space. Since homeomorphisms preserve the topology, these properties should be left unchanged (or invariant) under the action of a homeomorphism. This motivates the following definition.

Definition 3.3.12. A property (P) is said to be a *topological property* if whenever a topological space has property (P), then so does any space that is homeomorphic to it.

Much of what follows in this text is the study of various topological properties. We close this section with a brief discussion of open and closed functions.

Definitions 3.3.13. Let $(X, \mathscr{T})$ and $(Y, \mathscr{R})$ be two topological spaces. A function $f : X \to Y$ is said to be an *open function* if $f(U) \in \mathscr{R}$ whenever $U \in \mathscr{T}$. (That is, if the image of an open set is an open set.) Similarly, the function f is said to be a *closed function* if $f(A)$ is $\mathscr{R}$-closed whenever A is $\mathscr{T}$-closed. (That is, if the image of a closed set is a closed set.)

Example 3.3.14. Let $X = \{1, 2, 3\}$ and $Y = \{a, b, c\}$. Define a topology $\mathscr{T}$ on X and a topology $\mathscr{R}$ on Y by

$$\mathscr{T} = \{\emptyset, X, \{1\}, \{2\}, \{1, 2\}\} \quad \text{and} \quad \mathscr{R} = \{\emptyset, Y, \{a\}, \{a, b\}\}.$$

Define a function $f : X \to Y$ by means of the following table:

x	1	2	3
$f(x)$	a	a	b

(a) Is f a continuous function?

(b) Is f an open function?

(c) Is f a closed function?

Solution. (a) In order to show that f is continuous, we will show that $f^{-1}(U) \in \mathscr{T}$ for every $U \in \mathscr{R}$. Computing:

$$f^{-1}(\emptyset) = \emptyset, \quad f^{-1}(Y) = X, \quad f^{-1}(\{a\}) = \{1, 2\}, \quad f^{-1}(\{a, b\}) = X.$$

Since each of these sets is open in the topology on X, we conclude that f is a continuous function.

(b) We now show that $f(V) \in \mathscr{R}$ for every $V \in \mathscr{T}$:

$$f(\emptyset) = \emptyset, \quad f(X) = \{a, b\}, \quad f(\{1\}) = \{a\}, \quad f(\{2\}) = \{a\}, \quad f(\{1, 2\}) = \{a\}.$$

We have shown that the image of each open set in X is an open set in Y, and thus f is an open function.

(c) The $\mathscr{T}$-closed sets in X are X, $\emptyset$, $\{2,3\}$, $\{1,3\}$, and $\{3\}$. The $\mathscr{R}$-closed sets in Y are Y, $\emptyset$, $\{b,c\}$, and $\{c\}$. Observe that $\{3\}$ is $\mathscr{T}$-closed in X, but $f(\{3\}) = \{b\}$ is not $\mathscr{R}$-closed in Y. Thus, f is not a closed function.

In the exercises, we will explore the relationship between continuous, open, and closed functions, as well as their relationship to homeomorphisms.

Exercises

1. Let $(X, \mathscr{T})$ and $(Y, \mathscr{R})$ be topological spaces. Suppose $y_0 \in Y$ and define a function $f : X \to Y$ by $f(x) = y_0$ for all $x \in X$. Prove that f is $\mathscr{T}$-to-$\mathscr{R}$ continuous. (That is, prove that any constant function is continuous.)

2. Let $(X, \mathscr{T})$ be a topological space and define a function $f : X \to X$ by the formula $f(x) = x$ for all $x \in X$. Show that f is $\mathscr{T}$-to-$\mathscr{T}$ continuous.

3. Let X be any set with at least two elements. Let $\mathscr{R}$ be the discrete topology on X and let $\mathscr{T}$ be the indiscrete topology on X. Show that the function $f : X \to X$ defined by $f(x) = x$ for all $x \in X$ is not $\mathscr{T}$-to-$\mathscr{R}$ continuous.

4. Let $\mathscr{T}$ be the standard topology on $\mathbb{R}$ and let $\mathscr{R}$ be the discrete topology on $\mathbb{R}$. Show that the function $f : \mathbb{R} \to \mathbb{R}$ defined by $f(x) = x$ for all $x \in \mathbb{R}$ is not $\mathscr{T}$-to-$\mathscr{R}$ continuous. Is f $\mathscr{R}$-to-$\mathscr{T}$ continuous? Justify your answer.

5. Give an example of a function that is closed but not continuous.

6. Give an example of a function that is open but not continuous.

7. Give an example of a function that is closed but not open.

8. Give an example of a function that is continuous but neither open nor closed.

9. Give an example of a function that is both open and closed but fails to be continuous.

10. Let $(X, \mathscr{T})$ and $(Y, \mathscr{R})$ be topological spaces and suppose $f : X \to Y$ is a one-to-one and onto function. Prove that the following statements are equivalent:

    (a) f^{-1} is continuous.

    (b) f is an open function.

    (c) f is a closed function.

11. Let $(X, \mathscr{T})$ and $(Y, \mathscr{R})$ be topological spaces and suppose $f : X \to Y$ is a one-to-one and onto function. Prove that the following statements are equivalent:

    (a) f is a homeomorphism.

    (b) f and f^{-1} are both open functions.

    (c) f and f^{-1} are both closed functions.

12. Consider the space $X = \{1, 2, 3\}$ with topology $\mathscr{T} = \{\emptyset, \{1\}, \{1, 2\}, X\}$. Suppose $f : X \to X$ is a $\mathscr{T}$-to-$\mathscr{T}$ continuous function. If $f(2) = 1$ and $f(3) = 2$, then what is $f(1)$?

13. Give an example of a function that is one-to-one, onto, and continuous, but is not a homeomorphism.

14. Let $(X, \mathscr{T})$ and $(Y, \mathscr{R})$ be topological spaces and suppose $f : X \to Y$ is a $\mathscr{T}$-to-$\mathscr{R}$ continuous function. Suppose there exists some $\mathscr{R}$-to-$\mathscr{T}$ continuous function $g : Y \to X$ such that $f(g(x)) = x$ for all $x \in Y$ and $g(f(x)) = x$ for all $x \in X$. Prove that f is a homeomorphism.

15. Let $(X, \mathscr{T})$ and $(Y, \mathscr{R})$ be homeomorphic topological spaces. Can there be more than one homeomorphism between the two spaces? If the spaces are finite, how many homeomorphisms can there be? If X and Y are infinite spaces, can there be infinitely many homeomorphisms?

16. Let $D = \{(x, y) \mid x^2 + y^2 \leq 1\}$ and $S = \{(x, y) \mid |x| + |y| \leq 1\}$ be sets in $\mathbb{R}^2$. (Note that D is a disk and S is a square in the Cartesian plane.) Construct a homeomorphism between D and S. Assume $\mathbb{R}^2$ has the standard topology and give D and S the subspace topology.

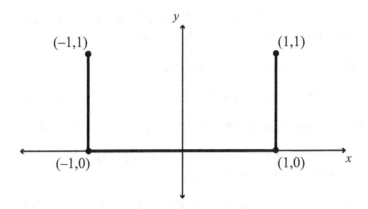

Figure 3.3.a: The above illustration represents the set U from Exercise 17. Construct a homeomorphism between this set and the line segment I.

17. Let I be the line segment $\{(x,0) \mid -1 \leq x \leq 1\}$ in $\mathbb{R}^2$. Construct a homeomorphism between I and the set

$$U = \{(-1, y) \mid 0 \leq y \leq 1\} \cup I \cup \{(1, y) \mid 0 \leq y \leq 1\}.$$

Assume $\mathbb{R}^2$ has the standard topology and each subset has the subspace topology. (See Figure 3.3.a.)

18. Let $(X, \mathscr{T}_d)$ be a metric space and suppose $\{x_n\}_{n=1}^{\infty}$ and $\{y_n\}_{n=1}^{\infty}$ are two sequences with terms in X having the property that $\lim\limits_{n \to \infty} d(x_n, y_n) = 0$. Prove that $\{x_n\}_{n=1}^{\infty}$ converges to $x_0 \in X$ if and only if $\{y_n\}_{n=1}^{\infty}$ converges to x_0.

19. Let $\mathbb{R}$ be the set of real numbers equipped with the standard topology. Suppose $f : \mathbb{R} \to \mathbb{R}$ is a continuous function and let $\{x_n\}_{n=1}^{\infty}$ be a sequence of real numbers that converges to $x_0 \in \mathbb{R}$. If $f(x_n) \leq x_n$ for each $n \in \mathbb{N}$, prove that $f(x_0) \leq x_0$.

3.4 More on Continuous Functions

In the previous section, we defined the notion of a continuous function between topological spaces. In some situations, however, it is not easy to verify the continuity of a function using the definition. In this section, we introduce some properties that are equivalent to continuity. That is, we will provide statements which are equivalent to the statement "The function f is continuous." These equivalent formulations of continuity will be useful in some of our later work. Before going on to the next theorem, the reader might wish to review the definitions of Sections 2.3 and 2.5.

Theorem 3.4.1. *Let $(X, \mathscr{T})$ and $(Y, \mathscr{R})$ be topological spaces and let $f : X \to Y$ be a function. The following statements are equivalent:*

(1) The function f is $\mathscr{T}$-to-$\mathscr{R}$ continuous.

(2) If A is an $\mathscr{R}$-closed subset of Y, then $f^{-1}(A)$ is a $\mathscr{T}$-closed subset of X.

(3) If $\mathscr{S}$ is a subbase for $\mathscr{R}$, then $f^{-1}(V) \in \mathscr{T}$ whenever $V \in \mathscr{S}$.

(4) For each $x \in X$, if $N_{f(x)}$ is a neighborhood of $f(x)$ in Y, then $f^{-1}\left(N_{f(x)}\right)$ is a neighborhood of x in X.

(5) For each $x \in X$, if $N_{f(x)}$ is a neighborhood of $f(x)$ in Y, then there is a neighborhood M_x of x in X such that $f(M_x) \subseteq N_{f(x)}$.

(6) If $A \subseteq X$, then $f\left(\overline{A}\right) \subseteq \overline{f(A)}$.

(7) If $B \subseteq Y$, then $\overline{f^{-1}(B)} \subseteq f^{-1}\left(\overline{B}\right)$.

Proof. Suppose statement (1) is true and let A be an $\mathscr{R}$-closed subset of Y. Then $A' \in \mathscr{R}$, and so $f^{-1}(A') \in \mathscr{T}$. But $f^{-1}(A') = [f^{-1}(A)]'$. Thus, $[f^{-1}(A)]' \in \mathscr{T}$, and so $f^{-1}(A)$ is $\mathscr{T}$-closed. This shows that (1) implies (2).

Now suppose that statement (2) is true and let $\mathscr{S}$ be a subbase for $\mathscr{R}$. Let $V \in \mathscr{S}$. Since $\mathscr{S}$ is a subbase for $\mathscr{R}$, it follows that $\mathscr{S} \subseteq \mathscr{R}$. Thus, $V \in \mathscr{R}$ and so V' is $\mathscr{R}$-closed. Hence, by assumption, $f^{-1}(V')$ is $\mathscr{T}$-closed. But $f^{-1}(V') = [f^{-1}(V)]'$. Consequently, $[f^{-1}(V)]'$ is $\mathscr{T}$-closed, and so $f^{-1}(V)$ is $\mathscr{T}$-open. We have shown (2) implies (3).

Assume that statement (3) is true. Let $x \in X$ and suppose N is a neighborhood of $f(x)$ in Y. Then $f(x) \in N$ and $N \in \mathscr{R}$. Let $\mathscr{S}$ be a subbase for $\mathscr{R}$ and let $\mathscr{B}$ be the base generated by $\mathscr{S}$. Since $N \in \mathscr{R}$, there is a subcollection Ω of $\mathscr{B}$ such that $N = \bigcup_{U \in \Omega} U$. Since $\mathscr{S}$ is a subbase, each U in Ω is the intersection of finitely many members of $\mathscr{S}$. That is, for $U \in \Omega$, there exists a finite collection of sets $\{W_1, W_2, \ldots, W_n\} \subset \mathscr{S}$ (which depends on U) such that

$$U = W_1 \cap W_2 \cap \cdots \cap W_n.$$

Observe that

$$f^{-1}(U) = f^{-1}(W_1 \cap \cdots \cap W_n) = f^{-1}(W_1) \cap \cdots \cap f^{-1}(W_n).$$

Because we have assumed that statement (3) is true, the set $f^{-1}(W_k)$ is $\mathscr{T}$-open for each $k \in \{1, \ldots, n\}$. Thus, $f^{-1}(U)$ is open for each U in Ω. But $N = \bigcup_{U \in \Omega} U$, and hence

$$f^{-1}(N) = f^{-1}\left(\bigcup_{U \in \Omega} U\right) = \bigcup_{U \in \Omega} f^{-1}(U).$$

Therefore, $f^{-1}(N)$ is a $\mathscr{T}$-open set. Since $x \in f^{-1}(N)$, we have shown that (3) implies (4).

Suppose now that (4) is true. Let $x \in X$ and suppose N is a neighborhood of $f(x)$. Then $f^{-1}(N)$ is a neighborhood of x. Let $M = f^{-1}(N)$. It follows that $f(M) = f(f^{-1}(N)) \subseteq N$, and so we have shown that (4) implies (5).

Now assume (5) is true and let $A \subseteq X$. We want to show that $f(\overline{A}) \subseteq \overline{f(A)}$. Suppose that this is not the case. Then there is a $y \in f(\overline{A})$ such that $y \notin \overline{f(A)}$. Choose $x \in \overline{A}$ such that $y = f(x)$. Since $y \notin \overline{f(A)}$, it must be that $x \notin A$. Thus, x must be a boundary point of A. Let $N = \left[\overline{f(A)}\right]'$. Then N is a neighborhood of y. We assumed that (5) was true, and so there must be a neighborhood M of x such that $f(M) \subseteq N$. We now have both that M is an open set containing x and that x is a boundary point of A. Therefore, there is some $z \in M \cap A$. It follows that $f(z) \in f(A) \subseteq \overline{f(A)}$. On the other hand, $f(z) \in f(M) \subseteq N$. But this is a contradiction, because $N = \left[\overline{f(A)}\right]'$ and it is not possible for $f(z)$ to be a member of both N and $\overline{f(A)}$. We arrived at a contradiction by assuming that (6) was false, and so we have shown that (5) implies (6).

Suppose that statement (6) is true. Let $B \subseteq Y$ and let $A = f^{-1}(B)$. By assumption, we have that $f(\overline{A}) \subseteq \overline{f(A)}$. That is,

$$f\left(\overline{f^{-1}(B)}\right) \subseteq \overline{f\left(f^{-1}(B)\right)} \subseteq \overline{B}.$$

Consequently,

$$f^{-1}\left[f\left(\overline{f^{-1}(B)}\right)\right] \subseteq f^{-1}(\overline{B}).$$

We know that $K \subseteq f^{-1}(f(K))$ whenever $K \subseteq X$, and so it follows that

$$\overline{f^{-1}(B)} \subseteq f^{-1}(\overline{B}).$$

This shows that (6) implies (7).

Assume now that statement (7) is true. We wish to show that f is continuous. Let V be an open set in $\mathscr{R}$ and let $A = V'$. Then $A = \overline{A}$. By (7), we know that

$$\overline{f^{-1}(A)} \subseteq f^{-1}(\overline{A}) = f^{-1}(A).$$

We always have $f^{-1}(A) \subseteq \overline{f^{-1}(A)}$, and so we conclude that $f^{-1}(A) = \overline{f^{-1}(A)}$. Thus, $f^{-1}(A)$ is closed. Because $[f^{-1}(V)]' = f^{-1}(V') = f^{-1}(A)$, we have that $f^{-1}(V)$ is open in $\mathscr{T}$, and therefore f is a continuous function.

We have shown

$$(1) \implies (2) \implies (3) \implies (4) \implies (5) \implies (6) \implies (7) \implies (1).$$

Q.E.D.

It is often convenient to think of a function $f : X \to Y$ as a surjection onto its image. It is important, therefore, to determine if a continuous function remains continuous when thought of in this way. The content of the next theorem settles the issue.

Theorem 3.4.2. *A continuous function is a continuous surjection onto its image.*

Proof. Let $(X, \mathscr{T})$ and $(Y, \mathscr{R})$ be topological spaces and suppose there is a continuous function $f : X \to Y$. Let $Z = f(X)$ and let $\mathscr{S}$ be the subspace topology induced on Z by $\mathscr{R}$. We are assuming that f is $\mathscr{T}$-to-$\mathscr{R}$ continuous. What we wish to show is that $f : X \to Z$ is $\mathscr{T}$-to-$\mathscr{S}$ continuous.

To that end, let $U \in \mathscr{S}$. By definition, there is a set $V \in \mathscr{R}$ such that $U = V \cap Z$. Thus,

$$f^{-1}(U) = f^{-1}(V \cap Z) = f^{-1}(V) \cap f^{-1}(Z) = f^{-1}(V) \cap X = f^{-1}(V).$$

By the continuity of f, we know $f^{-1}(V) \in \mathscr{T}$. Thus, $f^{-1}(U) \in \mathscr{T}$ and we have shown that f is $\mathscr{T}$-to-$\mathscr{S}$ continuous. Q.E.D.

Definition 3.1.5 introduced the notion of the restriction of a function. It is natural, in light of the previous theorem, to ask if the restriction of a continuous function to a subset remains continuous. The next theorem answers this question.

Theorem 3.4.3. *The restriction of a continuous function is continuous.*

Proof. Let $(X, \mathscr{T})$ and $(Y, \mathscr{R})$ be topological spaces and suppose that $f : X \to Y$ is a continuous function. Let $Z \subseteq X$ and let $\mathscr{S}$ be the topology induced on Z by $\mathscr{T}$. We assumed that f is $\mathscr{T}$-to-$\mathscr{R}$ continuous. Our goal is to show that $f|_Z : Z \to Y$ is $\mathscr{S}$-to-$\mathscr{R}$ continuous.

Let $U \in \mathscr{R}$. We will show that $(f|_Z)^{-1}(U) \in \mathscr{S}$. Observe the following equivalences:

$$
\begin{aligned}
x \in (f|_Z)^{-1}(U) &\iff [x \in Z \wedge f(x) \in U] \\
&\iff [x \in Z \wedge x \in f^{-1}(U)] \\
&\iff x \in f^{-1}(U) \cap Z.
\end{aligned}
$$

We conclude that $(f|_Z)^{-1}(U) = f^{-1}(U) \cap Z$. By the continuity of f on X, we know that $f^{-1}(U) \in \mathscr{T}$, and hence $f^{-1}(U) \cap Z$ is open in the induced topology $\mathscr{S}$ on Z. Therefore, $(f|_Z)^{-1}(U) \in \mathscr{S}$ and so $f|_Z$ is continuous. Q.E.D.

Definition 3.4.4. Let X and Y be sets and suppose $f : X \to Y$ is a function. If $Z \subseteq X$ and $g : Z \to Y$ is a function with the property that $g = f|_Z$, then f is called an *extension* of g over X relative to Y.

It is easy to see that extensions are not unique, as we demonstrate in the following example.

Example 3.4.5. Let $\mathbb{R}$ be the set of real numbers and let $Z = [-\frac{\pi}{2}, \frac{\pi}{2}]$. The restriction of the sine function to the interval Z is denoted by Sin. That is, the function $\text{Sin} : Z \to \mathbb{R}$ is defined by

$$\text{Sin}(x) = \sin|_Z(x),$$

for all $x \in Z$. (We still read Sin as "sine.") Define three different extensions of Sin over $\mathbb{R}$.

Solution. (a) Let $f(x) = \sin x$ for all $x \in \mathbb{R}$. Then f is an extension of Sin over $\mathbb{R}$. Note that f is continuous on $\mathbb{R}$.

(b) Define $g : \mathbb{R} \to \mathbb{R}$ as follows:

$$g(x) = \begin{cases} -1 & \text{if } x < -\frac{\pi}{2}, \\ \sin x & \text{if } x \in [-\frac{\pi}{2}, \frac{\pi}{2}], \\ 1 & \text{if } x > \frac{\pi}{2}. \end{cases}$$

Note that g is a continuous extension of Sin to $\mathbb{R}$. The reader is encouraged to draw the graph of g on $\mathbb{R}$.

(c) Now define $h : \mathbb{R} \to \mathbb{R}$ by:

$$h(x) = \begin{cases} -x & \text{if } x < -\frac{\pi}{2}, \\ \sin x & \text{if } x \in [-\frac{\pi}{2}, \frac{\pi}{2}], \\ x & \text{if } x > \frac{\pi}{2}. \end{cases}$$

The function h is an extension of Sin to $\mathbb{R}$. In this case, however, the extension is not continuous at the points $x = -\frac{\pi}{2}$ and $x = \frac{\pi}{2}$.

Example 3.4.6. Let $\mathbb{R}$ be the set of real numbers and let $Z = \mathbb{R} \backslash \{0\}$. Define $f : Z \to \mathbb{R}$ by $f(x) = \frac{1}{x}$. Define an extension of f over $\mathbb{R}$. Is it possible for this extension to be continuous?

Solution. Let $g : \mathbb{R} \to \mathbb{R}$ be defined by

$$g(x) = \begin{cases} 1/x & \text{if } x \neq 0, \\ 0 & \text{if } x = 0. \end{cases}$$

Certainly, g is an extension of f over $\mathbb{R}$, but it is not continuous. Observe that f itself is continuous over Z. It is not possible to define a continuous extension of f over $\mathbb{R}$, however, because if we let $h(x) = f(x)$ for all $x \neq 0$, then $\lim_{x \to 0} h(x)$ does not exist, regardless of what $h(0)$ is defined to be.

Question 3.4.7. Suppose that $(X, \mathscr{T})$ and $(Y, \mathscr{R})$ are topological spaces and that $(W, \mathscr{S})$ is a subspace of $(X, \mathscr{T})$. If $f : W \to Y$ is $\mathscr{S}$-to-$\mathscr{R}$ continuous, under what conditions does there exist a continuous extension of f over X?

Question 3.4.7 and related questions are answered in more advanced texts devoted to topology.

Exercises

1. Let $X = \{1,2,3\}$ and $Y = \{a,b,c\}$. Suppose that $\mathcal{T} = \{\emptyset, X, \{1,2\}, \{3\}\}$ and $\mathcal{R} = \{\emptyset, Y, \{a\}, \{b\}, \{a,b\}\}$.

   (a) Verify that $\mathcal{T}$ is a topology for X and $\mathcal{R}$ is a topology for Y.

   (b) Let f be defined by the table

x	1	2	3
$f(x)$	b	b	a

   Verify that f is $\mathcal{T}$-to-$\mathcal{R}$ continuous.

   (c) For each subset A of X, verify that $f(\overline{A}) \subseteq \overline{f(A)}$, where f is the function from (b). (*Note:* There are eight subsets of X.)

   (d) For each subset B of Y, verify that $\overline{f^{-1}(B)} \subseteq f^{-1}(\overline{B})$, where f is the function from (b). (*Note:* There are eight subsets of Y.)

2. Repeat Exercise 1, but with the function $f : X \to Y$ defined by the table

x	1	2	3
$f(x)$	a	a	c

3. Let $(X, \mathcal{T})$ and $(Y, \mathcal{R})$ be the topological spaces from Exercise 1 and let the function $f : X \to Y$ be defined by the table

x	1	2	3
$f(x)$	a	c	b

   Show that f is not $\mathcal{T}$-to-$\mathcal{R}$ continuous. Find at least one subset A of X for which $f(\overline{A}) \subseteq \overline{f(A)}$ is false and at least one subset B of Y for which $\overline{f^{-1}(B)} \subseteq f^{-1}(\overline{B})$ is false.

4. Explain how statement (5) of Theorem 3.4.1 is a generalization of the definition of continuity in Elementary Calculus.

5. Let $(X, \mathcal{T})$ and $(Y, \mathcal{R})$ be topological spaces. Show that $f : X \to Y$ is $\mathcal{T}$-to-$\mathcal{R}$ continuous if and only if $f^{-1}(B^{\circ}) \subseteq [f^{-1}(B)]^{\circ}$ for each $B \subseteq Y$. (Recall that B° denotes the interior of B. See Definition 2.5.1.)

6. Let $(X, \mathcal{T})$ be a topological space and let $(\mathbb{R}, \mathcal{R})$ be the set of real numbers equipped with the standard topology (generated by the collection of all open intervals). A function $f : X \to \mathbb{R}$ is said to be *lower semicontinuous*

provided that $f^{-1}[(a, \infty)] \in \mathscr{T}$ for each $a \in \mathbb{R}$. Similarly, f is said to be *upper semicontinuous* provided that $f^{-1}[(-\infty, b)] \in \mathscr{T}$ for each $b \in \mathbb{R}$.

(a) Prove that $f : X \to \mathbb{R}$ is $\mathscr{T}$-to-$\mathscr{R}$ continuous if and only if it is both lower semicontinuous and upper semicontinuous.

(b) Give an example of a lower semicontinuous function that is not continuous.

(c) Give an example of an upper semicontinuous function that is not continuous.

7. Using the definitions of upper and lower semicontinuity given in Exercise 6, prove:

(a) The function $f : X \to \mathbb{R}$ is lower semicontinuous if and only if $-f$ is upper semicontinuous.

(b) The function $f : X \to \mathbb{R}$ is lower semicontinuous if and only if for each $x \in X$ and each $a < f(x)$ there is a neighborhood N of x such that $a < f(t)$ whenever $t \in N$.

8. State and prove the analog of 7(b) for an upper semicontinuous function.

9. Let $(X, \mathscr{T})$ be a topological space and suppose $A \subseteq X$. Recall that the characteristic function of A is the function $\chi_A : X \to \mathbb{R}$ defined by

$$\chi_A(x) = \begin{cases} 1 & \text{if } x \in A, \\ 0 & \text{if } x \in X \setminus A. \end{cases}$$

(a) Prove that χ_A is lower semicontinuous if and only if A is an open set.

(b) Prove that χ_A is upper semicontinuous if and only if A is a closed set.

10. Let $\mathbb{R}$ have the standard topology and let $A = \mathbb{R} \setminus \{2\}$. Define $f : A \to \mathbb{R}$ by $f(x) = \dfrac{x^2 - 4}{x - 2}$ for each $x \in A$.

(a) Define a continuous extension of f over $\mathbb{R}$.

(b) Define a noncontinuous extension of f over $\mathbb{R}$.

(c) Sketch the graph of the extensions in (a) and (b).

11. Let $\mathbb{R}$ have the standard topology and let $B = \mathbb{R} \setminus \{3\}$. Define $g : B \to \mathbb{R}$ by $g(x) = \dfrac{x^3 - 27}{x - 3}$ for each $x \in B$.

(a) Define a continuous extension of g over $\mathbb{R}$.

(b) Define a noncontinuous extension of g over $\mathbb{R}$.

(c) Sketch the graph of the extensions in (a) and (b).

12. Let $(X, \mathscr{T})$ be a topological space and let $(\mathbb{R}, \mathscr{R})$ be the set of real numbers equipped with the standard topology (generated by the collection of all open intervals). Suppose $f : X \to \mathbb{R}$ and $g : X \to \mathbb{R}$ are $\mathscr{T}$-to-$\mathscr{R}$ continuous functions and let $A = \{x \mid g(x) = 0\}$. Prove the following:

   (a) If $S(x) = f(x) + g(x)$ for all $x \in X$, then S is continuous.

   (b) If $D(x) = f(x) - g(x)$ for all $x \in X$, then D is continuous.

   (c) If $P(x) = f(x) g(x)$ for all $x \in X$, then P is continuous.

   (d) If $Q(x) = f(x)/g(x)$ for all $x \in X \backslash A$, then Q is continuous (on $X \backslash A$).

   (e) If $A(x) = |f(x)|$ for all $x \in X$, then A is continuous.

   (f) If $M(x) = \max\{f(x), g(x)\}$ for all $x \in X$, then M is continuous.

   (g) If $m(x) = \min\{f(x), g(x)\}$ for all $x \in X$, then m is continuous.

13. Let $\mathbb{R}$ be the set of real numbers with the standard topology. Suppose $f : \mathbb{R} \to \mathbb{R}$ is continuous and define $g : \mathbb{R} \to \mathbb{R}^2$ by $g(x) = (x, f(x))$ for all $x \in \mathbb{R}$. Prove that g is continuous, where $\mathbb{R}^2$ is given the standard topology (generated by the Euclidean metric).

14. The *graph* of a function f is the set $\{(x, f(x)) : x$ is in the domain of $f\}$. Use Exercise 13 (above) to show that the graph of a continuous function from $\mathbb{R}$ to $\mathbb{R}$ is a closed subset of $\mathbb{R}^2$ (in the standard topology). (*Hint:* See Exercise 11 in Section 2.5.)

15. Let X be a nonempty set, let $(Y, \mathscr{R})$ be a topological space, and suppose $f : X \to Y$ is a function.

   (a) Let $\mathscr{T} = \{f^{-1}(U) \mid U \in \mathscr{R}\}$. Prove that $\mathscr{T}$ is a topology for X.

   (b) Prove that f is $\mathscr{T}$-to-$\mathscr{R}$ continuous.

   (c) Prove that if $\mathscr{S}$ is any topology for X that makes f into a $\mathscr{S}$-to-$\mathscr{R}$ continuous function, then $\mathscr{T} \subseteq \mathscr{S}$.

16. Let $(X, \mathscr{T})$, $(Y, \mathscr{R})$, and $(Z, \mathscr{S})$ be three topological spaces. Suppose also that $f : X \to Y$ and $g : Y \to Z$ are two functions. Prove the following:

   (a) If f and g are open, then so is $g \circ f$.

   (b) If f and g are closed, then so is $g \circ f$.

   (c) If $g \circ f$ is open and f is continuous and onto, then g is open.

   (d) If $g \circ f$ is closed and f is continuous and onto, then g is closed.

   (e) If $g \circ f$ is open and g is continuous and one-to-one, then f is open.

   (f) If $g \circ f$ is closed and g is continuous and one-to-one, then f is closed.

17. Let $X = \{a, b, c, d, e\}$, $Y = \{w, x, y, z\}$, and $Z = \{1, 2, 3\}$. Define a topology $\mathscr{T}$ on Y and a topology $\mathscr{R}$ on Z by

$$\mathscr{T} = \{\emptyset, Y, \{w, x\}, \{y, z\}\} \quad \text{and} \quad \mathscr{R} = \{\emptyset, Z, \{1, 2\}, \{2, 3\}, \{2\}\}.$$

Define functions $f : X \to Y$ and $g : X \to Z$ by

$$f(a) = f(c) = y, \quad f(b) = z, \quad f(d) = w, \quad f(e) = x,$$

and

$$g(a) = g(d) = 2, \quad g(b) = g(e) = 3, \quad g(c) = 1.$$

Construct the smallest topology on X that will make both f and g continuous.

18. Prove the *Pasting Lemma* (see Figures 3.4.a and 3.4.b): Let $(X, \mathscr{T})$ and $(Y, \mathscr{R})$ be topological spaces and let A and B be open subsets of X such that $X = A \cup B$. Suppose $g : A \to Y$ and $h : B \to Y$ are continuous functions such that $g(x) = h(x)$ for each $x \in A \cap B$. If $f : X \to Y$ is defined so that $f(a) = g(a)$ if $a \in A$ and $f(b) = h(b)$ if $b \in B$, then f is a continuous function from X to Y. (This is also true if A and B are assumed to be closed subsets of X. See Exercise 20.)

19. Let $X = \{a, b, c, d, e\}$ and $Y = \{1, 2, 3\}$ be sets. Define a topology on X by $\mathscr{T} = \{\emptyset, X, \{a, b, c, d\}, \{a, c, d\}, \{a, b\}, \{a\}, \{b\}\}$ and a topology on Y by $\mathscr{R} = \{\emptyset, Y, \{1, 2\}, \{3\}\}$. Let $A = \{a, b, c, d\}$ be given the subspace topology and define $g : A \to Y$ by

$$g(a) = 2, \quad g(b) = 3, \quad g(c) = 2, \quad g(d) = 2.$$

Let $B = \{c, d, e\}$ be given the subspace topology and define $h : B \to Y$ by

$$h(c) = h(d) = h(e) = 2.$$

Define a function f on $X = A \cup B$ as in Exercise 18.

   (a) Verify that $(X, \mathscr{T})$ and $(Y, \mathscr{R})$ are topological spaces.

   (b) Verify that g and h are continuous functions.

   (c) Verify that g and h agree on $A \cap B$.

   (d) Show that f is not continuous.

   (e) Explain why (d) does not contradict the Pasting Lemma.

20. The Pasting Lemma (as stated in Exercise 18) remains true if the sets A and B are assumed to be closed subsets of X (instead of open, as in Exercise 18). Prove the Pasting Lemma under the assumption that the sets A and B are closed in X.

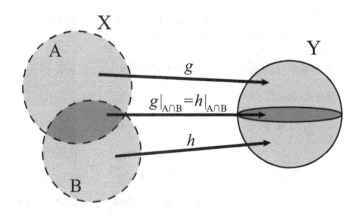

Figure 3.4.a: The Pasting Lemma (Part 1): In this illustration, A and B are open sets and g and h are both continuous functions that agree on $A \cap B$. If f is defined as in Exercise 18, then f is continuous on $A \cup B$.

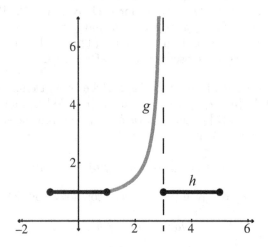

Figure 3.4.b: The Pasting Lemma (Part 2): In this illustration, let $A = (-1, 3)$ and let $B = [-1, 1] \cup [3, 5]$. Define g so that $g(x) = 1$ for $x \in (-1, 1)$ and $g(x) = \frac{1}{2} + \frac{1}{3-x}$ for $x \in [1, 3)$. Define h so that $h(x) = 1$ for all $x \in B$. Then g and h are continuous, but the function f defined in Exercise 18 is not. The Pasting Lemma does not apply because A is open and B is closed.

3.5 More on Homeomorphism

Recall that two topological spaces $(X, \mathscr{T})$ and $(Y, \mathscr{R})$ are said to be *homeomorphic* if there exists a function $h : X \to Y$ that is one-to-one and onto such that both h and h^{-1} are continuous. In such a case, the function h is called a *homeomorphism*.

We first defined homeomorphism in Section 3.3. Also in that section, we introduced the notion of a "topological property." We repeat the definition here, for completeness.

Definition 3.5.1. A property (P) is said to be a *topological property* if whenever a space has property (P), then so does any space that is homeomorphic to it.

We are now able describe the nature of the branch of mathematics known as topology: *Topology is the study of topological properties.*

As an example, we recall the notion of a Hausdorff space, which was introduced in Section 2.6. When a topological space is a Hausdorff space, we say it possesses the *Hausdorff property*. As we now show, the Hausdorff property is a topological property.

Theorem 3.5.2. *The Hausdorff property is a topological property.*

Proof. Suppose that $(X, \mathscr{T})$ is a Hausdorff space and suppose that $(Y, \mathscr{R})$ is a topological space that is homeomorphic to $(X, \mathscr{T})$. Our objective is to show that $(Y, \mathscr{R})$ is also a Hausdorff space.

By assumption, there is a one-to-one and onto function $h : X \to Y$ such that both h and h^{-1} are continuous. Let y_1 and y_2 be distinct elements in Y. We wish to find disjoint neighborhoods of these two points.

Because h is one-to-one, we know that $h^{-1}(y_1)$ and $h^{-1}(y_2)$ are distinct elements of X. Since $(X, \mathscr{T})$ is a Hausdorff space, there are sets U_1 and U_2 in $\mathscr{T}$ such that $h^{-1}(y_1) \in U_1$, $h^{-1}(y_2) \in U_2$, and $U_1 \cap U_2 = \emptyset$. Let $V_1 = h(U_1)$ and $V_2 = h(U_2)$. We have already seen that h is an open function. (See Exercise 11 in Section 3.3.) Thus, both V_1 and V_2 are $\mathscr{R}$-open sets. It is certainly the case that $y_1 \in V_1$ and $y_2 \in V_2$. It remains to show that $V_1 \cap V_2 = \emptyset$. Suppose to the contrary that $V_1 \cap V_2 \neq \emptyset$. Then

$$V_1 \cap V_2 \neq \emptyset \implies \text{There is some } y \in V_1 \cap V_2$$
$$\implies y \in V_1 \text{ and } y \in V_2$$
$$\implies h^{-1}(y) \in h^{-1}(V_1) \text{ and } h^{-1}(y) \in h^{-1}(V_2)$$
$$\implies h^{-1}(y) \in U_1 \cap U_2$$
$$\implies U_1 \cap U_2 \neq \emptyset.$$

This is a contradiction, and so it must be that $V_1 \cap V_2 = \emptyset$. Therefore, the topological space $(Y, \mathscr{R})$ is a Hausdorff space. Q.E.D.

For illustration, we introduce another topological property that we will study in greater detail later.

Definitions 3.5.3. A topological space $(X, \mathscr{T})$ is said to have the *fixed point property* (abbreviated f.p.p.) provided that if $f : X \to X$ is a continuous function, then there exists a point $x_0 \in X$ such that $f(x_0) = x_0$. In such a case, we call x_0 a *fixed point* of the function f.

The fixed point property is illustrated for $X = [0, 1]$ (with the standard subspace topology) in Figure 3.5.a.

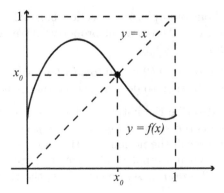

Figure 3.5.a: If X is the closed interval $[0,1]$ and $\mathscr{T}$ is the subspace topology induced by the standard topology on the set of real numbers, then $(X, \mathscr{T})$ has the fixed point property.

Theorem 3.5.4. *The fixed point property is a topological property.*

Proof. Suppose that $(X, \mathscr{T})$ and $(Y, \mathscr{R})$ are homeomorphic topological spaces and that $(X, \mathscr{T})$ has the f.p.p. By assumption, there is a one-to-one and onto function $h : X \to Y$ such that both h and h^{-1} are continuous.

Let $f : Y \to Y$ be any continuous function. Define a function $g : X \to X$ by the rule $g = h^{-1} \circ f \circ h$. Then g is continuous because it is the composition of continuous functions. Since $(X, \mathscr{T})$ has the f.p.p., there is some $x_0 \in X$ such that $g(x_0) = x_0$. Let $y_0 = h(x_0)$. Observe now that

$$
\begin{aligned}
g(x_0) = x_0 &\implies h^{-1} \circ f \circ h(x_0) = x_0 \\
&\implies f \circ h(x_0) = h(x_0) \\
&\implies f\big(h(x_0)\big) = h(x_0) \\
&\implies f(y_0) = y_0.
\end{aligned}
$$

Therefore, y_0 is a fixed point of f. Since every continuous function $f : Y \to Y$ has a fixed point, it follows that $(Y, \mathscr{R})$ has the f.p.p.   Q.E.D.

For the purposes of illustration, we introduce some other topological properties that are discussed in greater detail in more advanced texts.

Definition 3.5.5. Let $(X, \mathscr{T})$ be a topological space. A subset A of X is said to be *dense* in X if $\overline{A} = X$.

Definition 3.5.6. A topological space is called *separable* provided that it has a countable dense subset.

Definition 3.5.7. A topological space is said to be *second countable* if there is a countable base that generates the topology.

Theorem 3.5.8. *A second countable topological space is separable.*

Proof. Let $(X, \mathcal{T})$ be a second countable topological space. By assumption, there exists a countable base $\mathcal{B}$ that generates the topology $\mathcal{T}$. Choose a point z in each nonempty member of $\mathcal{B}$ and let A be the set of all points so chosen. Since $\mathcal{B}$ is countable, the set A is also countable. We claim that $\overline{A} = X$. Certainly, $\overline{A} \subseteq X$. Let $x \in X$ and let U be any $\mathcal{T}$-neighborhood of x. We wish to show U intersects A nontrivially. Because $\mathcal{B}$ generates $\mathcal{T}$, and because $U \in \mathcal{T}$, there is a $V \in \mathcal{B}$ such that $x \in V$ and $V \subseteq U$. Since V is a nonempty member of $\mathcal{B}$, there is a point $z \in V$ that was chosen to be a representative of V in A. Then $z \in V \cap A \subseteq U \cap A$, and so $U \cap A$ is nonempty. It follows that any open set containing x contains one or more points from A, which means that $x \in \overline{A}$. Therefore $X \subseteq \overline{A}$, and so $X = \overline{A}$.   Q.E.D.

Theorem 3.5.9. *The image of a separable topological space under a continuous function is separable.*

Proof. Let $(X, \mathcal{T})$ be a separable topological space. Suppose that $(Y, \mathcal{R})$ is a topological space and that $f : X \to Y$ is a continuous function. Without loss of generality, assume that f is an onto function. We wish to show that $(Y, \mathcal{R})$ is separable.

Since $(X, \mathcal{T})$ is separable, there is a countable set A in X such that $\overline{A} = X$. Let $B = f(A)$. The set B is countable because A is countable. By the continuity of f, we know that $f(\overline{A}) \subseteq \overline{f(A)}$. (See Theorem 3.4.1.) Thus,

$$Y = f(X) = f(\overline{A}) \subseteq \overline{f(A)} = \overline{B}.$$

Thus, $Y \subseteq \overline{B}$. Since it is also true that $\overline{B} \subseteq Y$, we have shown that $\overline{B} = Y$. Therefore, B is a countable dense subset of Y, and so $(Y, \mathcal{R})$ is a separable topological space.   Q.E.D.

Corollary 3.5.10. *Separability is a topological property.*

Proof. Let $(X, \mathcal{T})$ and $(Y, \mathcal{R})$ be homeomorphic topological spaces and suppose that $(X, \mathcal{T})$ is separable. By assumption, there is a function $h : X \to Y$ that is both continuous and onto. Thus, by Theorem 3.5.9, $(Y, \mathcal{R})$ is separable. Q.E.D.

Exercises

1. A topological space $(X, \mathcal{T})$ is said to be a T_1-*space* if for each pair of distinct points x and y in X, there are open sets U and V (not necessarily disjoint) such that $x \in U$ but $y \notin U$ and $y \in V$ but $x \notin V$. Prove that a space $(X, \mathcal{T})$ is T_1 if and only if each finite subset of X is closed.

2. A topological space $(X, \mathcal{T})$ is said to be *regular* if it is a T_1-space and if for each closed subset A of X and for each $a \notin A$ there are disjoint open sets U and V such that $a \in U$ and $A \subseteq V$. (See Figure 3.5.b.) Prove that regularity is a topological property.

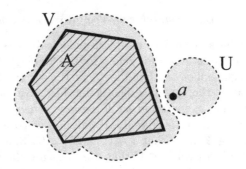

Figure 3.5.b: A topological space is called regular when any closed subset A can be separated from a point a not in A by means of disjoint open sets V and U, as illustrated above.

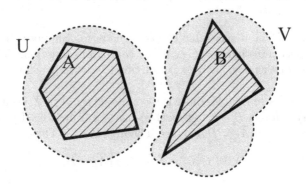

Figure 3.5.c: A topological space is called normal when any two disjoint closed subsets A and B can be separated by means of disjoint open sets U and V, as illustrated above.

3. Prove that a T_1-space $(X, \mathscr{T})$ is regular if and only if for each $x \in X$ and each open set U containing x, there is an open set W such that $x \in W$ and $\overline{W} \subseteq U$.

4. Prove that a T_1-space $(X, \mathscr{T})$ is regular if and only if for each $x \in X$ and each closed subset A of X with $x \notin A$, there exist open sets U and V such that $x \in U$, $A \subseteq V$, and $\overline{U} \cap \overline{V} = \emptyset$.

5. A topological space $(X, \mathscr{T})$ is said to be *normal* if it is a T_1-space and if for each pair A and B of disjoint closed subsets of X there are disjoint open sets U and V such that $A \subseteq U$ and $B \subseteq V$. (See Figure 3.5.c.) Prove that normality is a topological property.

6. Prove that a T_1-space $(X, \mathscr{T})$ is normal if and only if for each closed subset A and each set open set U such that $A \subseteq U$, there is an open set V such that $A \subseteq V$ and $\overline{V} \subseteq U$.

7. Prove that a T_1-space $(X, \mathcal{T})$ is normal if and only if for each pair A and B of disjoint closed subsets of X, there exist open sets U and V such that $A \subseteq U$, $B \subseteq V$, and $\overline{U} \cap \overline{V} = \emptyset$.

8. Prove that the closed interval $[0, 1]$ with the standard topology has the fixed point property (f.p.p.) (*Hint:* Use the Intermediate Value Theorem.)

9. Show that the open interval $(0, 1)$ with the standard topology lacks the fixed point property. That is, find a continuous function $f : (0, 1) \to (0, 1)$ such that $f(x) \neq x$ for any $x \in (0, 1)$.

10. Prove that the closed interval $[0, 1]$ and the open interval $(0, 1)$ are not homeomorphic. (*Hint:* See Exercises 8 and 9.)

11. Let $(Y, \mathcal{R})$ be a subspace of the topological space $(X, \mathcal{T})$. If there exists a continuous onto function $r : X \to Y$ such that $r(x) = x$ for all $x \in Y$, then the subspace $(Y, \mathcal{R})$ is called a *retract* of $(X, \mathcal{T})$. (See Figure 3.5.d.) Prove that if a topological space has the f.p.p., then so does any of its retracts.

12. Show the set of real numbers with the standard topology is separable.

13. Show the set of real numbers with the discrete topology is not separable.

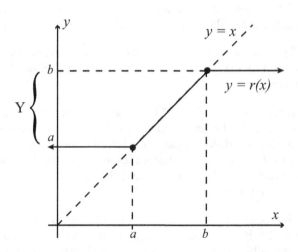

Figure 3.5.d: The subspace $Y = [a, b]$ is a retract of the set of real numbers because there exists a continuous function $r : \mathbb{R} \to [a, b]$ such that $r(x) = x$ for all $x \in [a, b]$. This property is illustrated in the graph of the function by containing the appropriate portion of the diagonal in the xy-plane.

14. Let $\mathbb{R}$ be the set of real numbers and let $X = \mathbb{R} \times \mathbb{R}$. Assign to X the topology $\mathcal{T}$ generated by the collection of all half-open rectangles

$$\{[a,b) \times [c,d) \mid a < b \text{ and } c < d\}.$$

Show that the topological space $(X, \mathcal{T})$ is separable.

15. Let $\mathbb{R}$ be the set of real numbers and let $\mathbb{R}^+$ be the set of non-negative real numbers. Let $X = \mathbb{R} \times \mathbb{R}^+$. Assign to X the topology $\mathcal{T}$ generated by the collection of neighborhoods $N_\epsilon(x, y)$ defined as follows:

(a) If $y > 0$, then for every $x \in \mathbb{R}$ and every ϵ such that $0 < \epsilon < y$,

$$N_\epsilon(x, y) = \{(r, s) \mid d((r, s), (x, y)) < \epsilon\}.$$

(b) If $y = 0$, then for every $x \in \mathbb{R}$ and every $\epsilon > 0$,

$$N_\epsilon(x, 0) = \{(x, 0)\} \cup \{(r, s) \mid d((r, s), (x, 0)) < \epsilon \text{ and } s > 0\}.$$

Show that $(X, \mathcal{T})$ is separable. (*Note:* d is the standard metric on $\mathbb{R}^2$.)

16. Give an example to show that a subspace of a separable space need not be separable. (*Hint:* Use the space $(X, \mathcal{T})$ of Exercise 14 or 15 and find a subset for which the induced topology is the discrete topology.)

17. Is the set of real numbers with the cofinite topology separable? (See Example 2.2.5.)

18. Let $(\mathbb{R}, \mathcal{T})$ be the set of real numbers where $\mathcal{T}$ is the indiscrete topology. (See Example 2.2.2.) Show that the set of even numbers is dense in $(\mathbb{R}, \mathcal{T})$.

19. Let $(X, \mathcal{T})$ be a topological space where X is a nonempty set and $\mathcal{T}$ is the indiscrete topology. Show that any nonempty subset of X is dense in X. (See Exercise 18.)

20. Let $X = \{(x, y) \mid x^2 + (y-1)^2 = 1\} \backslash \{(0, 2)\}$ be a "punctured circle" and let $Y = \{(x, 0) \mid x \in \mathbb{R}\}$ be the x-axis. Give X and Y the subspace topology induced by the standard topology on $\mathbb{R}^2$. Define a homeomorphism $h : X \to Y$. (*Hint:* Start by drawing a picture of X and Y. For $(x, y) \in X$, draw a ray starting at $(0, 2)$ and passing through (x, y). Let $h(x, y)$ denote the point where the ray intersects Y.)

21. Let $X = \{(x, y, z) \mid x^2 + y^2 + (z-1)^2 = 1\} \backslash \{(0, 0, 2)\}$ be a "punctured sphere" and let $Y = \{(x, y, 0) \mid (x, y) \in \mathbb{R} \times \mathbb{R}\}$ be the xy-plane. Give X and Y the subspace topology induced by the standard topology on $\mathbb{R}^3$. Define a homeomorphism $h : X \to Y$. (*Hint:* See Exercise 20.)

Chapter 4

Product Spaces

4.1 Products of Sets

We recall that if A and B are sets, then the set $\{(a,b) \mid (a \in A) \wedge (b \in B)\}$ is called the *Cartesian product* of A and B. The Cartesian product of A and B is denoted by the symbol $A \times B$. Similarly, if $\{A_1, A_2, \ldots, A_n\}$ is a finite collection of sets, then the set

$$\Big\{(x_1, x_2, \ldots, x_n) \;\Big|\; (x_1 \in A_1) \wedge (x_2 \in A_2) \wedge \cdots \wedge (x_n \in A_n)\Big\}$$

is called the Cartesian product of the collection $\{A_1, A_2, \ldots, A_n\}$ and is denoted by the symbol

$$\prod_{i=1}^{n} A_i = A_1 \times A_2 \times \cdots \times A_n.$$

Each $(x_1, x_2, \ldots, x_n)$ is called an *ordered n-tuple*.

We wish to generalize the idea of a Cartesian product to infinite families of sets. In order to do this, we take another look at the notion of ordered n-tuples. It is convenient to interpret $(x_1, x_2, \ldots, x_n)$ as a function x on the set $\{1, 2, \ldots, n\}$. The value of the function x at 1 is x_1, the value at 2 is x_2, the value at 3 is x_3, and so on. Of course, we require that for each i, the function value x_i is in A_i. To be more precise,

$$x : \{1, 2, \ldots, n\} \to \bigcup_{i=1}^{n} A_i,$$

where $x(i) = x_i$ and $x_i \in A_i$ for each $i \in \{1, 2, \ldots, n\}$. In this context, the Cartesian product $\prod_{i=1}^{n} A_i$ is the collection of all such functions. Specifically,

$$\prod_{i=1}^{n} A_i = \Big\{x \;\Big|\; x : \{1, 2, \ldots, n\} \to \bigcup_{i=1}^{n} A_i \text{ where } x(i) \in A_i \text{ for each } i\Big\}.$$

We take a moment to remind the reader that a set K is called an *index set* for a collection of sets $\mathscr{U}$ if $\mathscr{U} = \{X_\alpha \mid \alpha \in K\}$.

Definitions 4.1.1. Let K be an index set and for each $\alpha \in K$, let X_α be a set. The *Cartesian product* of the indexed collection $\{X_\alpha \mid \alpha \in K\}$ is defined to be the set

$$\prod_{\alpha \in K} X_\alpha = \left\{ f \mid f : K \to \bigcup_{\alpha \in K} X_\alpha \text{ where } f(\alpha) \in X_\alpha \text{ for each } \alpha \in K \right\}.$$

The value $f(\alpha)$ is called the αth *coordinate* of f.

In elementary mathematics, when we consider the Cartesian product $\mathbb{R} \times \mathbb{R}$, where $\mathbb{R}$ is the set of real numbers, we often consider the projections p_x and p_y of a point (a, b) onto the x-axis and the y-axis, respectively. To be precise, the projections $p_x : \mathbb{R} \times \mathbb{R} \to \mathbb{R}$ and $p_y : \mathbb{R} \times \mathbb{R} \to \mathbb{R}$ are functions defined by

$$p_x(a, b) = a \quad \text{and} \quad p_y(a, b) = b,$$

for all $(a, b) \in \mathbb{R} \times \mathbb{R}$. (See Figure 4.1.a.)

If we think of $\mathbb{R} \times \mathbb{R}$ as the Cartesian product indexed by the set $\{1, 2\}$, then

$$\mathbb{R} \times \mathbb{R} = \left\{ f \mid f : \{1, 2\} \to X_1 \cup X_2 \text{ where } f(1) \in X_1 \text{ and } f(2) \in X_2 \right\},$$

where $X_1 = \mathbb{R}$ and $X_2 = \mathbb{R}$. Thus, a point (a, b) in $\mathbb{R} \times \mathbb{R}$ can be interpreted as a function $f : \{1, 2\} \to X_1 \cup X_2$, where $f(1) = a$ and $f(2) = b$. If we think of X_1 as the x-axis and X_2 and the y-axis, and p_1 as the projection onto the x-axis and p_2 as the projection onto the y-axis, then we have

$$p_1(f) = p_1(a, b) = a = f(1) \quad \text{and} \quad p_2(f) = p_2(a, b) = b = f(2).$$

Thus, it is natural to give the following definition.

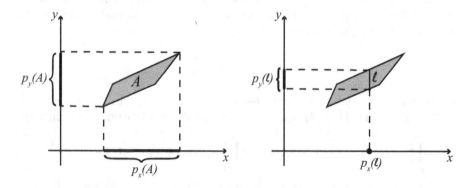

Figure 4.1.a: Projections onto the x-axis and y-axis in $\mathbb{R}^2$. Observe that the projection of the line segment ℓ onto the y-axis is an interval, but the projection onto the x-axis is a point.

Definition 4.1.2. Let K be an index set and for each $\alpha \in K$, let X_α be a set. For each $\beta \in K$, let the βth *projection function* (or *coordinate function*) be the function

$$p_\beta : \prod_{\alpha \in K} X_\alpha \to X_\beta$$

defined by $p_\beta(f) = f(\beta)$ for all $f \in \prod_{\alpha \in K} X_\alpha$.

Example 4.1.3. Let $K = \{u, v\}$ be an index set and consider the two sets $A_u = \{a, b, c, d\}$ and $A_v = \{x, y, z\}$.

(a) Describe $\prod_{\alpha \in K} A_\alpha$.

(b) Find the value of the projection function p_u at two elements of $\prod_{\alpha \in K} A_\alpha$.

(c) Find the value of the projection function p_v at two elements of $\prod_{\alpha \in K} A_\alpha$.

(d) Find $p_u^{-1}(\{b, c\})$.

(e) Find $p_v^{-1}(\{x, z\})$.

Solution. (a) Observe that $\displaystyle\prod_{\alpha \in K} A_\alpha = A_u \times A_v$ and

$$A_u \times A_v = \left\{ f \,\middle|\, f : \{u, v\} \to A_u \cup A_v \text{ where } f(u) \in A_u \text{ and } f(v) \in A_v \right\}.$$

Since there are 12 pairings of elements in A_u with elements in A_v, it follows that $A_u \times A_v$ contains 12 functions, say $\{f_1, \ldots, f_{12}\}$. We define the functions in $A_u \times A_v$ as follows:

$$f_1(u) = a \text{ and } f_1(v) = x, \qquad f_2(u) = a \text{ and } f_2(v) = y,$$
$$f_3(u) = a \text{ and } f_3(v) = z, \qquad f_4(u) = b \text{ and } f_4(v) = x,$$
$$f_5(u) = b \text{ and } f_5(v) = y, \qquad f_6(u) = b \text{ and } f_6(v) = z,$$
$$f_7(u) = c \text{ and } f_7(v) = x, \qquad f_8(u) = c \text{ and } f_8(v) = y,$$
$$f_9(u) = c \text{ and } f_9(v) = z, \qquad f_{10}(u) = d \text{ and } f_{10}(v) = x,$$
$$f_{11}(u) = d \text{ and } f_{11}(v) = y, \qquad f_{12}(u) = d \text{ and } f_{12}(v) = z.$$

(b) $p_u(f_5) = f_5(u) = b$ and $p_u(f_8) = f_8(u) = c$.

(c) $p_v(f_7) = f_7(v) = x$ and $p_v(f_{10}) = f_{10}(v) = x$.

(d) We compute the preimage:

$$p_u^{-1}(\{b, c\}) = \{f \mid p_u(f) \in \{b, c\}\} = \{f \mid f(u) \in \{b, c\}\}$$
$$= \{f \mid (f(u) = b) \vee (f(u) = c)\} = \{f_4, f_5, f_6, f_7, f_8, f_9\}.$$

(e) We compute the preimage:

$$p_v^{-1}(\{x, y\}) = \{f \mid p_v(f) \in \{x, y\}\} = \{f \mid f(v) \in \{x, y\}\}$$
$$= \{f \mid (f(v) = x) \vee (f(v) = y)\}$$
$$= \{f_1, f_2, f_4, f_5, f_7, f_8, f_{10}, f_{11}\}.$$

The following table may help to clarify the basic idea behind the solution to Example 4.1.3 because the members of the Cartesian product are displayed in a way similar to the Cartesian plane used in elementary mathematics.

$$A_u$$

$$A_v \begin{cases} \quad \end{cases}$$

		a	b	c	d
	x	$f_1 = (a, x)$	$f_4 = (b, x)$	$f_7 = (c, x)$	$f_{10} = (d, x)$
	y	$f_2 = (a, y)$	$f_5 = (b, y)$	$f_8 = (c, y)$	$f_{11} = (d, y)$
	z	$f_3 = (a, z)$	$f_6 = (b, z)$	$f_9 = (c, z)$	$f_{12} = (d, z)$

Example 4.1.4. Let $K = \{1, 2\}$ and let $A_1 = \mathbb{R}$ and $A_2 = \mathbb{R}$. We adopt the convention that $f = (a, b)$ means $f \in A_1 \times A_2$ is such that $f(1) = a$ and $f(2) = b$, where $a \in A_1$ and $b \in A_2$. Suppose $p_1 : A_1 \times A_2 \to A_1$ and $p_2 : A_1 \times A_2 \to A_2$ denote the projection functions. Find each of the following:

(a) $p_1(3, 7)$

(b) $p_2(-2, 5)$

(c) $p_1^{-1}(\{2\})$

(d) $p_1^{-1}(U)$, where U is the open interval $(-1, 3)$

(e) $p_2^{-1}(V)$, where V is the open interval $(1, 2)$

(f) $p_1^{-1}(U) \cap p_2^{-1}(V)$

Solution. (a) $p_1(3,7) = 3$ since $f = (3,7)$ implies that $p_1(f) = f(1) = 3$.

(b) $p_2(-2,5) = 5$ since $f = (-2,5)$ implies that $p_2(f) = f(2) = 5$.

(c) $p_1^{-1}(\{2\}) = \{f \mid p_1(f) \in \{2\}\} = \{f \mid f(1) = 2\} = \{(2,y) \mid y \in \mathbb{R}\}$.

Geometrically, the set $p_1^{-1}(\{2\})$ represents the vertical line passing through 2 on the x-axis.

(d) Observe that

$$p_1^{-1}(U) = p_1^{-1}\big[(-1,3)\big] = \{f \mid f(1) \in (-1,3)\}$$
$$= \{(x,y) \mid x \in (-1,3) \text{ and } y \in \mathbb{R}\}.$$

This set represents a vertical, unbounded strip, the intersection of which with the x-axis is the open interval $(-1,3)$.

(e) Note that

$$p_2^{-1}(V) = p_2^{-1}\big[(1,2)\big] = \{f \mid f(2) \in (1,2)\}$$
$$= \{(x,y) \mid x \in \mathbb{R} \text{ and } y \in (1,2)\}.$$

This set represents a horizontal, unbounded strip, the intersection of which with the y-axis is the open interval $(1,2)$.

(f) Recalling the answers from (d) and (e), we see
$$p_1^{-1}(U) \cap p_2^{-1}(V)$$

$$= \{(x,y) \mid x \in (-1,3) \text{ and } y \in \mathbb{R}\} \cap \{(x,y) \mid x \in \mathbb{R} \text{ and } y \in (1,2)\}$$
$$= \{(x,y) \mid -1 < x < 3 \text{ and } 1 < y < 2\}.$$

Note that this set is the interior of the rectangle in the Cartesian plane with vertices $(-1,1)$, $(-1,2)$, $(3,2)$, and $(3,1)$.

(See Figure 4.1.b for illustrations of the solutions to (d), (e), and (f).)

Theorem 4.1.5. Let K be an index set. For each $\alpha \in K$, let A_α be a set. If B is an arbitrary set, then

(a) $\left(\bigcap_{\alpha \in K} A_\alpha \right) \times B = \bigcap_{\alpha \in K} (A_\alpha \times B),$

(b) $\left(\bigcup_{\alpha \in K} A_\alpha \right) \times B = \bigcup_{\alpha \in K} (A_\alpha \times B).$

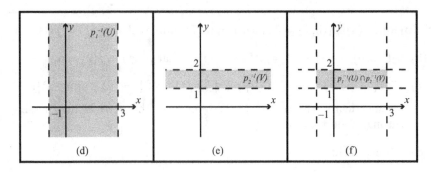

Figure 4.1.b: These graphs illustrate the solutions to parts (d), (e), and (f) of Example 4.1.4.

Proof. (a) The index set for the product $\left(\bigcap_{\alpha \in K} A_\alpha\right) \times B$ is the set $\{1, 2\}$. Thus,

$$f \in \left(\bigcap_{\alpha \in K} A_\alpha\right) \times B$$

$$\Longrightarrow f : \{1, 2\} \to \left(\bigcap_{\alpha \in K} A_\alpha\right) \cup B, \text{ where } f(1) \in \bigcap_{\alpha \in K} A_\alpha \text{ and } f(2) \in B$$

$$\Longrightarrow f : \{1, 2\} \to \bigcap_{\alpha \in K} (A_\alpha \cup B), \text{ where } f(1) \in A_\alpha \text{ for all } \alpha \in K \text{ and } f(2) \in B$$

$$\Longrightarrow f : \{1, 2\} \to A_\alpha \cup B, \text{ where } f(1) \in A_\alpha \text{ and } f(2) \in B \text{ for all } \alpha \in K$$

$$\Longrightarrow f \in A_\alpha \times B \text{ for every } \alpha \in K$$

$$\Longrightarrow f \in \bigcap_{\alpha \in K} (A_\alpha \times B).$$

This proves that $\left(\bigcap_{\alpha \in K} A_\alpha\right) \times B \subseteq \bigcap_{\alpha \in K} (A_\alpha \times B)$. The steps are reversible, and hence $\bigcap_{\alpha \in K} (A_\alpha \times B) \subseteq \left(\bigcap_{\alpha \in K} A_\alpha\right) \times B$. The proof of (b) is left as an exercise. (See Exercise 7.)   Q.E.D.

Exercises

1. Let $K = \{1, 2, 3\}$ and suppose $A_1 = \{a, b, c\}$, $A_2 = \{m, n\}$, and $A_3 = \{x, y, z\}$. How many members are in the set $\prod_{i \in K} A_i$? Describe three of these members.

2. Let $K = \{1, 2, 3, 4\}$ and let $A_i = \mathbb{R}$ for each $i \in K$. Let $f \in \prod_{i \in K} A_i$ be defined by $f(1) = -2$, $f(2) = 3$, $f(3) = 8$, and $f(4) = -\pi$. Let p_1, p_2, p_3, and p_4 be the four projection functions. Find $p_i(f)$ for each $i \in K$.

3. Repeat Exercise 2, but this time let $f(1) = 5$, $f(2) = 0$, $f(3) = -1$, and $f(4) = e$. (Recall that e is the base of the natural logarithm.)

4. Let $K = \mathbb{N}$ and let $A_\alpha = \mathbb{R}$ for each $\alpha \in K$. What is each member of $\prod_{\alpha \in K} A_\alpha$ called in elementary calculus?

5. For each $n \in \mathbb{N}$, let A_n be the closed interval $\left[-\frac{1}{n}, \frac{1}{n} \right]$. What property does each member of $\prod_{n \in \mathbb{N}} A_n$ possess? (*Hint:* See Exercise 4.)

6. The following statement, called the Axiom of Choice, cannot be proven using the standard axioms, but is usually assumed to be true.

    Axiom 4.1.6 (Axiom of Choice). *If X is a set and $\mathcal{T}$ is a nonempty collection of nonempty subsets of X, then there exists a "choice" function $f : \mathcal{T} \to X$ such that $f(A) \in A$ for each $A \in \mathcal{T}$. (That is, we can "choose" one element from each of the members of the collection $\mathcal{T}$.)*

    Let K be a nonempty set and for each $\alpha \in K$ let A_α be a set. Use Axiom 4.1.6 to prove $\prod_{\alpha \in K} A_\alpha \neq \emptyset$ if and only if $A_\alpha \neq \emptyset$ for each $\alpha \in K$.

7. Prove part (b) of Theorem 4.1.5.

8. Let K be a nonempty index set and for each $\alpha \in K$ let $\{A_\alpha, B_\alpha\}$ be a pair of sets. Prove each of the following:

    (a) $\left(\prod_{\alpha \in K} A_\alpha \right) \cap \left(\prod_{\alpha \in K} B_\alpha \right) = \prod_{\alpha \in K} (A_\alpha \cap B_\alpha)$.

    (b) $\left(\prod_{\alpha \in K} A_\alpha \right) \cup \left(\prod_{\alpha \in K} B_\alpha \right) \subseteq \prod_{\alpha \in K} (A_\alpha \cup B_\alpha)$.

9. Let K be a nonempty index set and for each $\alpha \in K$ let $\{A_\alpha, B_\alpha\}$ be a pair of sets. Prove the following:

    (a) If $A_\alpha \subseteq B_\alpha$ for each $\alpha \in K$, then $\prod_{\alpha \in K} A_\alpha \subseteq \prod_{\alpha \in K} B_\alpha$.

    (b) If $A_\alpha \neq \emptyset$ for each $\alpha \in K$ and $\prod_{\alpha \in K} A_\alpha \subseteq \prod_{\alpha \in K} B_\alpha$, then $A_\alpha \subseteq B_\alpha$ for all $\alpha \in K$.

    (c) Explain why we must assume $A_\alpha \neq \emptyset$ for each $\alpha \in K$ in order for the statement in (b) to be true. (*Hint:* See Exercise 6.)

10. Let K be a nonempty index set and for each $\alpha \in K$ let $\{A_\alpha, B_\alpha\}$ be a pair of sets such that $A_\alpha \subseteq B_\alpha$. Suppose that $B = \prod_{\alpha \in K} B_\alpha$. For each $\alpha \in K$, let $p_\alpha : B \to B_\alpha$ denote the αth projection function. Prove the following:

    (a) $\prod_{\alpha \in K} A_\alpha = \bigcap_{\alpha \in K} p_\alpha^{-1}(A_\alpha)$.

    (b) $B \setminus p_\alpha^{-1}(A_\alpha) = p_\alpha^{-1}(B_\alpha \setminus A_\alpha)$.

    (c) $B \setminus \left(\prod_{\alpha \in K} A_\alpha \right) = \bigcap_{\alpha \in K} p_\alpha^{-1}(B_\alpha \setminus A_\alpha)$.

11. Suppose $X_1 = \mathbb{R}$ and $X_2 = \mathbb{R}$ and let $X = X_1 \times X_2$. Let p_1 and p_2 be the projection functions defined by $p_1(a, b) = a$ and $p_2(a, b) = b$ for each point $(a, b) \in X$. Show for any $\epsilon > 0$, there exists a $\delta > 0$ such that, whenever (x, y) is inside the circle centered at (a, b) with radius δ, we have $p_1(x, y) = x \in (a - \epsilon, a + \epsilon)$ and $p_2(x, y) = y \in (b - \epsilon, b + \epsilon)$. What does this tell you about the two projection functions?

12. Let K be a nonempty index set. For each $\alpha \in K$, let X_α be a nonempty set and define $X = \prod_{\alpha \in K} X_\alpha$. Suppose $N \subseteq M \subseteq K$ and, for each $\alpha \in M$, let the set A_α be a nonempty subset of X_α. Define two subsets of X as follows:
$$B = \{f \mid f \in X \text{ and } f(\alpha) \in A_\alpha \text{ for } \alpha \in M\},$$
$$C = \{f \mid f \in X \text{ and } f(\alpha) \in A_\alpha \text{ for } \alpha \in N\}.$$
Is it true that $B \subseteq C$ or $C \subseteq B$? Justify your answer carefully.

13. Let K be a nonempty index set. For each $\alpha \in K$, let X_α be a nonempty set and define $X = \prod_{\alpha \in K} X_\alpha$. For each $\alpha \in K$, let $p_\alpha : X \to X_\alpha$ denote the αth projection function. Suppose $A \subseteq X$. Prove or disprove the following:

(a) $A \subseteq \prod_{\alpha \in K} p_\alpha(A)$,

(b) $A = \prod_{\alpha \in K} p_\alpha(A)$,

(c) If $f \in X \setminus A$, then $f(\alpha) \notin p_\alpha(A)$ for any $\alpha \in K$.

4.2   Product Spaces

Suppose now that K is an index set and, for each $\alpha \in K$, suppose the pair $(X_\alpha, \mathscr{T}_\alpha)$ is a topological space. Let $X = \prod_{\alpha \in K} X_\alpha$. We would like to define a topology on X in a meaningful way. It is natural to want each projection function $p_\alpha : X \to X_\alpha$ to be continuous. (See Exercise 11 in Section 4.1.) Thus, whatever topology $\mathscr{T}$ we assign to X, we want $p_\alpha^{-1}(V_\alpha) \in \mathscr{T}$ whenever $V_\alpha \in \mathscr{T}_\alpha$ (for each $\alpha \in K$). Consequently, we require $\mathscr{T}$ to be the smallest topology that contains the collection $\mathscr{S}$ of all sets of the form $p_\alpha^{-1}(V_\alpha)$, where $\alpha \in K$ and $V_\alpha \in \mathscr{T}_\alpha$. It is easy to see that $\mathscr{S}$ is not in general a topology, but we can use $\mathscr{S}$ as a subbase for the topology $\mathscr{T}$. That is, members of the base for the product topology will be of the form $p_{\alpha_1}^{-1}(V_{\alpha_1}) \cap \cdots \cap p_{\alpha_n}^{-1}(V_{\alpha_n})$, where $n \in \mathbb{N}$ and $V_{\alpha_k} \in \mathscr{T}_{\alpha_k}$ for all $k \in \{1, \ldots, n\}$.

Definitions 4.2.1. Let K be an index set and, for each $\alpha \in K$, let $(X_\alpha, \mathscr{T}_\alpha)$ be a topological space. Let $X = \prod_{\alpha \in K} X_\alpha$ and let $p_\alpha : X \to X_\alpha$ be the αth projection function. The unique topology $\mathscr{T}$ generated by the subbase of all sets of the form $p_\alpha^{-1}(V_\alpha)$, where $\alpha \in K$ and $V_\alpha \in \mathscr{T}_\alpha$, is called the *product topology* on X. The topological space $(X, \mathscr{T})$ is called a *product space* and for each $\alpha \in K$ the pair $(X_\alpha, \mathscr{T}_\alpha)$ is called the αth *coordinate space* (or *factor space*).

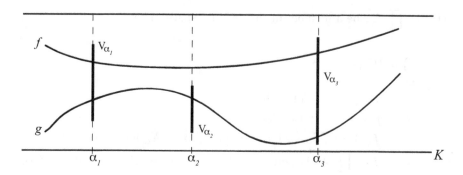

Figure 4.2.a: The functions f and g are elements of some member of the base for the product topology.

In Figure 4.2.a, the index set K is represented as a horizontal line and α_1, α_2, and α_3 are indices in K. The vertical line segments V_{α_1}, V_{α_2}, and V_{α_3} represent open sets in the respective coordinate spaces. The functions f and g are elements of a product space indexed by K. Note that $f(\alpha_1) \in V_{\alpha_1}$ and $f(\alpha_3) \in V_{\alpha_3}$. Consequently, $p_{\alpha_1}(f) \in V_{\alpha_1}$ and $p_{\alpha_3}(f) \in V_{\alpha_3}$. Thus, f is in $A = p_{\alpha_1}^{-1}(V_{\alpha_1}) \cap p_{\alpha_3}^{-1}(V_{\alpha_3})$, which is a member of the base for the product topology.

The function g is also in the set A. However, it is also true that $g(\alpha_2) \in V_{\alpha_2}$, and so we have $g \in p_{\alpha_1}^{-1}(V_{\alpha_1}) \cap p_{\alpha_2}^{-1}(V_{\alpha_2}) \cap p_{\alpha_3}^{-1}(V_{\alpha_3})$. This set is also a member of the base for the product topology. Note that f is not in this set.

The next theorem gives a useful characterization of elements in the base for the product topology.

Theorem 4.2.2. *For each α in an index set K, let $(X_\alpha, \mathscr{T}_\alpha)$ be a topological space and let p_α be the αth projection function from the product space onto $(X_\alpha, \mathscr{T}_\alpha)$. If $V_\alpha \in \mathscr{T}_\alpha$ for each $\alpha \in \{\alpha_1, \ldots, \alpha_n\} \subseteq K$, then*

$$\bigcap_{i=1}^{n} p_{\alpha_i}^{-1}(V_{\alpha_i}) = \prod_{\alpha \in K} A_\alpha,$$

where $A_\alpha = V_\alpha$ if $\alpha \in \{\alpha_1, \ldots, \alpha_n\}$ and $A_\alpha = X_\alpha$ if $\alpha \in K \backslash \{\alpha_1, \ldots, \alpha_n\}$.

Proof. Let $\Phi = \{\alpha_1, \ldots, \alpha_n\}$, so that

$$\bigcap_{i=1}^{n} p_{\alpha_i}^{-1}(V_{\alpha_i}) = \bigcap_{\alpha \in \Phi} p_\alpha^{-1}(V_\alpha).$$

Let $f \in \prod_{\alpha \in K} X_\alpha$ be a member of the product space. Then

$$f \in \bigcap_{\alpha \in \Phi} p_\alpha^{-1}(V_\alpha)$$

$\iff f \in p_\alpha^{-1}(V_\alpha)$ for each $\alpha \in \Phi$

$\iff p_\alpha(f) \in V_\alpha$ for each $\alpha \in \Phi$

$\iff f(\alpha) \in V_\alpha$ for each $\alpha \in \Phi$ [and $f(\alpha) \in X_\alpha$ for each $\alpha \in K \backslash \Phi$]

$\iff f \in \prod_{\alpha \in K} A_\alpha$, where $A_\alpha = V_\alpha$ if $\alpha \in \Phi$ and $A_\alpha = X_\alpha$ if $\alpha \in K \backslash \Phi$.

Q.E.D.

Theorem 4.2.2 tells us that each member of the base for the product topology on $\prod_{\alpha \in K} X_\alpha$ is of the form $\prod_{\alpha \in K} A_\alpha$, where each $A_\alpha \in \mathcal{T}_\alpha$ and $A_\alpha = X_\alpha$ for all but finitely many $\alpha \in K$.

The product topology is defined so that all projection functions are continuous. However, we can say even more about the projection functions.

Theorem 4.2.3. *The projection functions on a product space are continuous open functions onto their respective coordinate spaces.*

Proof. Let K be an index set and, for each $\alpha \in K$, let $(X_\alpha, \mathcal{T}_\alpha)$ be a topological space. Suppose $(X, \mathcal{T})$ is the product space. For $\beta \in K$, we wish to show that the projection $p_\beta : X \to X_\beta$ is onto, open, and continuous.

To show that p_β is onto, let $x_\beta \in X_\beta$. Suppose $f \in X$ and define $g \in X$ by

$$g(\gamma) = \begin{cases} f(\gamma) & \text{if } \gamma \neq \beta, \\ x_\beta & \text{if } \gamma = \beta. \end{cases}$$

Then, $p_\beta(g) = x_\beta$, and so p_β is onto. If $U \in \mathcal{T}_\beta$, then $p_\beta^{-1}(U)$ is a member of the subbase for the topology $\mathcal{T}$. Therefore, $p_\beta^{-1}(U)$ is an open set, and so p_β is continuous. Finally, if V is a member of the base for the topology, then $V = \prod_{\alpha \in K} A_\alpha$, where each $A_\alpha \in \mathcal{T}_\alpha$ (by Theorem 4.2.2). Then $p_\beta(V) = A_\beta$ is an element of $\mathcal{T}_\beta$, and so p_β is an open function.   Q.E.D.

Theorem 4.2.4. *Let K be an index set and, for each $\alpha \in K$, let $(X_\alpha, \mathcal{T}_\alpha)$ be a topological space. Let $(X, \mathcal{T})$ be the product space and suppose $(Y, \mathcal{B})$ is another topological space. A function $f : Y \to X$ is continuous if and only if the composition $p_\alpha \circ f : Y \to X_\alpha$ is continuous for each $\alpha \in K$.*

Proof. If f is continuous, then $p_\alpha \circ f$ is continuous for each $\alpha \in K$ because the composition of continuous functions is continuous (by Theorem 3.3.7). Conversely, suppose $p_\alpha \circ f$ is continuous for each $\alpha \in K$. By Theorem 3.4.1, in order to show that f is $\mathcal{B}$-to-$\mathcal{T}$ continuous, it suffices to show that $f^{-1}(V)$ is in $\mathcal{B}$ for any member V of a subbase for the topology $\mathcal{T}$. Recall that if V is a member of the subbase for the product topology $\mathcal{T}$, then $V = p_\alpha^{-1}(U_\alpha)$ for some $\alpha \in K$ and some $U_\alpha \in \mathscr{T}_\alpha$. Since $p_\alpha \circ f : Y \to X_\alpha$ is continuous, it follows that $(p_\alpha \circ f)^{-1}(U_\alpha) \in \mathscr{B}$. However,

$$(p_\alpha \circ f)^{-1}(U_\alpha) = f^{-1}(p_\alpha^{-1}(U_\alpha)) = f^{-1}(V).$$

Therefore, $f^{-1}(V) \in \mathscr{B}$, and so f is $\mathscr{B}$-to-$\mathscr{T}$ continuous. Q.E.D.

Theorem 4.2.4 is often reformulated in the following way.

Corollary 4.2.5. *Let $(X_\alpha, \mathscr{T}_\alpha)$ be a topological space for each $\alpha \in K$, where K is an index set, and let $(X, \mathscr{T})$ be the product space. For a topological space $(Y, \mathscr{B})$ and functions $f_\alpha : Y \to X_\alpha$ (for all $\alpha \in K$), define a function $F : Y \to X$ by assigning to each $y \in Y$ the element $F(y)$ in X defined to be*

$$F(y)(\alpha) = f_\alpha(y),$$

for all $\alpha \in K$. If f_α is continuous for each $\alpha \in K$, then F is continuous.

Proof. For each $\alpha \in K$ and each $y \in Y$, note that

$$(p_\alpha \circ F)(y) = p_\alpha(F(y)) = F(y)(\alpha) = f_\alpha(y).$$

Thus, for each $\alpha \in K$, the function $p_\alpha \circ F = f_\alpha$ is continuous. It follows that F is continuous, by Theorem 4.2.4. Q.E.D.

The function f_α in the preceding corollary is called the αth *component function* of the function F, where $\alpha \in K$.

A natural question to ask is the following: If each topological space in a collection has a certain property, when does the product space share the same property? The following theorem answers this question for the property of being a Hausdorff space. Recall that $(X, \mathscr{T})$ is a Hausdorff space if and only if for each pair of distinct points x and y in X, there are disjoint open sets U and V such that $x \in U$ and $y \in V$.

Theorem 4.2.6. *The product of Hausdorff spaces is a Hausdorff space.*

Proof. Let K be an index set and let $(X_\alpha, \mathscr{T}_\alpha)$ be a Hausdorff space for each $\alpha \in K$. Let $(X, \mathscr{T})$ be the product space.

Suppose f and g are distinct members of X. Then $f(\alpha) \neq g(\alpha)$ for some $\alpha \in K$. By assumption $(X_\alpha, \mathscr{T}_\alpha)$ is a Hausdorff space, and so there are disjoint open sets U_α and V_α in $\mathscr{T}_\alpha$ such that $f(\alpha) \in U_\alpha$ and $g(\alpha) \in V_\alpha$.

Let p_α denote the αth projection function. The projection functions are continuous (by Theorem 4.2.3), and so the sets $U = p_\alpha^{-1}(U_\alpha)$ and $V = p_\alpha^{-1}(V_\alpha)$ are open in $(X, \mathscr{T})$. Observe that $f \in U$ because $p_\alpha(f) = f(\alpha) \in U_\alpha$. Similarly, we note that $g \in V$. If we show that $U \cap V = \emptyset$, then we will have established that $(X, \mathscr{T})$ is a Hausdorff space. In order to show that U and V are disjoint, we assume to the contrary that there exists some $h \in U \cap V$. Then $h \in p_\alpha^{-1}(U_\alpha)$ and $h \in p_\alpha^{-1}(V_\alpha)$. Consequently,

$$h(\alpha) = p_\alpha(h) \in U_\alpha \cap V_\alpha.$$

This is a contradiction, since U_α and V_α were assumed to be disjoint. Therefore $U \cap V = \emptyset$, and so $(X, \mathscr{T})$ is a Hausdorff space. Q.E.D.

Theorem 4.2.7. *Let K be an index set and, for each $\alpha \in K$, suppose $(X_\alpha, \mathscr{T}_\alpha)$ is a topological space. If $(X, \mathscr{T})$ is the product space, and if $A_\alpha \subseteq X_\alpha$ for each $\alpha \in K$, then*

$$\overline{\prod_{\alpha \in K} A_\alpha} = \prod_{\alpha \in K} \overline{A_\alpha}.$$

Proof. For ease of notation, let $A = \prod_{\alpha \in K} A_\alpha$. Let $f \in \prod_{\alpha \in K} \overline{A_\alpha}$ and suppose $V \in \mathscr{T}$ is such that $f \in V$. We will show that $V \cap \overline{A}$ is nonempty. The set V is open in the product space, and so there is some member W of the base such that $f \in W$ and $W \subseteq V$. Since W is a member of the base for the product topology, there is a finite subset Φ of K such that

$$W = \left(\prod_{\alpha \in \Phi} U_\alpha \right) \times \left(\prod_{\alpha \in K \setminus \Phi} X_\alpha \right),$$

where $U_\alpha \in \mathscr{T}_\alpha$ for each $\alpha \in \Phi$. (See Theorem 4.2.2.)

Recall that $f(\alpha) = p_\alpha(f)$. By assumption, $p_\alpha(f) \in \overline{A_\alpha}$ for all $\alpha \in K$. But $f \in W$ implies that $p_\alpha(f) \in U_\alpha$ for every $\alpha \in \Phi$. Thus, for each $\alpha \in \Phi$, it must be the case that $U_\alpha \cap A_\alpha \neq \emptyset$. It follows that $W \cap \prod_{\alpha \in K} A_\alpha \neq \emptyset$. Because $W \subseteq V$, we conclude that $V \cap A \neq \emptyset$, and therefore $f \in \overline{A}$. We have now proved that $\prod_{\alpha \in K} \overline{A_\alpha} \subseteq \overline{A}$.

Conversely, let $g \in \overline{A}$. For $\alpha \in K$, let U_α be any member of $\mathscr{T}_\alpha$ containing $g(\alpha)$. Then, $p_\alpha^{-1}(U_\alpha)$ is a member of the subbase for the product topology and so is a member of $\mathscr{T}$. Because $g \in p_\alpha^{-1}(U_\alpha)$, it follows that $p_\alpha^{-1}(U_\alpha) \cap \prod_{\alpha \in K} A_\alpha \neq \emptyset$. Thus, $U_\alpha \cap A_\alpha \neq \emptyset$. This shows that $g(\alpha) \in \overline{A_\alpha}$. Since this is true for each $\alpha \in K$, we conclude that $g \in \prod_{\alpha \in K} \overline{A_\alpha}$, and so $\overline{A} \subseteq \prod_{\alpha \in K} \overline{A_\alpha}$. Q.E.D.

Exercises

1. Let $X_1 = \mathbb{R}$ and $X_2 = \mathbb{R}$, where $\mathbb{R}$ is given the standard topology. In each of the following cases, identify $p_1^{-1}(U_1)$, $p_2^{-1}(U_2)$, and $p_1^{-1}(U_1) \cap p_2^{-1}(U_2)$. Illustrate each set by means of a diagram. (Represent $X_1 \times X_2$ as the usual Cartesian plane.)

   (a) $U_1 = (-1, 3)$ and $U_2 = (1, 4)$.
   (b) $U_1 = (1, 5)$ and $U_2 = (-1, 2) \cup (3, 6)$.
   (c) $U_1 = (-3, 0) \cup (1, 3)$ and $U_2 = (-1, 5)$.
   (d) $U_1 = (-4, -1) \cup (0, 3)$ and $U_2 = (-\infty, 0) \cup (1, 4)$.

2. Let $\mathbb{R}$ denote the set of real numbers equipped with the standard topology. In this exercise, we consider the product space $\prod_{\alpha \in \mathbb{R}} X_\alpha$, where the index set is $\mathbb{R}$ and $X_\alpha = \mathbb{R}$ for each $\alpha \in \mathbb{R}$. Consider the functions f, g, and h defined by

$$f(x) = \frac{x}{2} + 3, \quad g(x) = \sin x, \quad h(x) = \begin{cases} 1/x & \text{if } x \neq 0, \\ 1 & \text{if } x = 0. \end{cases}$$

    (a) Sketch the graphs of f, g, and h in a Cartesian coordinate plane.

    (b) Explain why the graphs of f, g, and h represent three points in the product $\prod_{\alpha \in \mathbb{R}} X_\alpha$.

    (c) What is the $(-3)^{\text{th}}$ coordinate of f?

    (d) What is the $(\pi/6)^{\text{th}}$ coordinate of g?

    (e) What is the $(-4)^{\text{th}}$ coordinate of h?

    (f) What is the $(0)^{\text{th}}$ coordinate of h?

3. As in Exercise 2, let $X_\alpha = \mathbb{R}$ for each $\alpha \in \mathbb{R}$, where $\mathbb{R}$ is given the standard topology. Draw the index set $\mathbb{R}$ as a horizontal line, the way you would draw the x-axis in elementary mathematics. Draw the sets X_2 and X_π as lines perpendicular to the x-axis at $x = 2$ and $x = \pi$, respectively. Let U_2 be the open interval $(1,3)$ in X_2 and let U_π be the open interval $(-1,2)$ in X_π.

    (a) Sketch a member of $p_2^{-1}(U_2)$.

    (b) Sketch a member of $p_\pi^{-1}(U_\pi)$.

    (c) Sketch a member of $p_2^{-1}(U_2) \cap p_\pi^{-1}(U_\pi)$.

4. Let K be an index set. For each $\alpha \in K$, let $(X_\alpha, \mathscr{T}_\alpha)$ be a topological space and suppose $A_\alpha \subseteq X_\alpha$. Denote the product space by $(X, \mathscr{T})$. Prove that $\prod_{\alpha \in K} A_\alpha$ is dense in X if and only if A_α is dense in X_α (in the topology $\mathscr{T}_\alpha$) for each $\alpha \in K$. (See Definition 3.5.5.)

5. Let $X_1 = \mathbb{R}$ and $X_2 = \mathbb{R}$, where $\mathbb{R}$ is given the standard topology. The product $X_1 \times X_2$ can be represented geometrically by the Cartesian plane, where X_1 and X_2 are the x-axis and y-axis, respectively. Find a closed subset A of the product space $X_1 \times X_2$ such that $p_1(A)$ and $p_2(A)$ are each not closed.

6. Let K be an infinite index set and, for each $\alpha \in K$, let $(X_\alpha, \mathscr{T}_\alpha)$ be a topological space where X_α has at least two points and $\mathscr{T}_\alpha$ is the discrete topology. Let $(X, \mathscr{T})$ be the product space. Prove that $\mathscr{T}$ is not the discrete topology on X.

7. Let K be an infinite index set and, for each $\alpha \in K$, let $(X_\alpha, \mathscr{T}_\alpha)$ be a topological space such that $\mathscr{B}_\alpha$ is a base for $\mathscr{T}_\alpha$. Let $(X, \mathscr{T})$ be the product space and suppose $\mathscr{S} = \{p_\alpha^{-1}(U_\alpha) \mid \alpha \in K \text{ and } U_\alpha \in \mathscr{B}_\alpha\}$. Prove that $\mathscr{S}$ is a subbase for $\mathscr{T}$.

8. Let K be an infinite index set and, for each $\alpha \in K$, let $(X_\alpha, \mathscr{T}_\alpha)$ be a topological space. If $(X, \mathscr{T})$ is the product space, and if U is a nonempty set in $\mathscr{T}$, prove that $p_\alpha(U) = X_\alpha$ for all but finitely many $\alpha \in K$.

9. Let $X_1 = \mathbb{R}$ and $X_2 = \mathbb{R}$, where $\mathbb{R}$ is given the standard topology. Suppose that $f : X_1 \times X_2 \to X_1 \times X_2$ is a continuous function. Let $\{(x_n, y_n)\}_{n \in \mathbb{N}}$

be a sequence in $X_1 \times X_2$ that converges to the point (x_0, y_0). Suppose for each $n \in \mathbb{N}$ we have $f(x_n, y_n) = (r_n, s_n)$ where $r_n \leq x_n$ and $s_n \leq y_n$. Prove the following:

(a) $\lim_{n \to \infty} x_n = x_0$ and $\lim_{n \to \infty} y_n = y_0$.

(b) $\lim_{n \to \infty}(r_n, s_n) = f(x_0, y_0)$.

(c) If $f(x_0, y_0) = (r, s)$, then $r \leq x_0$ and $s \leq y_0$.

Note: In this problem, we have adopted the standard convention that $f(x, y)$ denotes $f\big((x, y)\big)$ for each $(x, y) \in X_1 \times X_2$.

10. (a) Show that $\mathbb{R}^n$ with the standard metric topology (introduced in Section 2.8) is homeomorphic to $\prod_{i=1}^n \mathbb{R}$ with the product topology.

(b) Show $\prod_{n \in \mathbb{N}} \left(-\frac{1}{n}, \frac{1}{n} \right)$ is not open in the product topology on $\prod_{n \in \mathbb{N}} \mathbb{R}$.

4.3 More on Product Spaces

In this section, we introduce several theorems that will be needed for some of our future work.

Theorem 4.3.1. *Let K be an index set and, for each $\alpha \in K$, let $(X_\alpha, \mathcal{T}_\alpha)$ be a topological space with subspace $(Y_\alpha, \mathcal{R}_\alpha)$. Let $(X, \mathcal{T})$ and $(Y, \mathcal{R})$ be the corresponding product spaces. If $\mathcal{S}$ is the subspace topology on Y inherited as a subset of X, then $\mathcal{R} = \mathcal{S}$.*

Proof. The collection $\mathcal{B} = \{Y \cap U \mid U \text{ is a member of the base for } \mathcal{T}\}$ is a base for the induced subspace topology $\mathcal{S}$. We need only show that $\mathcal{B}$ is also a base for the product topology $\mathcal{R}$ on Y.

Begin by assuming that $V \in \mathcal{B}$. Then $V = Y \cap U$, where U is a member of the base for $\mathcal{T}$. Because U is a member of the base for the product topology on X, it follows that $U = \prod_{\alpha \in K} A_\alpha$, where $A_\alpha \in \mathcal{T}_\alpha$ for each $\alpha \in K$ and there exists a finite subset Φ of K such that $A_\alpha = X_\alpha$ for all $\alpha \in K \backslash \Phi$. Consequently,

$$V = Y \cap U = Y \cap \left(\prod_{\alpha \in K} A_\alpha \right) = \prod_{\alpha \in K} (Y \cap A_\alpha).$$

Observe that $Y \cap A_\alpha \in \mathcal{R}_\alpha$, because $\mathcal{R}_\alpha$ is the subspace topology on Y_α induced by $\mathcal{T}_\alpha$. Furthermore, $Y_\alpha \cap A_\alpha = Y_\alpha$ for each $\alpha \in K \backslash \Phi$. Thus, V is a member of the base for the product topology $\mathcal{R}$ on Y.

Now, conversely, assume that W is a member of the base for the topology $\mathcal{R}$. Then there is a finite subset Ψ of K such that $W = \prod_{\alpha \in K} W_\alpha$, where $W_\alpha \in \mathcal{R}_\alpha$ for all $\alpha \in \Psi$ and $W_\alpha = Y_\alpha$ for all $\alpha \in K \backslash \Psi$. By the definition of $\mathcal{R}_\alpha$, there is a $U_\alpha \in \mathcal{T}_\alpha$ such that $W_\alpha = Y_\alpha \cap U_\alpha$. If $\alpha \in K \backslash \Psi$, then let $U_\alpha = X_\alpha$.

The set $\prod_{\alpha\in K} U_\alpha$ is a member of the base for the product topology $\mathscr{T}$, and so $Y \cap \left(\prod_{\alpha\in K} U_\alpha\right)$ is a member of the collection $\mathscr{B}$. Finally, we observe that

$$Y \cap \left(\prod_{\alpha\in K} U_\alpha\right) = \prod_{\alpha\in K} W_\alpha = W.$$

(See Exercise 2 at the end of this section.) Q.E.D.

Theorem 4.3.2. *Let K be an index set and, for each $\alpha \in K$, let $(X_\alpha, \mathscr{T}_\alpha)$ be a topological space. Denote the product space by $(X, \mathscr{T})$. Let $f \in X$ and $\beta \in K$. For each $\alpha \in K$, define a set Y_α according to the rule*

$$Y_\alpha = \begin{cases} \{f(\alpha)\} & \text{if } \alpha \neq \beta, \\ X_\beta & \text{if } \alpha = \beta. \end{cases}$$

If $Y_{f,\beta} = \prod_{\alpha\in K} Y_\alpha$ is given the subspace topology $\mathscr{R}$ induced by $\mathscr{T}$, then the subspace $(Y_{f,\beta}, \mathscr{R})$ of $(X, \mathscr{T})$ is homeomorphic to $(X_\beta, \mathscr{T}_\beta)$.

Proof. We first observe that the subspace topology $\mathscr{R}$ on $Y_{f,\beta}$ is the same as the product topology on $\prod_{\alpha\in K} Y_\alpha$, by Theorem 4.3.1. Let $F : X_\beta \to Y_{f,\beta}$ be defined so that, for each $x_\beta \in X_\beta$, the element $F(x_\beta)$ in $\prod_{\alpha\in K} Y_\alpha$ is given by the rule

$$F(x_\beta)(\gamma) = \begin{cases} f(\gamma) & \text{if } \gamma \neq \beta, \\ x_\beta & \text{if } \gamma = \beta. \end{cases}$$

We want to show that F is one-to-one and onto, and that both F and F^{-1} are continuous. Suppose that $F(x_\beta) = F(x'_\beta)$. Then, $F(x_\beta)(\beta) = F(x'_\beta)(\beta)$, from which it follows that $x_\beta = x'_\beta$. This shows that F is one-to-one.

Now suppose that $x \in Y_{f,\beta}$. Then, by definition, $x(\beta) \in X_\beta$ and $x(\gamma) = f(\gamma)$ for all $\gamma \neq \beta$. Because $x(\beta) \in X_\beta$, it follows that $F\big(x(\beta)\big)$ is an element of $Y_{f,\beta}$ and that

$$F\big(x(\beta)\big)(\gamma) = \begin{cases} f(\gamma) & \text{if } \gamma \neq \beta, \\ x(\beta) & \text{if } \gamma = \beta. \end{cases}$$

Consequently, $F\big(x(\beta)\big) = x$, and so F is onto.

For each $\alpha \in K$, let p_α denote the αth coordinate function. Note that, for each $x_\beta \in X_\beta$,

$$(p_\alpha \circ F)(x_\beta) = F(x_\beta)(\alpha) = \begin{cases} f(\alpha) & \text{if } \alpha \neq \beta, \\ x_\beta & \text{if } \alpha = \beta. \end{cases}$$

Thus, $p_\alpha \circ F$ is constant if $\alpha \neq \beta$ and is the identity if $\alpha = \beta$. It follows that $p_\alpha \circ F$ is continuous for all $\alpha \in K$, and so F is continuous, by Theorem 4.2.4.

We conclude the proof by observing that $F^{-1} : Y_{f,\beta} \to X_\alpha$ is the restriction of the projection function $p_\beta : X \to X_\beta$ to the subspace $Y_{f,\beta}$ of X, and so is continuous. We have shown that F is a one-to-one and onto function such that both F and F^{-1} are continuous. Therefore, F is a homeomorphism between $(Y_{f,\beta}, \mathscr{R})$ and $(X_\beta, \mathscr{T}_\beta)$.   Q.E.D.

Theorem 4.3.2 is often used to show that if a product space has a certain property, then each of its factors also has that property. As an example, we state and prove the converse of Theorem 4.2.6.

Theorem 4.3.3. *If a product space is a Hausdorff space, then each coordinate space is a Hausdorff space.*

Proof. Let K be an index set and, for each $\alpha \in K$, let $(X_\alpha, \mathscr{T}_\alpha)$ be a topological space. Denote the product space by $(X, \mathscr{T})$. We assume that $(X, \mathscr{T})$ is a Hausdorff space and show that $(X_\alpha, \mathscr{T}_\alpha)$ is a Hausdorff space for each $\alpha \in K$.

Let $\alpha \in K$ and $f \in X$. Define $Y_{f,\alpha}$ as in Theorem 4.3.2 and give it the subspace topology $\mathscr{R}$. Because $(Y_{f,\alpha}, \mathscr{R})$ is a subspace of the Hausdorff space $(X, \mathscr{T})$, it too must be a Hausdorff space. By Theorem 4.3.2, the factor $(X_\alpha, \mathscr{T}_\alpha)$ is homeomorphic to $(Y_{f,\alpha}, \mathscr{R})$, and so is also a Hausdorff space.   Q.E.D.

For the purposes of the next theorems, we introduce a new definition.

Definition 4.3.4. If X is a set, then the set $\Delta = \{(x,x) \mid x \in X\}$ is called the *diagonal* of $X \times X$.

Theorem 4.3.5. *A topological space $(X, \mathscr{T})$ is a Hausdorff space if and only if the diagonal Δ is closed in the product space $(X \times X, \mathscr{R})$.*

Proof. Assume $(X, \mathscr{T})$ is a Hausdorff space. We will demonstrate that Δ is closed by showing that its complement Δ' is open. Let $(x,y) \in \Delta'$. It follows that x and y are distinct elements of X. Because X is a Hausdorff space, there exist disjoint $\mathscr{T}$-open sets U and V such that $x \in U$ and $y \in V$.

Let p_1 and p_2 be the projection functions. Then $W = p_1^{-1}(U) \cap p_2^{-1}(V)$ is a member of $\mathscr{R}$. Certainly $(x,y) \in W$. We claim that $W \subseteq \Delta'$. To see this, observe that

$$(a,b) \in W \implies [(a,b) \in p_1^{-1}(U)] \wedge [(a,b) \in p_2^{-1}(V)]$$
$$\implies [p_1(a,b) \in U] \wedge [p_2(a,b) \in V]$$
$$\implies (a \in U) \wedge (b \in V).$$

Since $U \cap V = \emptyset$, we conclude that $a \neq b$, and so $(a,b) \in \Delta'$. Therefore, for each point (x,y) in Δ' there exists a neighborhood W of (x,y) such that $W \subseteq \Delta'$. We conclude that Δ' is open, and so Δ is closed.

Conversely, assume that Δ is closed in the product space $(X \times X, \mathscr{R})$. Let x and y be distinct points in X. It follows that $(x,y) \in \Delta'$. Thus, there exists a basic open set $W \in \mathscr{R}$ such that $(x,y) \in W$ and $W \subseteq \Delta'$. Because W is a basic set in the product topology, we can find $\mathscr{T}$-open sets U and V such that $W = p_1^{-1}(U) \cap p_2^{-2}(V)$. Consequently $p_1(x,y) \in U$ and $p_2(x,y) \in V$. That is,

$x \in U$ and $y \in V$. The sets U and V are known to be open in $\mathscr{T}$. Thus, it remains only to show that $U \cap V = \emptyset$. Observe that

$$z \in U \cap V \implies \left[(z,z) \in p_1^{-1}(U)\right] \wedge \left[(z,z) \in p_2^{-1}(V)\right]$$
$$\implies (z,z) \in p_1^{-1}(U) \cap p_2^{-1}(V)$$
$$\implies (z,z) \in W$$
$$\implies W \cap \Delta \neq \emptyset.$$

This contradicts the fact that $W \subseteq \Delta'$, and so $U \cap V = \emptyset$. Therefore, $(X, \mathscr{T})$ is a Hausdorff space. Q.E.D.

The next theorem is a useful consequence of Theorem 4.3.5.

Theorem 4.3.6. *Let $(X, \mathscr{T})$ be a topological space and let $(Y, \mathscr{R})$ be a Hausdorff space. If the functions $f : X \to Y$ and $g : X \to Y$ are continuous, then the set $\{x \mid f(x) = g(x)\}$ is closed in X.*

Proof. Define a function $h : X \to Y \times Y$ by $h(x) = \big(f(x), g(x)\big)$. Since f and g are assumed to be continuous, the function h is continuous, by Corollary 4.2.5. Let Δ be the diagonal in $Y \times Y$. By Theorem 4.3.5, the set Δ is closed in $Y \times Y$. Consequently, the continuity of h implies that $h^{-1}(\Delta)$ is closed in X. It remains only to observe that $h^{-1}(\Delta) = \{x \mid f(x) = g(x)\}$. Q.E.D.

Corollary 4.3.7. *Let $(X, \mathscr{T})$ be a topological space and let $(Y, \mathscr{R})$ be a Hausdorff space. Suppose that $f : X \to Y$ and $g : X \to Y$ are continuous functions. If the set $\{x \mid f(x) = g(x)\}$ is dense in X, then $f = g$.*

Proof. Let $B = \{x \mid f(x) = g(x)\}$. By Theorem 4.3.6, we know that B is closed in X. Thus, $\overline{B} = B$. On the other hand, we assumed that B was dense in X, and so $\overline{B} = X$. Therefore, $B = X$, and so $f(x) = g(x)$ for all $x \in X$. Q.E.D.

Theorem 4.3.8. *Let K be an index set. For each $\alpha \in K$, let $(X_\alpha, \mathscr{T}_\alpha)$ be a topological space, and let $(X, \mathscr{T})$ be the product space. If $f \in X$, then the set*

$$B_f = \big\{g \in X \mid \text{the set } \{\alpha \mid g(\alpha) \neq f(\alpha)\} \text{ is finite}\big\}$$

is dense in X.

Proof. We need only show that if $U \in \mathscr{T}$ is nonempty, then $U \cap B_f \neq \emptyset$. We know that U contains some member of the base for the product topology. Thus, there exists a finite set of indices $\{\alpha_1, \ldots, \alpha_n\}$ such that $\bigcap_{i=1}^{n} p_{\alpha_i}^{-1}(V_{\alpha_i}) \subseteq U$, where V_{α_i} is a nonempty set in $\mathscr{T}_{\alpha_i}$ for each $i \in \{1, \ldots, n\}$.

For each $i \in \{1, \ldots, n\}$, choose some $x_{\alpha_i} \in V_{\alpha_i}$. Now, define $h \in X$ as follows:

$$h(\alpha) = \begin{cases} f(\alpha) & \text{if } \alpha \notin \{\alpha_1, \ldots, \alpha_n\}, \\ x_\alpha & \text{if } \alpha \in \{\alpha_1, \ldots, \alpha_n\}. \end{cases}$$

We certainly have that $h \in B_f$, because $h(\alpha) = f(\alpha)$ for all but finitely many values of α. It is also clear that $h \in p_{\alpha_i}^{-1}(V_{\alpha_i})$ for each $i \in \{1, \ldots, n\}$. Therefore, $h \in U \cap B_f$, and so $U \cap B_f \neq \emptyset$. Q.E.D.

Exercises

1. Let K be an index set and, for each $\alpha \in K$, let $(X_\alpha, \mathscr{T}_\alpha)$ be a topological space. Let $(X, \mathscr{T})$ denote the product space. Suppose that $A_\alpha \in \mathscr{T}_\alpha$ is nonempty for each $\alpha \in K$. If $\prod_{\alpha \in K} A_\alpha$ is open, prove that $A_\alpha = X_\alpha$ for all but finitely many values of α.

2. Verify the claim made at the end of the proof of Theorem 4.3.1. That is, show that
$$Y \cap \left(\prod_{\alpha \in K} U_\alpha \right) = \prod_{\alpha \in K} W_\alpha = W.$$

3. Let $(X, \mathscr{T})$ and $(Y, \mathscr{R})$ be topological spaces and let $(X \times Y, \mathscr{P})$ be the product space. Suppose $f : X \to Y$ is a function. The *graph* of f is the set
$$G_f = \{(x, f(x)) \mid x \in X\}.$$
Define a function $g : X \to G_f$ by $g(x) = (x, f(x))$. Prove that f is continuous if and only if g is a homeomorphism, where G_f is given the subspace topology induced by $\mathscr{P}$.

4. Let K be an index set and, for each $\alpha \in K$, let $(X_\alpha, \mathscr{T}_\alpha)$ be a topological space. Let $(X, \mathscr{T})$ denote the product space. Suppose that A_α is a subset of X_α for each $\alpha \in K$. Prove that $\prod_{\alpha \in K} A_\alpha$ is closed if and only if A_α is closed for all $\alpha \in K$.

5. Let K be an infinite index set and, for each $\alpha \in K$, let $(X_\alpha, \mathscr{T}_\alpha)$ be a topological space. Let $(X, \mathscr{T})$ denote the product space. Suppose that A_α is a subset of X_α for each $\alpha \in K$. Prove or disprove the following:
$$\left(\prod_{\alpha \in K} A_\alpha \right)^\circ = \prod_{\alpha \in K} A_\alpha^\circ.$$

6. Prove or disprove the following: The image of a Hausdorff space under a continuous function is a Hausdorff space. (The image is given the subspace topology in the codomain.)

Chapter 5

Connectedness

5.1 Introduction to Connectedness

In the present chapter, we introduce a very useful topological property. We recall from elementary calculus that if the value of a continuous function is positive at x_1 and negative at x_2, then it must assume the value zero somewhere between x_1 and x_2. This important fact, which is used to prove the Intermediate Value Theorem, follows from a topological property that intervals possess. (Although it is not usually phrased this way in elementary calculus.) In this section, we define this topological property, which is known as "connectedness," and give several examples of interesting and familiar applications.

Definition 5.1.1. Let $(X, \mathscr{T})$ be a topological space and suppose A and B are subsets of X. If $\overline{A} \cap B = \emptyset$ and $A \cap \overline{B} = \emptyset$, then we say that A and B are *separated*. That is, A and B are separated if no point in one set belongs to the closure of the other.

Example 5.1.2. Give an example of a topological space with two subsets that are disjoint but not separated.

Solution. Let $\mathbb{R}$ be the set of real numbers with the standard topology. Let A and B denote the rational and irrational numbers, respectively. Certainly, A and B are disjoint. It is well-known that every irrational number is a boundary point of A and every rational number is a boundary point of B. In fact, both A and B are dense in $\mathbb{R}$, which means $\overline{A} = \mathbb{R}$ and $\overline{B} = \mathbb{R}$. Therefore,

$$\overline{A} \cap B = \mathbb{R} \cap B = B \quad \text{and} \quad A \cap \overline{B} = A \cap \mathbb{R} = A.$$

Since neither A nor B is empty, we have both $\overline{A} \cap B \neq \emptyset$ and $A \cap \overline{B} \neq \emptyset$. Therefore, A and B are disjoint, but not separated.

In Example 5.1.2, to show that A and B were not separated, it would have been sufficient to show either $\overline{A} \cap B \neq \emptyset$ or $A \cap \overline{B} \neq \emptyset$. It was not necessary to show both. An alternative solution would be to take $A = (-\infty, 0]$ and $B = (0, \infty)$. These sets are disjoint and $A \cap \overline{B} = \{0\}$, which is nonempty. Thus, A and B are disjoint sets that are not separated. Observe, however, that $\overline{A} \cap B = \emptyset$.

Definitions 5.1.3. A topological space $(X, \mathcal{T})$ is said to be *connected* if it is not the union of two nonempty separated subsets. A subset Y of X is said to a *connected set* if $(Y, \mathcal{R})$ is a connected space, where $\mathcal{R}$ is the subspace topology on Y.

For future applications of this concept, we will find it convenient to have alternate conditions that characterize connectedness (or the lack of it). Some of these conditions are stated in the next theorem. (We recall that a subset of a set X is said to be a *proper subset* if it is neither X nor $\emptyset$.)

Theorem 5.1.4. *For $(X, \mathcal{T})$ a topological space, the following are equivalent:*

(a) *The space $(X, \mathcal{T})$ is not connected.*

(b) *X is the union of two nonempty disjoint closed sets.*

(c) *X is the union of two nonempty disjoint open sets.*

(d) *There exists a proper subset of X which is both open and closed.*

(e) *There exists a continuous function f from X onto a two point discrete space.*

Proof. $(a \Rightarrow b)$ Suppose $(X, \mathcal{T})$ is not connected. Then there are two nonempty separated sets A and B such that $X = A \cup B$. The assumption is that A and B are separated, and so

$$A \cap B \subseteq \overline{A} \cap B = \emptyset.$$

It follows that A and B are disjoint. It remains to show that A and B are both closed sets. Since A and B are disjoint sets such that $X = A \cup B$, we know that $A = B'$. Consequently, since $\overline{A} \cap B = \emptyset$, it follows that $\overline{A} \subseteq B' = A$. This implies that $A = \overline{A}$, and so A is a closed set. A similar argument shows that B is also a closed set.

$(b \Rightarrow c)$ Suppose $X = A \cup B$, where A and B are nonempty disjoint closed sets. Then $A = B'$ and $B = A'$, and so A and B are also two nonempty disjoint open sets with union X.

$(c \Rightarrow d)$ Let A and B be nonempty disjoint open sets such that $X = A \cup B$. Then $B = A'$ and $A = B'$, and so A and B are both closed sets. Since $X = A \cup B$ and neither A nor B is empty, it follows that A and B are both proper subsets of X. Therefore, each of A and B is a proper subset of X that is both open and closed.

$(d \Rightarrow e)$ Assume that A is a proper subset of X which is both open and closed. If $B = A'$, then B is also both open and closed. The fact that A is a proper subset of X implies that $B = A'$ is nonempty. Now, let $Y = \{a, b\}$ be a two point set equipped with the discrete topology. Define a function $f : X \to Y$ by

$$f(x) = \begin{cases} a & \text{if } x \in A, \\ b & \text{if } x \in B. \end{cases}$$

Then f is a continuous function from X onto Y.

$(e \Rightarrow a)$ Assume there exists a space $Y = \{a, b\}$ with the discrete topology and a function $f : X \to Y$ that is both continuous and onto. Let $A = f^{-1}(\{a\})$ and $B = f^{-1}(\{b\})$. Because f maps X onto $\{a, b\}$,

$$A \cup B = f^{-1}(\{a\}) \cup f^{-1}(\{b\}) = f^{-1}(\{a, b\}) = X.$$

By the continuity of f, both sets A and B are closed. This means that $\overline{A} = A$ and $\overline{B} = B$. Consequently,

$$\overline{A} \cap B = A \cap B = f^{-1}(\{a\}) \cap f^{-1}(\{b\}) = \emptyset.$$

A similar argument shows that $A \cap \overline{B} = \emptyset$. Therefore, A and B are separated, and so the space $(X, \mathscr{T})$ is not connected. Q.E.D.

We will use condition (e) of Theorem 5.1.4 frequently, and so it is useful to give the following definition.

Definition 5.1.5. Let $(X, \mathscr{T})$ be a topological space and suppose that $\{a, b\}$ is a two point set which is given the discrete topology. If $g : X \to \{a, b\}$ is a function which is both onto and continuous, then g is said to *split* X. (See Figure 5.1.a.)

Theorem 5.1.6. *The image of a connected space under a continuous function is connected.*

Proof. Let $(X, \mathscr{T})$ and $(Y, \mathscr{R})$ be topological spaces and let $f : X \to Y$ be a continuous function from X onto Y. Suppose that $(X, \mathscr{T})$ is connected but $(Y, \mathscr{R})$ is not connected. By assumption, there exists a function g that splits Y. Consequently, the composition $g \circ f$ splits X. This is a contradiction, because $(X, \mathscr{T})$ is connected. Therefore, $(Y, \mathscr{R})$ is connected. Q.E.D.

Corollary 5.1.7. *Connectedness is a topological property.*

Proof. Assume that $(X, \mathscr{T})$ and $(Y, \mathscr{R})$ are homeomorphic topological spaces. Suppose $(X, \mathscr{T})$ is connected. By assumption, there exists a homeomorphism $h : X \to Y$. A homeomorphism is necessarily both continuous and onto. Thus $(Y, \mathscr{R})$ is the image of $(X, \mathscr{T})$ under the continuous function h, and so $(Y, \mathscr{R})$ is also connected, by Theorem 5.1.6. Q.E.D.

Theorem 5.1.8. *Let $(X, \mathscr{T})$ be a topological space and let A be a connected subset of X. If $g : X \to \{a, b\}$ splits X, then $A \subseteq g^{-1}(\{a\})$ or $A \subseteq g^{-1}(\{b\})$.*

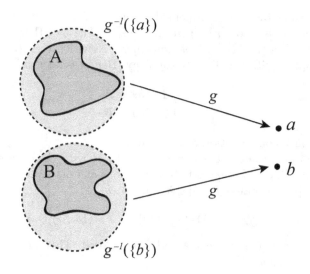

Figure 5.1.a: The continuous function g splits the space $X = A \cup B$.

Proof. Because g splits X (by assumption), we have $X = g^{-1}(\{a\}) \cup g^{-1}(\{b\})$. If the conclusion of the theorem is not satisfied, then we have $A \cap g^{-1}(\{a\}) \neq \emptyset$ and $A \cap g^{-1}(\{b\}) \neq \emptyset$. But, in that case, the restriction $g|_A$ splits A, which contradicts the assumption that A is connected. $\hfill$ Q.E.D.

See Figure 5.1.b for an illustration of Theorem 5.1.8.

Theorem 5.1.9. *Let $(X, \mathcal{T})$ be a topological space and let K be an index set. For each $\alpha \in K$, let A_α be a connected subset of $(X, \mathcal{T})$. If the sets A_α and A_β are not separated for all pairs $\{\alpha, \beta\} \subseteq K$, then $\bigcup_{\alpha \in K} A_\alpha$ is connected.*

Proof. Let $A = \bigcup_{\alpha \in K} A_\alpha$ and suppose that A is not connected. By Theorem 5.1.4, there exists a function $g : A \to \{a, b\}$ that splits A. Thus, there exist distinct x and y in A such that $x \in g^{-1}(\{a\})$ and $y \in g^{-1}(\{b\})$. By the definition of A, there must be some $\alpha \in K$ such that $x \in A_\alpha$ and some $\beta \in K$ such that $y \in A_\beta$. From Theorem 5.1.8, it follows that $A_\alpha \subseteq g^{-1}(\{a\})$ and $A_\beta \subseteq g^{-1}(\{b\})$. Note that $g^{-1}(\{a\})$ and $g^{-1}(\{b\})$ are closed because g is continuous and $\{a\}$ and $\{b\}$ are closed. We conclude that $\overline{A_\alpha} \subseteq g^{-1}(\{a\})$ and $\overline{A_\beta} \subseteq g^{-1}(\{b\})$. Since g splits A, it follows that $\overline{A_\alpha} \cap A_\beta = \emptyset$ and $A_\alpha \cap \overline{A_\beta} = \emptyset$. This contradicts the assumption that A_α and A_β are not separated. Therefore, we must reject the supposition that A is not connected. $\hfill$ Q.E.D.

Theorem 5.1.10. *Suppose A and B are subsets of a topological space such that $A \subseteq B \subseteq \overline{A}$. If A is connected, then so is B. In particular, the closure of a connected set is connected.*

Proof. Let $(X, \mathcal{T})$ be a topological space containing the sets A and B, where A is a subset of B. Suppose that A is connected, but B is not connected. Since B

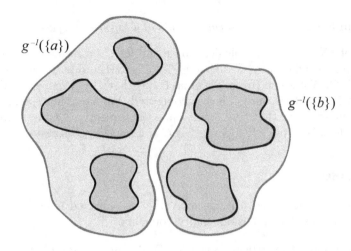

$g^{-1}(\{a\})$

$g^{-1}(\{b\})$

Figure 5.1.b: Each connected subset of X is contained within either $g^{-1}(\{a\})$ or $g^{-1}(\{b\})$. (See Theorem 5.1.8.)

is not connected, there exists a function $g : B \to \{a, b\}$ that splits B. Because A is connected and $A \subseteq B$, it follows from Theorem 5.1.8 that $A \subseteq g^{-1}(\{a\})$ or $A \subseteq g^{-1}(\{b\})$. Without loss of generality, assume $A \subseteq g^{-1}(\{a\})$.

The function g is continuous and $\{a\}$ is closed in the discrete topology. Thus, the set $g^{-1}(\{a\})$ is closed in $(X, \mathscr{T})$. It follows (from our assumption on A) that $\overline{A} \subseteq g^{-1}(\{a\})$, and hence $B \subseteq g^{-1}(\{a\})$ (since $B \subseteq \overline{A}$). From this we conclude that $g(B) \subseteq \{a\}$, which contradicts the assumption that $g : B \to \{a, b\}$ is an onto function. Consequently, we must reject the assumption that B is not connected. Q.E.D.

If a topological space $(X, \mathscr{T})$ is not connected, it is natural to try to obtain some information about the various pieces into which the space can be separated. Singletons (single point sets) are connected. So, for each $c \in X$, we have at least one connected subset of X that contains c. We wish to consider a maximal connected subset of X that contains c. To that end, we provide a definition.

Definition 5.1.11. Let $(X, \mathscr{T})$ be a topological space and let $c \in Y$, where Y is a subset of X. The *(connected) component* of Y corresponding to c is the set

$$C(Y, c) = \bigcup_{A \in \mathscr{C}} A,$$

where $\mathscr{C}$ is the collection of all connected subsets of Y that contain c.

By Theorem 5.1.9, each component is connected. (See also Exercise 3 at the end of this section.) It is not hard to show that components corresponding to different points of Y are either equal or disjoint. (See Exercise 10 at the end of this section.)

Theorem 5.1.12. *Connected components of a topological space are closed sets.*

Proof. Let $(X, \mathscr{T})$ be a topological space and let C be a component of X corresponding to some $x_0 \in C$. That is, let $C = C(X, x_0)$. It suffices to show that $\overline{C} \subseteq C$. We know (from Theorem 5.1.9) that C is connected. Hence, by Theorem 5.1.10, its closure $\overline{C}$ is also connected. We have thus established that $\overline{C}$ is a connected set that contains x_0, and consequently $\overline{C}$ is a member of the union that makes up C. It follows that $\overline{C} \subseteq C$, and so C is closed.   Q.E.D.

Exercises

1. Prove that a topological space $(X, \mathscr{T})$ is connected if and only if every proper subset of X has nonempty boundary.

2. Prove that a topological space $(X, \mathscr{T})$ is connected if and only if for every pair of distinct points x and y in X, there is a connected set C such that $\{x, y\} \subseteq C$ and $C \subseteq X$.

3. Let $(X, \mathscr{T})$ be a topological space and let K be an index set. For each $\alpha \in K$, let A_α be a connected subset of X. Prove that if $\bigcap_{\alpha \in K} A_\alpha \neq \emptyset$, then $\bigcup_{\alpha \in K} A_\alpha$ is connected.

4. Let $(X, \mathscr{T})$ be a topological space and, for each $i \in \mathbb{N}$, let A_i be a connected subset of X. If $A_i \cap A_{i+1} \neq \emptyset$ for each $i \in \mathbb{N}$, prove that $\bigcup_{i \in \mathbb{N}} A_i$ is connected.

5. Prove that a topological space $(X, \mathscr{T})$ is not connected if and only if there is a proper subset A of X such that $\overline{A} \cap \overline{A'} = \emptyset$.

6. Let $(X, \mathscr{T})$ be a topological space and suppose $(Y, \mathscr{R})$ is a subspace. Prove that Y is not connected if and only if there exist sets U and V in $\mathscr{T}$ such that $U \cap V = \emptyset$, $U \cap Y \neq \emptyset$, $V \cap Y \neq \emptyset$, and $Y \subseteq U \cup V$.

7. Let $(X, \mathscr{T})$ be a topological space, Y a connected subset of X, and K an index set. For each $\alpha \in K$, let A_α be a connected subset of X such that $Y \cap A_\alpha \neq \emptyset$. Prove that $Y \cup \left(\bigcup_{\alpha \in K} A_\alpha \right)$ is connected.

8. Suppose that $(X, \mathscr{T})$ is a connected topological space. If $\mathscr{T}$ is the discrete topology, what can we say about X?

9. Let $(X, \mathscr{T})$ be a topological space and let A be a connected subset of X. If B is a subset of X such that $A \cap B \neq \emptyset$ and $A \cap B' \neq \emptyset$, prove that $A \cap \partial B \neq \emptyset$. (Recall that ∂B denotes the *boundary* of B.)

10. Let $(X, \mathscr{T})$ be a topological space. Define a relation $\sim$ on X by setting $x \sim y$ if and only if there is a connected subset E of X such that $\{x, y\}$ is a subset of E. Prove that $\sim$ is an equivalence relation.

11. Let $(X, \mathscr{T})$ and $(X, \mathscr{R})$ be topological spaces such that $\mathscr{T} \subseteq \mathscr{R}$.

    (a) If $(X, \mathscr{T})$ is connected, must $(X, \mathscr{R})$ be connected?
    (b) If $(X, \mathscr{R})$ is connected, must $(X, \mathscr{T})$ be connected?

    Justify your answers by giving a proof or a counterexample.

12. Let $(X, \mathscr{T})$ be a topological space and suppose A is a subset of X. Prove that $\overline{A}$ is connected if and only if A is not the union of two nonempty sets B and C where $\overline{B} \cap \overline{C} = \emptyset$.

13. Let $(X, \mathscr{T})$ be a connected, metrizable topological space. Prove that X is uncountable. (See Definition 2.8.8 for the definition of a metrizable space.)

5.2   Products of Connected Spaces

It is easy to see that if the product of topological spaces is connected, then each coordinate space must be connected. This is because each factor X_α is the image of the product $\prod_{\alpha \in K} X_\alpha$ under the continuous projection function p_α. In this section, we prove that the converse is true also. We begin with the simpler case, where we have only two spaces.

Theorem 5.2.1. *Let $(X_1, \mathscr{T}_1)$ and $(X_2, \mathscr{T}_2)$ be topological spaces and let $(X, \mathscr{T})$ be the product space. The product space $(X, \mathscr{T})$ is connected if and only if each of the spaces $(X_1, \mathscr{T}_1)$ and $(X_2, \mathscr{T}_2)$ is connected.*

Proof. Let $p_1 : X \to X_1$ and $p_2 : X \to X_2$ be the projection functions. Suppose that $(X, \mathscr{T})$ is connected. The projection functions are continuous and onto (by Theorem 4.2.3). Thus, X_1 and X_2 are connected (by Theorem 5.1.6).

Conversely, assume that $(X_1, \mathscr{T}_1)$ and $(X_2, \mathscr{T}_2)$ are connected. We wish to show that the product space $(X, \mathscr{T})$ is connected. Suppose to the contrary that $(X, \mathscr{T})$ is not connected. Then there exists a function $G : X \to \{a, b\}$ that splits X. Let (x_1, x_2) be a point in X and assume, without loss of generality, that $G(x_1, x_2) = a$.

Let $B_1 = \{(u, x_2) \mid u \in X_1\}$. As a subspace of $(X, \mathscr{T})$, the space B_1 is homeomorphic to $(X_1, \mathscr{T}_1)$, by Theorem 4.3.2, and so must be connected. Consequently, the function $G|_{B_1}$ must be constant, for otherwise it would split B_1. Since $(x_1, x_2) \in B_1$ and $G(x_1, x_2) = a$, we must have that $G(u, x_2) = a$ for all $(u, x_2) \in B_1$.

Next, let (y_1, y_2) be an arbitrary point in the product space X and define a set $B_2 = \{(y_1, v) \mid v \in X_2\}$. As a subspace of $(X, \mathscr{T})$, the space B_2 is homeomorphic to $(X_2, \mathscr{T}_2)$, and so must be connected. Consequently, the function $G|_{B_2}$ must be constant. Now observe that $(y_1, x_2) \in B_1 \cap B_2$. It follows that

$$G(y_1, y_2) = G(y_1, x_2) = G(x_1, x_2) = a.$$

The point (y_1, y_2) was arbitrary in X, however, and so we have shown that G is constant on X. This is a contradiction since G splits X, and hence is onto $\{a, b\}$. Therefore, we reject the supposition that $(X, \mathscr{T})$ is not connected.   Q.E.D.

The proof of the general case is more involved, but it makes use of the same idea.

Theorem 5.2.2. *A product space is connected if and only if each coordinate space is connected.*

Proof. Let K be an index set and, for each $\alpha \in K$, let $(X_\alpha, \mathscr{T}_\alpha)$ be a topological space. Denote the product space by $(X, \mathscr{T})$.

Suppose first that the product space $(X, \mathscr{T})$ is connected and let $\alpha \in K$ be given. If $p_\alpha : X \to X_\alpha$ is the αth projection function, then p_α is continuous and onto, by Theorem 4.2.3. Thus, $X_\alpha = p_\alpha(X)$ is the image of a connected space under a continuous function, and so is also connected, by Theorem 5.1.6.

Now, conversely, we assume that $(X_\alpha, \mathscr{T}_\alpha)$ is connected for each $\alpha \in K$. Our objective is to show that the product space $(X, \mathscr{T})$ is connected. Assume to the contrary that it is not connected. Then there is a function $G : X \to \{a, b\}$ that splits X. Let $f \in X$ and assume, without loss of generality, that $G(f) = a$. Let

$$B_f = \{h \in X \mid h(\alpha) = f(\alpha) \text{ for all but finitely many } \alpha \in K\}.$$

Then, by Theorem 4.3.8, the set B_f is dense in $(X, \mathscr{T})$. We will show that $G(h) = a$ for all $h \in B_f$.

Let $h \in B_f$. Then there exists a finite set $\{\alpha_1, \dots, \alpha_n\} \subseteq K$ such that

$$h(\alpha) = f(\alpha) \text{ if } \alpha \in K \setminus \{\alpha_1, \dots, \alpha_n\}.$$

For each $k \in \{1, \dots, n\}$, define a set

$$B_{f,\alpha_1,\dots,\alpha_k} = \{\phi \in X \mid \phi(\alpha) = f(\alpha) \text{ if } \alpha \in K \setminus \{\alpha_1, \dots, \alpha_k\}\}.$$

Observe that $h \in B_{f,\alpha_1,\dots,\alpha_n}$. We will proceed by means of an inductive argument on $k \in \{1, \dots, n\}$.

First consider $k = 1$. As a subspace of $(X, \mathscr{T})$, the space B_{f,α_1} is homeomorphic to $(X_{\alpha_1}, \mathscr{T}_{\alpha_1})$, by Theorem 4.3.2. Consequently, B_{f,α_1} is connected, and so $G|_{B_{f,\alpha_1}}$ is constant. (For otherwise, the function $G|_{B_{f,\alpha_1}}$ would split B_{f,α_1}.) Because $f \in B_{f,\alpha_1}$ and $G(f) = a$, we conclude that $G(\phi) = a$ for all $\phi \in B_{f,\alpha_1}$.

Now let $1 \le k < n$ and suppose that $G|_{B_{f,\alpha_1,\dots,\alpha_k}}$ is constant. Arguing as before, since $f \in B_{f,\alpha_1,\dots,\alpha_k}$ and $G(f) = a$, it follows that $G(\phi) = a$ for all $\phi \in B_{f,\alpha_1,\dots,\alpha_k}$.

Let $\psi \in B_{f,\alpha_1,\dots,\alpha_k,\alpha_{k+1}}$ be arbitrary. Define an element $g \in X$ as follows:

$$g(\alpha) = \begin{cases} \psi(\alpha) & \text{if } \alpha \in \{\alpha_1, \dots, \alpha_k\}, \\ f(\alpha) & \text{if } \alpha \in K \setminus \{\alpha_1, \dots, \alpha_k\}. \end{cases}$$

Then $g \in B_{f,\alpha_1,\dots,\alpha_k}$, and so $G(g) = a$. Next, define a set

$$B_{g,\alpha_{k+1}} = \{\phi \in X \mid \phi(\alpha) = g(\alpha) \text{ if } \alpha \in K \setminus \{\alpha_{k+1}\}\}.$$

Once again making use of Theorem 4.3.2, we see that $B_{g,\alpha_{k+1}}$ is homeomorphic to $(X_{\alpha_{k+1}}, \mathscr{T}_{\alpha_{k+1}})$. Thus, $B_{g,\alpha_{k+1}}$ is a connected subset of X, and so $G|_{B_{g,\alpha_{k+1}}}$

is constant. Because $g \in B_{g,\alpha_{k+1}}$ and $G(g) = a$, we conclude that $G(\phi) = a$ for all ϕ in the set $B_{g,\alpha_{k+1}}$.

Finally, we observe that $\psi \in B_{g,\alpha_{k+1}}$, and consequently $G(\psi) = a$. Since ψ was an arbitrary element in $B_{f,\alpha_1,...,\alpha_k,\alpha_{k+1}}$, it follows that $G(\psi) = a$ for all $\psi \in B_{f,\alpha_1,...,\alpha_k,\alpha_{k+1}}$. Hence, by the Principle of Mathematical Induction, G is constant on $B_{f,\alpha_1,...,\alpha_n}$ and $G(\psi) = a$ for all $\psi \in B_{f,\alpha_1,...,\alpha_n}$. Because $h \in B_{f,\alpha_1,...,\alpha_n}$, we conclude that $G(h) = a$. The choice of h in B_f was arbitrary, and so we conclude that $G(h) = a$ for all $h \in B_f$.

To complete the proof, we define a function $F : X \to \{a, b\}$ by $F(h) = a$ for all $h \in X$. Then F and G agree on B_f, a dense subset of X. By Corollary 4.3.7, we conclude that $F(h) = G(h)$, and consequently $G(h) = a$, for all $h \in X$. This is a contradiction because the range of G is the set $\{a, b\}$. Thus, we must reject the assumption that $(X, \mathscr{T})$ is not connected. Q.E.D.

Exercises

1. Let $X = \{a, b, c, d, e\}$ and $Y = \{x, y, z\}$ be two sets. Define a topology $\mathscr{T}$ on X by $\mathscr{T} = \{\emptyset, X, \{a, b\}, \{c, d, e\}\}$ and a topology $\mathscr{R}$ on Y by $\mathscr{R} = \{\emptyset, Y\}$. Both $(X, \mathscr{T})$ and $(Y, \mathscr{R})$ are topological spaces. Are X and Y connected? Is their product space connected? Justify your answers.

2. Let $X = \{a, b, c, d, e\}$ and $Y = \{x, y, z\}$ be two sets. Define a topology $\mathscr{T}$ on X by $\mathscr{T} = \{\emptyset, X, \{a, b, c\}, \{c, d, e\}, \{c\}\}$ and a topology $\mathscr{R}$ on Y by $\mathscr{R} = \{\emptyset, Y\}$. Both $(X, \mathscr{T})$ and $(Y, \mathscr{R})$ are topological spaces. Are X and Y connected? Is their product space connected? Justify your answers.

3. Let K be an index set and for each $\alpha \in K$ let $(X_\alpha, \mathscr{T}_\alpha)$ be a topological space. Show that the product space $(X, \mathscr{T})$ is not connected if $(X_\beta, \mathscr{T}_\beta)$ is not connected for some $\beta \in K$.

4. Let $(X, \mathscr{T})$ and $(Y, \mathscr{R})$ be connected topological spaces. If A and B are proper subsets of X and Y, respectively, then prove that $(X \times Y) \backslash (A \times B)$ is a connected subset of the product space.

5. Write a detailed proof similar to that of Theorem 5.2.1 for the following:

   Theorem 5.2.3. *For each $i \in \{1, 2, 3\}$, let $(X_i, \mathscr{T}_i)$ be a topological space. The product space $(X, \mathscr{T})$ is connected if and only if $(X_i, \mathscr{T}_i)$ is connected for each $i \in \{1, 2, 3\}$.*

   When writing your proof, let $K = \{1, 2, 3\}$ and consider each element x of the product space as a function $x : K \to \bigcup_{i \in K} X_i$, where $x(i) \in X_i$ for each $i \in K$.

5.3   Connected Subsets of the Real Line

In this section, we show that the only connected subsets of the real line are singletons and intervals. We need to recall some definitions and an axiom from elementary calculus. If S is a set of real numbers, then a real number b is said to be an *upper bound* of S if $x \leq b$ for all $x \in S$. If S has an upper bound b, then any number $b' > b$ is also an upper bound, so S has an infinite number of upper bounds. If c is an upper bound of S and $c \leq b$ for any upper bound b of S, then c is called the *least upper bound* or the *supremum* of S and we write $c = \sup(S)$. Similarly, a number b is called a *lower bound* of S if $x \geq b$ for all $x \in S$. If c is a lower bound of S and $c \geq b$ for any lower bound b, then c is called the *greatest lower bound* or the *infimum* of S and we write $c = \inf(S)$.

We now state a very important axiom.

Axiom 5.3.1 (Axiom of Completeness). *If S is a nonempty set of real numbers that has an upper bound, then S has a least upper bound.*

Example 5.3.2. Let $A = \{1, 3, 5\}$, $B = [-2, 4)$, $C = \{1/n \mid n \in \mathbb{N}\}$, and $D = (-2, \infty)$.

(a) If possible, find three upper bounds for each set.

(b) If possible, find three lower bounds for each set.

(c) Find the supremum (the least upper bound) of each set that has an upper bound.

(d) Find the infimum (the greatest lower bound) of each set that has a lower bound.

Solution.   (a) The numbers 6, 8, and 17 are upper bounds of each A, B, and C. The set D has no upper bound.

(b) The numbers $-3, -5, -23$ are lower bounds of each set A, B, C, and D.

(c) $\sup A = 5$, $\sup B = 4$, $\sup C = 1$; the set D has no upper bound, and hence no supremum.

(d) $\inf A = 1$, $\inf B = -2$, $\inf C = 0$, $\inf D = -2$.

Example 5.3.2 illustrates the fact that the supremum of a set may or may not be in the set. (The same can be said of the infimum of a set.)

Theorem 5.3.3. *A set of real numbers with at least two members is connected if and only if it is an interval.*

Proof. We begin by making an observation about intervals. A set of real numbers I is an interval if and only if the following is true: If $\{x, y\} \subseteq I$ and r is any real number between x and y, then $r \in I$.

Assume that S is a connected set of real numbers. We wish to show that S is an interval. Suppose to the contrary that S is not an interval. Then there exist real numbers x and y in S and a real number r such that $x < r < y$, but $r \notin S$. Let $U = (-\infty, r) \cap S$ and $V = (r, \infty) \cap S$. We know that $x \in U$ and $y \in V$, by construction. It follows that U and V are nonempty subsets of S that are open in the subspace topology on S. Certainly, we have that $U \cap V = \emptyset$. We also have that $S = U \cup V$, because $r \notin S$. Therefore, S is not connected, which is a contradiction. Consequently, we reject our assumption and conclude that S is an interval.

Next, we show that all intervals are connected. Begin by assuming that S is an open interval. Suppose S is not connected. Then there are nonempty sets U and V, open in the subspace topology on S, such that $U \cap V = \emptyset$ and $U \cup V = S$. Let $x \in U$ and $y \in V$ and assume (without loss of generality) that $x < y$.

Let $A = \{z \in U \mid z < y\}$. We know that $x \in A$, and so $A \neq \emptyset$. By the Axiom of Completeness, A has a supremum. Let $c = \sup(A)$. Then $x \le c$, because $c = \sup(A)$ and $x \in A$. Since y is an upper bound of A, and since c is the least upper bound of A, it follows that $c \le y$. Therefore, $x \le c \le y$, and so c is between x and y. Because S is an interval that contains both x and y, it follows that $c \in S$. Consequently, we have that either $c \in U$ or $c \in V$. We will show that both cases lead to a contradiction.

Suppose $c \in U$. The set U is open in the subspace topology on S, and so there exists a set $\widetilde{U}$ that is open in the standard topology on $\mathbb{R}$ such that $U = \widetilde{U} \cap S$. Because S is an open interval, it follows that U is open in the standard topology on $\mathbb{R}$. Thus, since $c \in U$, there exists an $\epsilon > 0$ such that $(c - \epsilon, c + \epsilon) \subseteq U$. Pick any $s \in (c, c + \epsilon)$. Recall that c is the least upper bound of the set $\{z \in U \mid z < y\}$. Then, since $s \in U$ and $s > c$, it must be that $y \le s$, and consequently $y < c + \epsilon$. But we have already demonstrated that $c \le y$, and so $y \in [c, c + \epsilon) \subseteq U$. Thus, $y \in U \cap V$, contradicting the fact that U and V are disjoint. Therefore, $c \notin U$.

Suppose instead that $c \in V$. Then, as before, there exists some $\delta > 0$ such that $(c - \delta, c + \delta) \subseteq V$. We have that $A \subseteq U$ and $(c - \delta, c) \subseteq V$. Hence, $A \cap (c - \delta, c) = \emptyset$. If $t \in (c - \delta, c)$, then $z \le t$ for each $z \in A$, and so t is an upper bound for A such that $t < c$. This contradicts the fact that c is the least upper bound of A. Therefore, c cannot be in either U or V. We must reject the assumption that S is not connected and conclude that any open interval in $\mathbb{R}$ is connected.

Suppose that $I = (a, b)$ is an open interval, where a and b are real numbers such that $a < b$. We have established that I is connected. Observe that

$$I \subseteq (a, b] \subseteq [a, b] = \overline{I} \quad \text{and} \quad I \subseteq [a, b) \subseteq [a, b] = \overline{I}$$

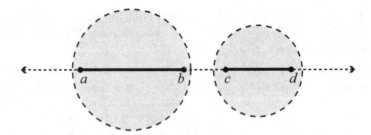

Figure 5.3.a: This image illustrates Theorem 5.3.3. The set $S = [a, b] \cup [c, d]$ is the union of two disjoint intervals. Since there exist disjoint open sets in $\mathbb{R}$ containing $[a, b]$ and $[c, d]$, it follows that $[a, b]$ and $[c, d]$ are both open in the subspace topology. Thus, S is the union of two disjoint nonempty open sets, and so S is not connected, by Theorem 5.1.4

Consequently, by Theorem 5.1.10, we have that the intervals $(a, b]$, $[a, b)$, and $[a, b]$ are connected. Similarly, if $I = (-\infty, b)$ or $I = (a, \infty)$, then

$$I \subseteq (-\infty, b] = \overline{I} \quad \text{or} \quad I \subseteq [a, \infty) = \overline{I}.$$

Therefore, all intervals (of finite or infinite length) are connected.   Q.E.D.

See Figure 5.3.a for an illustration of Theorem 5.3.3.

We can now prove a theorem that most students have used many times in their elementary mathematics classes.

Theorem 5.3.4 (Intermediate Value Theorem). *Let a and b be real numbers such that $a < b$. Suppose $f : [a, b] \to \mathbb{R}$ is a continuous function such that $f(a) \neq f(b)$. For each w between $f(a)$ and $f(b)$, there exists at least one number $c \in (a, b)$ such that $f(c) = w$.*

Proof. The interval $[a, b]$ is connected and f is continuous. Thus, the set $f([a, b])$ is connected (by Theorem 5.1.6), and so is an interval (by Theorem 5.3.3). Since $f(a)$ and $f(b)$ are in the interval $f([a, b])$, and since w is between $f(a)$ and $f(b)$, it must be that w is in $f([a, b])$. Therefore, there is some number $c \in [a, b]$ such that $f(c) = w$. Because $w \neq f(a)$ and $w \neq f(b)$, it follows that $c \neq a$ and $c \neq b$, and so $c \in (a, b)$.   Q.E.D.

See Figure 5.3.b for an illustration of Theorem 5.3.4.

The Intermediate Value Theorem has many applications, two of which are illustrated in the following examples. Other applications can be found in the exercises at the end of this section.

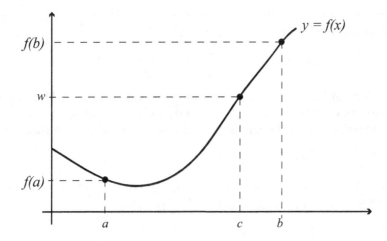

Figure 5.3.b: If f is a continuous function on $[a, b]$ and w is a number between $f(a)$ and $f(b)$, then there is a (not necessarily unique) number c between a and b for which $f(c) = w$.

Example 5.3.5. Show that $\sqrt{2}$ is a real number between 1 and 2.

Solution. The number $\sqrt{2}$ is a solution to the equation $x^2 = 2$. Let $f(x) = x^2$. Then $f : [1, 2] \to \mathbb{R}$ is a continuous function such that $f(1) = 1$ and $f(2) = 4$. Since $2 \in (1, 4)$, there is a number c between 1 and 2 such that $f(c) = 2$. That is, there is a number $c = \sqrt{2}$ in the interval $(1, 2)$ such that $c^2 = 2$.

Example 5.3.6. Suppose $p(x)$ is a degree n polynomial with real coefficients. If n is an odd integer, prove that $p(x)$ has at least one real zero.

Solution. Let $p(x) = a_0 x^n + a_1 x^{n-1} + a_2 x^{n-2} + \cdots + a_{n-1} x + a_n$, where $a_0 \neq 0$ and $a_k \in \mathbb{R}$ for each $k \in \{0, 1, \ldots, n\}$. We wish to show that the equation $p(x) = 0$ has at least one real solution. Since polynomial functions are continuous, it will suffice to show that there is a number $c > 0$ such that $p(c) > 0$ and $p(-c) < 0$. Without loss of generality, we may assume that $a_0 = 1$. Thus, we may write

$$p(x) = x^n \left(1 + \frac{a_1}{x} + \frac{a_2}{x^2} + \cdots + \frac{a_{n-1}}{x^{n-1}} + \frac{a_n}{x^n} \right),$$

whenever $x \neq 0$. The basic idea is to choose a positive number c so large that each of the fractions above, when evaluated at c and $-c$, will be very small in absolute value. In such a case, if we let

$$h(x) = 1 + \frac{a_1}{x} + \frac{a_2}{x^2} + \cdots + \frac{a_{n-1}}{x^{n-1}} + \frac{a_n}{x^n},$$

then $h(c)$ and $h(-c)$ will both be positive. Because n is an odd integer, we will have that $c^n > 0$ and $(-c)^n < 0$, and so $p(c) > 0$ and $p(-c) < 0$.

It remains to find the number c. Let c be any number such that

$$c > \max\left\{1, 2n|a_1|, 2n|a_2|, \ldots, 2n|a_{n-1}|, 2n|a_n|\right\}.$$

Let $k \in \{1, \ldots, n\}$ be given. Since $c > 1$, we have that $c^k \geq c$. It follows that $c^k > 2n|a_k|$, and consequently $\frac{|a_k|}{c^k} < \frac{1}{2n}$. Thus,

$$-\frac{1}{2n} < \frac{a_k}{c^k} < \frac{1}{2n} \quad \text{and} \quad -\frac{1}{2n} < \frac{a_k}{(-c)^k} < \frac{1}{2n}.$$

Summing over k in $\{1, \ldots, n\}$, we obtain

$$-\frac{1}{2} < \frac{a_1}{c} + \frac{a_2}{c^2} + \cdots + \frac{a_{n-1}}{c^{n-1}} + \frac{a_n}{c^n} < \frac{1}{2}$$

and

$$-\frac{1}{2} < \frac{a_1}{(-c)} + \frac{a_2}{(-c)^2} + \cdots + \frac{a_{n-1}}{(-c)^{n-1}} + \frac{a_n}{(-c)^n} < \frac{1}{2}.$$

We conclude that $\frac{1}{2} < h(c) < \frac{3}{2}$ and $\frac{1}{2} < h(-c) < \frac{3}{2}$. Hence,

$$p(c) = c^n h(c) > 0 \quad \text{and} \quad p(-c) = (-c)^n h(-c) < 0.$$

Therefore, by the Intermediate Value Theorem, there is a real number x_0 in $(-c, c)$ such that $p(x_0) = 0$. That is, x_0 is a real zero of $p(x)$.

Exercises

1. Let $A = \{-5, 3, 6\}$, $B = \left\{\frac{n}{n+1} \;\middle|\; n \in \mathbb{N}\right\}$, $C = (-\infty, 3]$, and $D = (-2, \infty)$.

   (a) If possible, find three upper bounds for each set.

   (b) If possible, find three lower bounds for each set.

   (c) Find the supremum (least upper bound) for each set that has an upper bound.

   (d) Find the infimum (greatest lower bound) for each set that has a lower bound.

2. Use the Axiom of Completeness (Axiom 5.3.1) to prove that any nonempty set of real numbers which has a lower bound must have a greatest lower bound (an *infimum*).

3. Use the Axiom of Completeness (Axiom 5.3.1) to prove that the set $\mathbb{N}$ of positive integers does not have an upper bound.

4. Let the universe be the set $\mathbb{Q}$ of rational numbers. Show that $\mathbb{Q}$ does not satisfy the Axiom of Completeness. (*Hint:* Consider the subset of $\mathbb{Q}$ defined by $S = \{x \in \mathbb{Q} \mid (x > 0) \wedge (x^2 < 2)\}$.)

5. Let $f(x) = x^2 + 5x + 2$. Show that the number 8 is between $f(0)$ and $f(2)$. Find a number c in the interval $(0, 2)$ such that $f(c) = 8$.

6. Let $g(x) = \dfrac{x}{x^2 + 2}$. Show that the number $\frac{4}{33}$ is between $g(0)$ and $g(1)$. Find a number c in the interval $(0, 1)$ such that $g(c) = \frac{4}{33}$.

7. Let $h(x) = |9 - x^2|$. Show that the number 8 is between $h(-2)$ and $h(5)$. Find a number c in the interval $(-2, 5)$ such that $h(c) = 8$.

8. Prove that if n is any positive integer, then the equation $x^n - 5$ has a unique positive solution.

9. Prove that if n is any positive integer, then any nonnegative real number has a unique nonnegative n^{th} root.

10. Let C be a circle and let P be any point inside the circle that is distinct from the center of the circle. Let a and b be the minimum and maximum distances from P to the circle, respectively. Show that there is a point on C with distance from P equal to $\frac{1}{2}(a + b)$.

11. Let K be the region bounded by an ellipse E and let P be a point on the ellipse. Show that there is a line through P that divides K into two regions of equal area.

12. Use the Intermediate Value Theorem (Theorem 5.3.4) to prove the Mean Value Theorem for Integrals:

    Theorem 5.3.7 (Mean Value Theorem for Integrals). *If $a < b$, and if the function $f : [a, b] \to \mathbb{R}$ is continuous, then there is a number c between a and b such that*

    $$f(c) = \frac{1}{b - a} \int_a^b f(x)\, dx.$$

    (*Hint:* Use the fact that numbers x_0 and x_1 can be found in $[a, b]$ such that $f(x_0) \le f(x) \le f(x_1)$ for all x in $[a, b]$.)

13. Let $f(x) = \dfrac{1}{x}$ for all $x \neq 0$.

    (a) Show that $\frac{1}{4}$ is between $f(-1)$ and $f(2)$, but there is no c in the interval $(-1, 2)$ such that $f(c) = \frac{1}{4}$.

    (b) Does this contradict the Intermediate Value Theorem? Explain.

    (c) Let $a < 0 < b$. Show that for each number w between $f(a)$ and $f(b)$ there is no c in the interval (a, b) such that $f(c) = w$.

14. Let $f(x) = \dfrac{x}{x^2 - 9}$ for all $|x| \neq 3$.

    (a) Show that $\frac{1}{7}$ is between $f(-1)$ and $f(4)$, but there is no c in the interval $(-1, 4)$ such that $f(c) = \frac{1}{7}$.

    (b) Does this contradict the Intermediate Value Theorem? Explain.

    (c) Let $a < -3$ and $3 < b$. Show that for each number w between $f(a)$ and $f(b)$ there is a c in the interval (a, b) such that $f(c) = w$. Thus, the conclusion of the Intermediate Value Theorem is true even though the hypothesis is not satisfied. Explain.

15. Show that any line segment in the Cartesian plane is connected.

16. A *polygonal path* is the union of finitely many line segments $I_1, I_2, \ldots, I_n$, where $n \in \mathbb{N}$, such that $I_k \cap I_{k+1}$ consists of a common endpoint for each $k \in \{1, 2, \ldots, n-1\}$. Show that any polygonal path in the Cartesian plane is connected.

17. (a) Explain why the Cartesian plane, with the standard topology, is a connected set.

    (b) Let C be the Cartesian plane and let A be any countable subset of C. Prove that $C \backslash A$ is a connected set. (*Hint:* Use Exercise 16 from above and the Pigeonhole Principle from Section 1.4.)

18. Let $A = \{(0, 0)\}$, $B = \{(0, 1)\}$, and $C = \{(0, y) \mid -1 \leq y \leq 1\}$. Suppose $S = \{(x, \sin(\pi/x)) \mid 0 < x \leq 1\}$. (See Figure 5.3.c.[1])

    (a) Show that S is a connected subset of the Cartesian plane.

    (b) Let $D = S \cup A$, $E = S \cup A \cup B$, and $F = S \cup C$. Show that D, E, and F are connected subsets of the Cartesian plane.

19. Let a and b be real numbers such that $a < b$ and let $f : [a, b] \to \mathbb{R}$ be a continuous function such that $f(a) < a$ and $b < f(b)$. Prove that there is a number x_0 in the interval (a, b) such that $f(x_0) = x_0$. (See Figure 5.3.d and also Exercise 8 in Section 3.5.)

---

[1] Image generated by Mathematica.

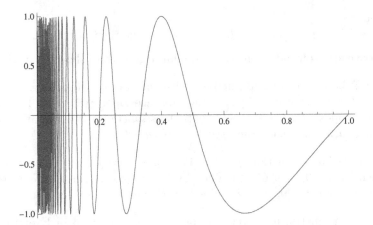

Figure 5.3.c: The graph of $y = \sin(\pi/x)$ over $(0,1]$ represents the set S in Exercise 18.

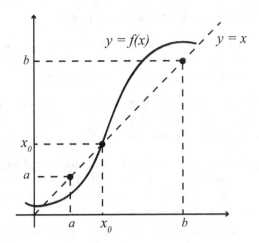

Figure 5.3.d: Illustration to accompany Exercise 19. Note that $f(a) < a$ and $f(b) > b$.

20. Let S, D, E, and F be as in Exercise 18. Show that S and E do not have the fixed point property (f.p.p.), but that both D and F do have the f.p.p.

   Hint: Each set is given the subspace topology induced by the standard topology on $\mathbb{R} \times \mathbb{R}$. To show that D has the f.p.p., assume there is a continuous function $f : D \to D$ such that $f(x,y) \neq (x,y)$ for all $x \in D$. Let

$$U = \{(x,y) \in D \mid f(x,y) \text{ is to the left of } (x,y)\}$$

and
$$V = \{(x, y) \in D \mid f(x, y) \text{ is to the right of } (x, y)\}.$$

Arrive at a contradiction. Use a similar argument to show F has the f.p.p.

21. Let $\mathbb{R}$ be the set of real numbers with the standard topology, let the set $C = \{(x, y) \mid x^2 + y^2 = 1\}$ have the subspace topology induced from $\mathbb{R} \times \mathbb{R}$, and suppose $f : C \to \mathbb{R}$ is a continuous function. Prove that there is a point (a, b) on C such that $f(a, b) = f(-a, -b)$.

    Hint: Define a function $g : C \to \mathbb{R}$ by $g(x, y) = f(x, y) - f(-x, -y)$ and then let $h : [0, 1] \to C$ be defined by $h(t) = (\cos \pi t, \sin \pi t)$. Now consider the function $F : [0, 1] \to \mathbb{R}$ defined by $F = g \circ h$. Also, see Exercise 22.

22. Give a geometric interpretation of the proof provided in Exercise 21 by sketching the graph of the function f in a three-dimensional coordinate system. It should be a continuous closed curve over a circle. For (x, y) in C, show that $g(x, y) = f(x, y) - f(-x, -y)$ is the difference between the functional values of f at two antipodal points and explain what happens when you rotate a diameter through an angle π.

23. Let a and b be real numbers such that $a < b$ and suppose $h : [a, b] \to [a, b]$ is a homeomorphism. Prove that either $h(a) = a$ and $h(b) = b$ or $h(a) = b$ and $h(b) = a$. (So h sends endpoints to endpoints.)

24. A sequence $\{C_n\}_{n=1}^{\infty}$ of sets is said to be a *nested sequence* if $C_{n+1} \subseteq C_n$ for all $n \in \mathbb{N}$. Give an example of a nested sequence $\{C_n\}_{n=1}^{\infty}$ of connected subsets of the Cartesian plane such that $\bigcap_{n=1}^{\infty} C_n$ is not connected.

25. (a) Let $(X, \mathcal{T})$ and $(Y, \mathcal{R})$ be topological spaces and suppose $f : X \to Y$ is a homeomorphism. Suppose that A is a nonempty subset of X. Prove that $(A, \mathcal{T}')$ is homeomorphic to $(f(A), \mathcal{R}')$, where $\mathcal{T}'$ and $\mathcal{R}'$ are the subspace topologies on A in X and $f(A)$ in Y, respectively.

    (b) Let $\mathbb{R}$ be the set of real numbers with $\mathcal{T}$ the standard topology. Let the product space be denoted by $(\mathbb{R} \times \mathbb{R}, \mathcal{P})$. We know that both $(\mathbb{R}, \mathcal{T})$ and $(\mathbb{R} \times \mathbb{R}, \mathcal{P})$ are connected. Prove that these two spaces are not homeomorphic.

    Hint: Assume there is a homeomorphism $h : \mathbb{R} \to \mathbb{R} \times \mathbb{R}$. Let x_0 be an arbitrary real number and arrive at a contradiction by applying (a) to $\mathbb{R} \backslash \{x_0\}$ and $\mathbb{R} \times \mathbb{R} \backslash \{h(x_0)\}$.

Chapter 6

Compactness

6.1 Introduction to Compactness

If a set of real numbers is finite, then we can be sure that we can find its largest (or smallest) element. If a collection of neighborhoods of a point x has finitely many members, then the intersection of those neighborhoods is also a neighborhood of x. We can see that finiteness plays an important role in mathematics. Naturally, however, sets and collections are often infinite. Thus, we need a vehicle that allows us to go from the infinite case to the finite case. Compactness is one of the tools that is often used to accomplish this goal. In order to formally define compactness, we need a number of preliminary definitions.

Definitions 6.1.1. Let $(X, \mathscr{T})$ be a topological space and let $Y \subseteq X$. Let $\mathscr{C}$ be a collection of subsets of X. We call $\mathscr{C}$ a *cover* of Y if $Y \subseteq \bigcup_{B \in \mathscr{C}} B$. If $\mathscr{S}$ is a cover of Y such that $\mathscr{S} \subseteq \mathscr{C}$, then $\mathscr{S}$ is called a *subcover of Y derived from $\mathscr{C}$* (or simply a *subcover* of Y, if there is no ambiguity). If the subcover $\mathscr{S}$ contains finitely many sets, we say that it is a *finite subcover*. If $\mathscr{U}$ is a cover of Y and $\mathscr{U} \subseteq \mathscr{T}$, then $\mathscr{U}$ is called an *open cover* of Y.

See Figure 6.1.a for an illustration of Definitions 6.1.1. We are now ready to introduce compactness.

Definition 6.1.2. Let $(X, \mathscr{T})$ be a topological space. A subset A of X is said to be *compact* if for each open cover $\mathscr{U}$ of A there is a finite subcover derived from $\mathscr{U}$. The topological space $(X, \mathscr{T})$ is said to be a *compact space* if X itself is compact.

Example 6.1.3. Let $\mathbb{R}$ be the set of real numbers with the standard topology. Show that the open interval $(0, 1)$ is not a compact subset of $\mathbb{R}$.

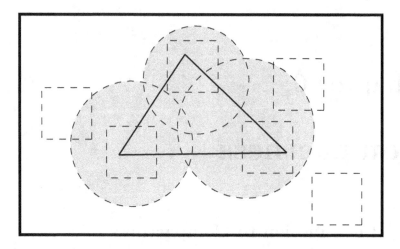

Figure 6.1.a: The set X is the triangle together with its interior. The interior of each square and each circle is an open set and the collection of these open sets is an open cover of X. If we remove all square open sets, the union of the circular open sets (the shaded region) contains X, and therefore the set of circular open sets is a subcover of X derived from the collection of all circular open sets and square open sets.

Solution. Let $\mathcal{U} = \left\{ (\frac{1}{n}, 1) \mid n \in \mathbb{N} \right\}$. We first show that $\mathcal{U}$ is a cover of $(0, 1)$. That is, we show that

$$(0,1) \subseteq \bigcup_{J \in \mathcal{U}} J = \bigcup_{n \in \mathbb{N}} \left(\frac{1}{n}, 1 \right).$$

Let $x \in (0, 1)$. Because $x > 0$ and $\lim_{n \to \infty} \frac{1}{n} = 0$, there exists some $N \in \mathbb{N}$ so large that $\frac{1}{N} < x$. It follows that $x \in (\frac{1}{N}, 1)$ and $(\frac{1}{N}, 1) \in \mathcal{U}$. Since this is true for all $x \in (0, 1)$, we have

$$(0,1) \subseteq \bigcup_{J \in \mathcal{U}} J.$$

Since each member of $\mathcal{U}$ is an open interval, we conclude that $\mathcal{U}$ is an open cover of $(0, 1)$.

We now show that no finite subcollection of sets from $\mathcal{U}$ forms a cover of $(0, 1)$. Suppose $\{J_1, J_2, \ldots, J_k\}$ is a finite collection of members of $\mathcal{U}$. For each $i \in \{1, 2, \ldots, k\}$, we know that $J_i = (\frac{1}{n_i}, 1)$ for some $n_i \in \mathbb{N}$. Let $M = \max\{n_1, n_2, \ldots, n_k\}$ and let $x = \frac{1}{2M}$. Certainly, it is the case that $x \in (0, 1)$. However, $x < \frac{1}{n_i}$ for all $i \in \{1, 2, \ldots, k\}$, and so $x \notin \bigcup_{i=1}^{k} J_i$. Consequently, the collection $\{J_1, J_2, \ldots, J_k\}$ is not a cover of $(0, 1)$.

We have shown that no finite subcollection of sets from $\mathscr{U}$ can be a cover of $(0, 1)$. Therefore, the set $(0, 1)$ is not a compact subset of $\mathbb{R}$ (when $\mathbb{R}$ is given the standard topology).

In the next section, we will show that the closed interval $[0, 1]$ *is* compact.

Example 6.1.4. Let $(X, \mathscr{T})$ be a topological space and let Y be a finite subset of X. Show that Y is compact.

Solution. Let $Y = \{x_1, x_2, \ldots, x_n\}$ and let $\mathscr{U}$ be an open cover of Y. By the definition of a cover, it must be that $Y \subseteq \bigcup_{U \in \mathscr{U}} U$, and so each x_i belongs to at least one member of $\mathscr{U}$. For each $i \in \{1, 2, \ldots, n\}$, choose one member of $\mathscr{U}$ containing x_i and denote it U_i. Then $Y \subseteq U_1 \cup U_2 \cup \cdots \cup U_n$, and so the collection $\{U_1, U_2, \ldots, U_n\}$ is a finite subcover of X derived from $\mathscr{U}$. Therefore, Y is compact.

We now show that compactness is a topological property.

Theorem 6.1.5. *The image of a compact set under a continuous function is compact.*

Proof. Let $(X, \mathscr{T})$ and $(Y, \mathscr{R})$ be topological spaces and suppose $f : X \to Y$ is a continuous function. Let A be a compact subset of X. We wish to show $f(A)$ is a compact subset of Y. Let $\mathscr{U}$ be an open cover of $f(A)$. For each $x \in A$, we know that $f(x) \in f(A)$, and consequently $f(x) \in U$ for some $U \in \mathscr{U}$. Thus, the collection $\mathscr{V} = \{f^{-1}(U) \mid U \in \mathscr{U}\}$ is an open cover of A. Since A is compact, there is a finite subcover

$$\{f^{-1}(U_1), f^{-1}(U_2), \ldots, f^{-1}(U_n)\}$$

derived from $\mathscr{V}$. We claim that $\{U_1, U_2, \ldots, U_n\}$ is a finite subcover of $f(A)$ derived from $\mathscr{U}$. Certainly, $\{U_1, U_2, \ldots, U_n\} \subseteq \mathscr{U}$. It remains only to show that $f(A) \subseteq U_1 \cup \cdots \cup U_n$.

If $y \in f(A)$, then $y = f(x)$ for some $x \in A$. But $x \in A$ implies that $x \in f^{-1}(U_k)$ for some $k \in \{1, 2, \ldots, n\}$. Thus, $f(x) \in U_k$, and so $y \in \bigcup_{i=1}^{n} U_i$. The choice of $y \in f(A)$ was arbitrary, and so $f(A) \subseteq \bigcup_{i=1}^{n} U_i$. Therefore, $\{U_1, U_2, \ldots, U_n\}$ is a finite subcover of $f(A)$ derived from $\mathscr{U}$, and so $f(A)$ is compact in Y. Q.E.D.

Corollary 6.1.6. *Compactness is a topological property.*

Proof. Let $(X, \mathscr{T})$ and $(Y, \mathscr{R})$ be homeomorphic topological spaces and suppose that $(X, \mathscr{T})$ is a compact space. By assumption, there exists a homeomorphism $h : X \to Y$. It particular, the function h is continuous with $h(X) = Y$. Thus, Y is a compact set, by Theorem 6.1.5, and so $(Y, \mathscr{R})$ is a compact space. Q.E.D.

Theorem 6.1.7. *Let $(X, \mathscr{T})$ be a topological space with $(Y, \mathscr{R})$ a subspace. A set $A \subseteq Y$ is compact with respect to $\mathscr{R}$ if and only if it is compact with respect to $\mathscr{T}$.*

Proof. Suppose that A is compact with respect to $\mathscr{R}$. Let $\mathscr{U}$ be a cover of A where each member of $\mathscr{U}$ is open in $\mathscr{T}$. Let $\mathscr{V} = \{Y \cap U \mid U \in \mathscr{U}\}$. Then each member of $\mathscr{V}$ is open in $\mathscr{R}$. Since $A \subseteq Y$ and $A \subseteq \bigcup_{U \in \mathscr{U}} U$, it follows that

$$A = Y \cap A \subseteq Y \cap \left(\bigcup_{U \in \mathscr{U}} U \right) = \bigcup_{U \in \mathscr{U}} (Y \cap U) = \bigcup_{V \in \mathscr{V}} V.$$

Therefore, $\mathscr{V}$ is an $\mathscr{R}$-open cover of A. By assumption, A is compact in $\mathscr{R}$, and so there exists a finite subcover of A derived from $\mathscr{V}$, say

$$\{Y \cap U_1, Y \cap U_2, \ldots, Y \cap U_n\}.$$

Then $\{U_1, U_2, \ldots, U_n\}$ is a finite $\mathscr{T}$-open subcover of A derived from $\mathscr{U}$. Thus, A is compact with respect to $\mathscr{T}$.

The converse is left as an exercise. (See Exercise 5.)   Q.E.D.

Corollary 6.1.8. *Let $(X, \mathscr{T})$ be a topological space with $(Y, \mathscr{R})$ a subspace. Then $(Y, \mathscr{R})$ is a compact space if and only if Y is a compact subset of X.*

Proof. The topological space $(Y, \mathscr{R})$ is a compact space precisely when Y is compact with respect to $\mathscr{R}$. By Theorem 6.1.7, the subset Y of X is compact with respect to $\mathscr{R}$ if and only if it is compact with respect to $\mathscr{T}$. Therefore, $(Y, \mathscr{R})$ is a compact space if and only if Y is a compact subset of X.   Q.E.D.

Theorem 6.1.9. *A closed subset of a compact space is compact.*

Proof. Let $(X, \mathscr{T})$ be a compact space and suppose A is a closed subset of X. Let $\mathscr{U}$ be an open cover of A. Since A is closed, the complement A' is open. Thus, $\mathscr{U} \cup \{A'\}$ is an open cover of X. By assumption, X is compact, and so there is a finite subcover of X derived from $\mathscr{U} \cup \{A'\}$. Denote this finite subcover of X by $\{V_1, V_2, \ldots, V_n\}$.

Because $A \subseteq X$, it follows that $\{V_1, V_2, \ldots, V_n\}$ is a finite subcover of A. One of these sets might be A'. Then $\{V_1, V_2, \ldots, V_n\} \setminus \{A'\}$ is a finite cover of A, all elements of which are members of $\mathscr{U}$. Therefore, $\mathscr{U}$ admits a finite subcover of A, and so A is a compact subset of X.   Q.E.D.

Theorem 6.1.10. *A compact subset of a Hausdorff space is closed.*

Proof. Suppose $(X, \mathscr{T})$ is a Hausdorff space and let $A \subseteq X$ be a compact set. Let $x \in A'$. We will show that A' is open by constructing an open neighborhood about x that is disjoint from A. By the Hausdorff assumption on X, for each $a \in A$, there exist disjoint open sets U_a and V_a such that $x \in U_a$ and $a \in V_a$. The collection of sets $\mathscr{V} = \{V_a \mid a \in A\}$ is an open cover of A. Since A is compact, there is a finite subcover of A derived from $\mathscr{V}$, say $\{V_{a_1}, V_{a_2}, \ldots, V_{a_n}\}$. Consider the corresponding collection $\{U_{a_1}, U_{a_2}, \ldots, U_{a_n}\}$ of neighborhoods of x

and let $U = U_{a_1} \cap U_{a_2} \cap \ldots \cap U_{a_n}$. Then U is open (as the intersection of *finitely* many open sets) and $x \in U$.

We claim that $U \cap A = \emptyset$. To see this, suppose there exists some $a \in A$ such that $a \in U$. Then $a \in U_{a_i}$ for each $i \in \{1, 2, \ldots, n\}$, and consequently $a \notin V_{a_i}$ for any $i \in \{1, 2, \ldots, n\}$. This means that $a \notin \bigcup_{i=1}^{n} V_{a_i}$. But $\{V_{a_1}, V_{a_2}, \ldots, V_{a_n}\}$ is a cover of A, and so $a \notin A$. This is a contradiction. Therefore, $U \cap A = \emptyset$, and so U is an open set such that $x \in U$ and $U \subseteq A'$. We conclude that A' is open, and hence A is closed. $\hspace{2cm}$ Q.E.D.

Observe that, by Theorems 6.1.9 and 6.1.10, a subset of a compact Hausdorff space is compact if and only if it is closed.

The proof of Theorem 6.1.10 requires the construction of an open neighborhood of x that is disjoint from A. For each $a \in A$, we found disjoint open sets U_a and V_a such that $x \in U_a$ and $a \in V_a$. It certainly is always the case that $A \subseteq \bigcup_{a \in A} V_a$ and $x \in \bigcap_{a \in A} U_a$ and that $\bigcup_{a \in A} V_a$ is disjoint from $\bigcap_{a \in A} U_a$. This is not sufficient to show that A is closed, however, because there is no reason to assume that the (possibly infinite) intersection $\bigcap_{a \in A} U_a$ is open. We need the compactness assumption on A to go from a case where we have an infinite number of sets to a case where we have a finite number of sets. (Because the finite intersection of open sets is necessarily open.)

The proofs of the next two theorems employ the same technique to use compactness to reduce an infinite number of sets to a finite number. (Refer to Exercises 2 and 5 in Section 3.5 for the relevant definitions.)

Theorem 6.1.11. *Any compact Hausdorff space is regular.*

Proof. Let $(X, \mathscr{T})$ be a compact Hausdorff space. Suppose A is a closed set and $x \notin A$. We wish to find disjoint open sets U and V containing x and A, respectively. By Theorem 6.1.9, the set A is compact. For each $a \in A$, there exist disjoint open sets U_a and V_a such that $x \in U_a$ and $a \in V_a$. The collection $\mathscr{V} = \{V_a \mid a \in A\}$ is an open cover of A. Since A is compact, there is a finite subcover of A derived from $\mathscr{V}$, say $\{V_{a_1}, V_{a_2}, \ldots, V_{a_n}\}$. Consider the corresponding collection $\{U_{a_1}, U_{a_2}, \ldots, U_{a_n}\}$ of neighborhoods of x. Let $U = U_{a_1} \cap U_{a_2} \cap \ldots \cap U_{a_n}$ and $V = V_{a_1} \cup V_{a_2} \cup \cdots \cup V_{a_n}$. Then U and V are open sets such that $x \in U$ and $A \subseteq V$.

We claim that $U \cap V = \emptyset$. To see this, suppose that there exists some $a \in U \cap V$. Then $a \in U$, and so $a \in U_{a_i}$ for each $i \in \{1, 2, \ldots, n\}$. It follows that $a \notin V_{a_i}$ for any $i \in \{1, 2, \ldots, n\}$, and consequently $a \notin \bigcup_{i=1}^{n} V_{a_i} = V$. This contradicts the assumption that $a \in U \cap V$. We conclude that $U \cap V = \emptyset$, and so X is regular. $\hspace{2cm}$ Q.E.D.

Theorem 6.1.12. *Any compact Hausdorff space is normal.*

The proof uses the same technique as the proof of Theorem 6.1.11. The details are left as an exercise. (See Exercise 12.)

Exercises

1. Let X be an infinite set and let $\mathscr{T}$ be the discrete topology on X. Prove that the space $(X, \mathscr{T})$ is not compact.

2. Let $\mathbb{R}$ be the set of real numbers and let $\mathscr{T}$ be the standard topology on $\mathbb{R}$. Show that the space $(\mathbb{R}, \mathscr{T})$ is not compact.

3. Let $\mathbb{R}$ be the set of real numbers and let $\mathscr{T}$ be the cofinite topology on $\mathbb{R}$. That is, let $\mathscr{T} = \{V \mid V \subseteq \mathbb{R} \text{ and } V' \text{ is finite}\} \cup \{\emptyset\}$. Show that $(\mathbb{R}, \mathscr{T})$ is compact.

4. Let $(X, \mathscr{T})$ be a topological space and suppose that $\mathscr{C}$ is a finite collection of compact subsets of X. Prove that $\bigcup_{A \in \mathscr{C}} A$ is compact.

5. Let $(X, \mathscr{T})$ be a topological space with $(Y, \mathscr{R})$ a subspace. If a set $A \subseteq Y$ is compact with respect to $\mathscr{T}$, show that it is compact with respect to $\mathscr{R}$. (This completes the proof of Theorem 6.1.7.)

6. Give an example of a topological space that has a closed subset that is not compact.

7. Give an example of a topological space that has a compact subset that is neither open nor closed.

8. Give an example of a topological space that has a compact subset that is open.

9. Prove that a topological space $(X, \mathscr{T})$ is compact if and only if there is a base $\mathscr{B}$ for the topology $\mathscr{T}$ such that every cover $\mathscr{U}$ of X with $\mathscr{U} \subseteq \mathscr{B}$ admits a finite subcover derived from $\mathscr{U}$.

10. Let $(X, \mathscr{T})$ be a compact space, let $(Y, \mathscr{R})$ be a Hausdorff space, and suppose $f : X \to Y$ is a one-to-one, onto, and continuous function. Prove that f is a homeomorphism.

11. Give an example of a noncompact Hausdorff space $(X, \mathscr{T})$ and a compact Hausdorff space $(Y, \mathscr{R})$ for which there is a one-to-one, onto, continuous function that is not a homeomorphism.

12. Prove Theorem 6.1.12.

13. A collection $\mathscr{C}$ of sets has the *finite intersection property* if the intersection of finitely many sets from $\mathscr{C}$ is nonempty; that is, if $\bigcap_{B \in \mathscr{F}} B \neq \emptyset$ for each finite subcollection $\mathscr{F}$ of $\mathscr{C}$. Let $(X, \mathscr{T})$ be a topological space and prove that X is compact if and only if $\bigcap_{B \in \mathscr{C}} B \neq \emptyset$ for each collection $\mathscr{C}$ of closed subsets of X that has the finite intersection property.

14. A topological space is said to be *locally compact* if each of its points has a neighborhood with compact closure.

(a) Show that any compact space is locally compact.

(b) Show that the set of real numbers with the standard topology is locally compact, but not compact.

(c) Show that the set of natural numbers with the discrete topology is locally compact, but not compact.

15. Let $(X, \mathscr{U})$ be a locally compact Hausdorff space (see Exercise 14) and suppose ∞ is an element not in X. Let $Y = X \cup \{\infty\}$ and let $\mathscr{V}$ be the collection of subsets V of Y for which any one of the following is true:

(a) V is open in X, or

(b) V contains ∞ and $Y \setminus V$ is compact in X, or

(c) $V = Y$.

Prove that $\mathscr{V}$ is a topology on Y.

16. Let $(X, \mathscr{U})$ be a locally compact Hausdorff space and let $(Y, \mathscr{V})$ be the topological space defined in Exercise 15. Prove that $(Y, \mathscr{V})$ is a compact space. This space $(Y, \mathscr{V})$ is called the *one-point compactification* of $(X, \mathscr{U})$.

17. Let $(X, \mathscr{U})$ be the set of natural numbers with $\mathscr{U}$ the discrete topology. Show that the one-point compactification of $(X, \mathscr{U})$ is homeomorphic to the set $K = \{0\} \cup \{\frac{1}{n} \mid n \in \mathbb{N}\}$ with the subspace topology $\mathscr{T}$ induced by the standard topology on the set of real numbers.

18. Let $X = \{(x, y) \mid x^2 + (y - 1)^2 = 1\}$ and $Y = \{(x, 0) \mid x \in \mathbb{R}\}$. Give X and Y the subspace topology induced by the standard topology on $\mathbb{R}^2$. Let $Z = Y \cup \{\infty\}$ be the one-point compactification of Y. Prove that X and Z are homeomorphic. (*Hint:* See Exercise 20 in Section 3.5)

19. Let $X = \{(x, y, z) \mid x^2 + y^2 + (z - 1)^2 = 1\}$ and $Y = \{(x, y, 0) \mid (x, y) \in \mathbb{R} \times \mathbb{R}\}$. Give X and Y the subspace topology induced by the standard topology on $\mathbb{R}^3$. Let $Z = Y \cup \{\infty\}$ be the one-point compactification of Y. Prove that X and Z are homeomorphic. (*Hint:* See Exercise 21 in Section 3.5.)

6.2   Compactness in the Space of Real Numbers

Let $\mathbb{R}$ be the set of real numbers with the standard topology. We shall show that a subset of $\mathbb{R}$ is compact if and only if it is closed and bounded. We will use this fact to prove a very important theorem from elementary calculus. We begin with the following theorem.

Theorem 6.2.1. *If a and b are real numbers with $a < b$, then the closed interval $[a, b]$ is compact in $\mathbb{R}$.*

Proof. Let $\mathscr{U}$ be an arbitrary open cover of $[a, b]$. The basic idea is to consider the "largest" subset $[a, c]$ of $[a, b]$ that can be covered by finitely many members of $\mathscr{U}$ and show that $c = b$. Define a set

$$A = \{x \in [a, b] \mid [a, x] \text{ can be covered by finitely many elements of } \mathscr{U}\}.$$

We have that $a \in A$, and hence A is nonempty. Also, b is an upper bound for A. Thus, by the Axiom of Completeness, A has a least upper bound, say c. Certainly, $c \in [a, b]$. Thus, we can choose a member V of $\mathscr{U}$ such that $c \in V$. Since V is open, there is some $d \in (a, c)$ such that $(d, c) \subseteq V$. Because $a < d < c$, it must be that $d \in A$. Therefore, a finite number of members of $\mathscr{U}$ cover $[a, d]$, say $\{U_1, \ldots, U_n\}$. Observe that $(d, c) \subseteq V$ and $c \in V$. Consequently, because $V \in \mathscr{U}$, we conclude that $\{U_1, \ldots, U_n, V\}$ is a finite subcollection of $\mathscr{U}$ that covers $[a, c]$. It follows that $c \in A$.

It remains to show that $c = b$. Suppose to the contrary that $c < b$. Recall that V was chosen in $\mathscr{U}$ so that $c \in V$. Since V is open, we can find some $\epsilon > 0$ such that $(c - \epsilon, c + \epsilon) \subseteq V$ and $c + \epsilon < b$. By choosing ϵ sufficiently small, we can insure that $c + \epsilon \in V$. Consequently, the collection of sets $\{U_1, \ldots, U_n, V\}$ is a finite subcollection of $\mathscr{U}$ that covers $[a, c + \epsilon]$. It follows that $c + \epsilon \in A$. This is a contradiction, since c is the least upper bound of A. We must reject the assumption that $c < b$ and conclude that $c = b$.   Q.E.D.

We are now ready to prove the main theorem of this section.

Theorem 6.2.2 (Heine-Borel Theorem). *A subset of the real numbers (equipped with the standard topology) is compact if and only if it is closed and bounded.*

Proof. Assume that A is a set of real numbers that is closed and bounded. Since A is bounded, there is a closed bounded interval $[a, b]$ such that $A \subseteq [a, b]$. By Theorem 6.2.1, the closed interval $[a, b]$ is compact. Thus, A is a closed subset of a compact set, and hence is itself compact, by Theorem 6.1.9.

Conversely, suppose that A is compact. Observe that A is a compact subset of the Hausdorff space $\mathbb{R}$, and consequently is closed, by Theorem 6.1.10. It remains to show that A is bounded. Let $\mathscr{U} = \{(x - 1, x + 1) \mid x \in A\}$. It is evident that $\mathscr{U}$ is an open cover of A. Since A is compact, it follows that there is a finite subcover derived from $\mathscr{U}$, say

$$\{(x_1 - 1, x_1 + 1), (x_2 - 1, x_2 + 1), \ldots, (x_n - 1, x_n + 1)\}.$$

Let $m = \min\{x_1 - 1, \ldots, x_n - 1\}$ and $M = \max\{x_1 + 1, \ldots, x_n + 1\}$. Suppose $x \in A$ is arbitrary. Then $x \in (x_i - 1, x_i + 1)$ for some $i \in \{1, \ldots, n\}$, and so $m < x < M$. Since this is true for all numbers $x \in A$, it follows that A is bounded.   Q.E.D.

The following is a very important theorem from elementary calculus. Among other things, this theorem is used to prove that the Riemann integral $\int_a^b f(x)\, dx$ exists whenever $f : [a, b] \to \mathbb{R}$ is continuous.

Theorem 6.2.3 (Extreme Value Theorem). *If a function $f : [a, b] \to \mathbb{R}$ is continuous, then there are numbers x_1 and x_2 in $[a, b]$ such that $f(x_1) \le f(x) \le f(x_2)$ for all $x \in [a, b]$.*

Proof. By Theorem 6.2.1, the closed interval $[a, b]$ is compact. Thus, the image $f([a, b])$ is compact, by Theorem 6.1.5. From Theorem 6.2.2, it follows that $f([a, b])$ is closed and bounded. Let $m = \inf f([a, b])$ and $M = \sup f([a, b])$ be the greatest lower bound and least upper bound of $f([a, b])$, respectively. Then, for all x in $[a, b]$, we have $m \le f(x) \le M$.

It remains to show that there exist numbers x_1 and x_2 in $[a, b]$ such that $f(x_1) = m$ and $f(x_2) = M$. Let $\epsilon > 0$ be given and note that there must be some $x \in [a, b]$ such that $m \le f(x) < m + \epsilon$. If this were not the case, then $m + \epsilon$ would be a lower bound of $f([a, b])$ greater than the greatest lower bound. Since each open set containing m must contain a set $(m - \epsilon, m + \epsilon)$ for some $\epsilon > 0$, this shows that each open neighborhood of m intersects $f([a, b])$. Therefore, m is in the closure of $f([a, b])$. However, $f([a, b])$ is known to be closed, and so $m \in \overline{f([a, b])} = f([a, b])$. Thus, there exists a number $x_1 \in [a, b]$ such that $f(x_1) = m$.

The proof that $M = f(x_2)$ for some $x_2 \in [a, b]$ is similar and is left as an exercise. (See Exercise 3 at the end of this section.) Q.E.D.

For a different proof of Theorem 6.2.3, see Example 1.4.3. (See also Exercise 4 below.)

Exercises

1. Give an example of a continuous function $f : (0, 1) \to \mathbb{R}$ that is not bounded.

2. Give an example of a bounded function $f : [0, 1] \to \mathbb{R}$ for which

$$\inf f([0, 1]) < f(x) < \sup f([0, 1])$$

   for all $x \in [0, 1]$.

3. Complete the proof of Theorem 6.2.3 by showing that there is a number x_2 in $[a, b]$ such that $f(x_2) = M$, the least upper bound (supremum) of the set $f([a, b])$.

4. Give a proof similar to the argument in Example 1.4.3 to show that if f is continuous on $[a, b]$, then there is a number $x_1 \in [a, b]$ such that $f(x_1) = \inf f([a, b])$, the greatest lower bound (infimum) of the set $f([a, b])$.

5. Let $\mathbb{R}$ be the set of real numbers with the standard topology. Let U and V be open subsets of $\mathbb{R}$ and show that $U \cap V$ is compact if and only if $U \cap V = \emptyset$.

6. (*Cantor's Intersection Theorem*) Let B_1 be a nonempty closed and bounded set of real numbers and suppose that

$$B_1 \supseteq B_2 \supseteq B_3 \supseteq \cdots$$

is a nested sequence of nonempty closed sets. Prove that $\bigcap_{i=1}^{\infty} B_i \neq \emptyset$.

7. Let A be a compact set of real numbers and let $\mathscr{U}$ be an open cover of A. Prove that there exists a number $\epsilon > 0$ such that if $\{x,y\} \subseteq A$ and $|x - y| < \epsilon$, then there is some $U \in \mathscr{U}$ such that $\{x,y\} \subseteq U$.

8. Let $\mathbb{R}$ be the set of real numbers and define $d : \mathbb{R} \times \mathbb{R} \to \mathbb{R}$ by

$$d(x,y) = \begin{cases} 1 & \text{if } x \neq y, \\ 0 & \text{if } x = y. \end{cases}$$

Then d is a metric for $\mathbb{R}$ and the topology $\mathscr{T}_d$ generated by d is the discrete topology. (See Examples 2.7.3 and 2.7.9.) Let $A = \{x \mid 0 < x < 1\}$. Show that A is closed and bounded, but not compact, in the space $(\mathbb{R}, \mathscr{T}_d)$.

9. Let $(X, \mathscr{T})$ be a compact space and let $\mathbb{R}$ be the set of real numbers with the standard topology. Prove that if $f : X \to \mathbb{R}$ is a continuous function, then f is bounded. That is, show there is a number $B > 0$ such that $|f(x)| < B$ for all $x \in X$.

6.3   The Product of Compact Spaces

Let K be an index set and, for each $\alpha \in K$, let $(X_\alpha, \mathscr{T}_\alpha)$ be a topological space. Let $(X, \mathscr{T})$ be the product space. An important theorem in topology, known as *Tychonoff's Theorem*, states that the product space is compact if and only if $(X_\alpha, \mathscr{T}_\alpha)$ is compact for each $\alpha \in K$. We will not prove Tychonoff's Theorem in its full generality in this text, but we will provide a weakened version. As a first step, we prove one half of Tychonoff's Theorem.

Theorem 6.3.1. *If a product space is compact, then each coordinate space is compact.*

Proof. Let K be an index set and, for each $\alpha \in K$, let $(X_\alpha, \mathscr{T}_\alpha)$ be a topological space. Let $(X, \mathscr{T})$ be the product space. We suppose that $(X, \mathscr{T})$ is compact and show that $(X_\alpha, \mathscr{T}_\alpha)$ is compact for each $\alpha \in K$.

For each $\alpha \in K$, let p_α be the αth projection function. Observe that p_α is continuous and $p_\alpha(X) = X_\alpha$. It follows that X_α is the image of a compact set under a continuous function, and hence is compact, by Theorem 6.1.5. Q.E.D.

The proof of the converse is much more difficult. In this text, we will prove only the case where K is finite. We begin by considering the case when K has only two elements.

Theorem 6.3.2. *If $(X_1, \mathcal{T}_1)$ and $(X_2, \mathcal{T}_2)$ are compact topological spaces, then the product space $(X_1 \times X_2, \mathcal{T})$ is compact.*

Proof. In light of Exercise 9 in Section 6.1, we need only show that every open cover of $X_1 \times X_2$ consisting of sets from a base for the product topology $\mathcal{T}$ admits a finite subcover. Each member of the base for the product topology on $X_1 \times X_2$ is of the form $U \times V$, where $U \in \mathcal{T}_1$ and $V \in \mathcal{T}_2$. Thus, let M be an index set and let $\mathcal{C} = \{U_\beta \times V_\beta \mid \beta \in M\}$ be a cover of $X_1 \times X_2$ by elements belonging to the base for the product topology. We will show that $\mathcal{C}$ admits a finite subcover of $X_1 \times X_2$.

Let $x \in X_1$. The set $\{x\} \times X_2$, equipped with the subspace topology, is homeomorphic to $(X_2, \mathcal{T}_2)$, by Theorem 4.3.2. It follows that $\{x\} \times X_2$ is a compact subset of $X_1 \times X_2$. Consequently, since $\{x\} \times X_2$ is covered by $\mathcal{C}$, there exists a finite subcover of $\{x\} \times X_2$ derived from $\mathcal{C}$. Thus, there exists a finite subset N_x of M such that $\{U_\beta \times V_\beta \mid \beta \in N_x\}$ forms a cover for $\{x\} \times X_2$. Note that N_x depends on x. We require only sets that have nonempty intersection with $\{x\} \times X_2$. Hence, we may assume that $x \in U_\beta$ for all $\beta \in N_x$ (by discarding any extra sets, if necessary). Now, let $U(x) = \bigcap_{\beta \in N_x} U_\beta$. Then $U(x)$ is an open set such that $x \in U(x)$ and $\{U(x) \times V_\beta \mid \beta \in N_x\}$ is a finite cover of $\{x\} \times X_2$.

Proceeding in the above fashion for each $x \in X_1$, we obtain a collection of sets $\mathcal{U} = \{U(x) \mid x \in X_1\}$. Observe that $\mathcal{U}$ is an open cover of X_1. Consequently, by the compactness of X_1, we obtain a finite subcover, say $\{U(x_1), U(x_2), \ldots, U(x_n)\}$, where $x_i \in X_1$ for each $i \in \{1, 2, \ldots, n\}$. Let $N = N_{x_1} \cup N_{x_2} \cup \cdots \cup N_{x_n}$. Then N is a subset of M, the index set. Furthermore, since N_{x_i} has finitely many members for each $i \in \{1, 2, \ldots, n\}$, so does N. It follows that $\{U_\beta \times V_\beta \mid \beta \in N\}$ is a finite collection of open sets, each of which belongs to $\mathcal{C}$. Thus, to show that this is a finite subcover derived from $\mathcal{C}$, it remains only to show that $X_1 \times X_2 \subseteq \bigcup_{\beta \in N} (U_\beta \times V_\beta)$.

Let $(a, b) \in X_1 \times X_2$. Then $a \in X_1$, and so we have that $a \in U(x_i)$ for some index $i \in \{1, 2, \ldots, n\}$. We know that $\{U_\beta \times V_\beta \mid \beta \in N_{x_i}\}$ is a cover for $\{x_i\} \times X_2$. Choose $\gamma \in N_{x_i}$ such that $b \in V_\gamma$. Since $a \in U(x_i)$ and $U(x_i) = \bigcap_{\beta \in N_{x_i}} U_\beta$, it follows that $a \in U_\gamma$. Thus $(a, b) \in U_\gamma \times V_\gamma$. Finally, we observe that

$$\gamma \in N_{x_i} \subseteq N_{x_1} \cup \ldots \cup N_{x_n} = N.$$

Therefore, $(a, b) \in \bigcup_{\beta \in N} (U_\beta \times V_\beta)$. Since the choice of (a, b) in $X_1 \times X_2$ was arbitrary, we conclude that $X_1 \times X_2 \subseteq \bigcup_{\beta \in N} (U_\beta \times V_\beta)$. Q.E.D.

We wish to prove that the product of n compact spaces is compact, where $n \in \mathbb{N}$. To that end, we need the following theorem.

Theorem 6.3.3. *Let $\{(X_i, \mathcal{T}_i) \mid i \in \{1, 2, \ldots, n\}\}$ be a finite collection of topological spaces. Suppose*

$$X = \prod_{i=1}^{n} X_i, \quad Y = \prod_{i=1}^{n-1} X_i, \quad \text{and} \quad Z = Y \times X_n.$$

If $(X, \mathcal{T})$, $(Y, \mathcal{S})$, and $(Z, \mathcal{R})$ are the corresponding product spaces, then $(X, \mathcal{T})$ and $(Z, \mathcal{R})$ are homeomorphic.

Proof. Before proving the theorem, we note that the points of the space X are n-tuples; that is, functions with domain $\{1, 2, \ldots, n\}$ and such that $x(i) \in X_i$ for each $i \in \{1, 2, \ldots, n\}$. The points in Z are ordered pairs (z_1, z_2), where z_1 is an $(n-1)$-tuple with $z_1(i) \in X_i$ for each $i \in \{1, 2, \ldots, n-1\}$ and $z_2 \in X_n$.

Define a function $h : X \to Z$ by

$$h(x_1, x_2, \ldots, x_{n-1}, x_n) = \big((x_1, x_2, \ldots, x_{n-1}), x_n\big),$$

where $x_i \in X_i$ for each $i \in \{1, 2, \ldots, n\}$. Certainly, h is one-to-one and onto. Also, if $A_i \subseteq X_i$ for each $i \in \{1, 2, \ldots, n\}$, then

$$h(A_1 \times A_2 \times \cdots \times A_{n-1} \times A_n) = (A_1 \times A_2 \times \cdots \times A_{n-1}) \times A_n.$$

Observe that members of the base for the product topology for $(X, \mathcal{T})$ and $(Z, \mathcal{R})$ are of the form $U_1 \times U_2 \times \cdots \times U_{n-1} \times U_n$ and $(U_1 \times U_2 \times \cdots \times U_{n-1}) \times U_n$, respectively, where $U_i \in \mathcal{T}_i$ for each $i \in \{1, 2, \ldots, n\}$. It follows that the image under h of members of the base for $\mathcal{T}$ are members of the base for $\mathcal{R}$, and so both h and h^{-1} are continuous. Thus, h is a homeomorphism.   Q.E.D.

Theorem 6.3.4. *If $(X_i, \mathcal{T}_i)$ is a compact topological space for each i in the index set $\{1, 2, \ldots, n\}$, then the product space $(X, \mathcal{T})$ is compact.*

Proof. We give a proof by induction. By Theorem 6.3.2, the theorem is true for the case $n = 2$. Assume the theorem is true for $n = k$; that is, assume that the product space $(X_1 \times \cdots \times X_k, \mathcal{S})$ is compact. Then, by Theorem 6.3.2, the product space $\big((X_1 \times \cdots \times X_k) \times X_{k+1}, \mathcal{R}\big)$ is compact, where $\mathcal{R}$ is the product topology. By Theorem 6.3.3, the product spaces $\big((X_1 \times \cdots \times X_k) \times X_{k+1}, \mathcal{R}\big)$ and $(X_1 \times \cdots \times X_k \times X_{k+1}, \mathcal{R}')$ are homeomorphic. Thus, $(X_1 \times \cdots \times X_k \times X_{k+1}, \mathcal{R}')$ is compact. Therefore, $(X, \mathcal{T})$ is compact, by the Principle of Mathematical Induction.   Q.E.D.

We introduced Euclidean space in Section 2.8. The base for the topology on $\mathbb{R}^n$ was the collection of all open balls. It is easy to see that this is the same as the product topology for the Cartesian product of n copies of the real line. (See Exercise 10 in Section 4.2.) We are now ready to prove the Heine-Borel Theorem (Theorem 6.2.2) for $\mathbb{R}^n$.

Theorem 6.3.5. *A subset of $\mathbb{R}^n$ is compact if and only if it is closed and bounded.*

Proof. Let A be a compact subset of $\mathbb{R}^n$. Since $\mathbb{R}^n$ is a Hausdorff space, we know that A is closed by Theorem 6.1.10. We next show that A is bounded. For each $i \in \mathbb{N}$, let $V_i = \prod_{k=1}^{n} (-i, i)$, the product of n copies of the interval $(-i, i)$, which is also denoted $(-i, i)^n$. For each $i \in \mathbb{N}$, the set V_i is a basic open set in $\mathbb{R}^n$. Furthermore,

$$V_1 \subseteq V_2 \subseteq V_3 \subseteq \cdots \quad \text{and} \quad \bigcup_{i=1}^{\infty} V_i = \mathbb{R}^n.$$

Consequently, $A \subseteq \bigcup_{i=1}^{\infty} V_i$, and so $\{V_i \mid i \in \mathbb{N}\}$ is an open cover of A. Since A is compact, there is a finite subcover of A derived from $\{V_i \mid i \in \mathbb{N}\}$, say $\{V_{i_1}, V_{i_2}, \ldots, V_{i_N}\}$. We may assume $i_1 < i_2 < \cdots < i_N$ (by reordering the sets, if necessary). Because the sets are nested, it follows that $V_{i_1} \subseteq V_{i_2} \subseteq \cdots \subseteq V_{i_N}$. Thus,

$$A \subseteq V_{i_1} \cup \cdots \cup V_{i_N} = V_{i_N} = \prod_{k=1}^{n}(-i_N, i_N).$$

Hence, if $(x_1, \ldots, x_n) \in A$, we have that $|x_j| < i_N$ for each $j \in \{1, \ldots, n\}$. Therefore,

$$\|(x_1, \ldots, x_n)\| = \sqrt{x_1^2 + \cdots + x_n^2} < \sqrt{n \cdot i_N{}^2} = i_N \sqrt{n},$$

and so A is bounded.

Conversely, suppose that A is closed and bounded. For each $i \in \{1, 2, \ldots, n\}$, let p_i denote the ith projection function. If $B_i = p_i(A)$, then B_i is a bounded set of real numbers, because projections do not increase distances. (See Exercises 3 and 4.) For each $i \in \{1, 2, \ldots, n\}$, let $L_i > 0$ be a positive real number such that $-L_i < x < L_i$ for all $x \in B_i$. Then,

$$A \subseteq \prod_{i=1}^{n} B_i \subseteq \prod_{i=1}^{n}[-L_i, L_i].$$

(See Exercise 5.) By Theorem 6.2.1, each closed interval $[-L_i, L_i]$ is compact. Thus, the product $\prod_{i=1}^{n}[-L_i, L_i]$ is compact, by Theorem 6.3.4. The set A is a closed subset of this compact set, and hence A is a compact set, by Theorem 6.1.9. $\qquad$ Q.E.D.

Exercises

1. Give an example of a closed subset of $\mathbb{R}^3$ that is not compact.

2. Give an example of a bounded subset of $\mathbb{R}^3$ that is not compact.

3. Let $x = (x_1, \ldots, x_n)$ and $y = (y_1, \ldots, y_n)$ be two points in $\mathbb{R}^n$ and let $i \in \{1, \ldots, n\}$. If p_i denotes the ith projection function, show that

$$d_1\big(p_i(x), p_i(y)\big) = |x_i - y_i| \leq d_n(x, y),$$

   where d_1 and d_n denote the metrics on $\mathbb{R}$ and $\mathbb{R}^n$, respectively. (See Section 2.8 for the relevant definitions.)

4. Let A be a bounded subset of $\mathbb{R}^n$ and let $i \in \{1, \ldots, n\}$. If p_i denotes the ith projection function, show that the set $B_i = p_i(A)$ is a bounded set of real numbers.

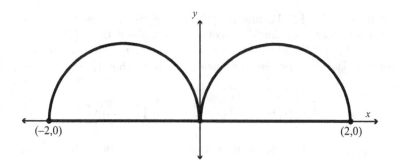

Figure 6.3.a: This graph represents the set $X = A \cup B \cup C$ in Exercise 8.

5. Suppose that $A \subseteq \mathbb{R}^n$. For each $i \in \{1, \ldots, n\}$, let p_i be the ith projection function on $\mathbb{R}^n$ and let $B_i = p_i(A)$. If for each $i \in \{1, \ldots, n\}$ the set C_i is a subset of $\mathbb{R}$ such that $B_i \subseteq C_i$, then show that $A \subseteq \prod_{i=1}^{n} C_i$.

6. Let $(X, \mathscr{T})$ and $(Y, \mathscr{R})$ be two topological spaces. Suppose that A and B are compact subsets of X and Y, respectively. If W is an open set in $X \times Y$ such that $A \times B \subseteq W$, prove that there exist sets $U \in \mathscr{T}$ and $V \in \mathscr{R}$ such that $A \times B \subseteq U \times V$ and $U \times V \subseteq W$.

7. Let A be a nonempty subset of the metric space $(X, \mathscr{T}_d)$ and suppose $x \in X$. The *distance from x to A* is denoted $d(x, A)$ and is defined by

$$d(x, A) = \inf\{d(x, y) \mid y \in A\}.$$

If A is compact, show there is a point $y_0 \in A$ such that $d(x, y_0) = d(x, A)$. Give an example to show this need not be true if A is not compact.

8. Let $A = \{(x, 0) \mid x \in [-2, 2]\}$,

$$B = \{(x, y) \mid (x - 1)^2 + y^2 = 1 \text{ and } y > 0\},$$

and

$$C = \{(x, y) \mid (x + 1)^2 + y^2 = 1 \text{ and } y > 0\}.$$

Now, let $X = A \cup B \cup C$ and $Y = \{(x, 0) \mid x \in \mathbb{R}\}$. Give X and Y the subspace topology induced by the standard topology on $\mathbb{R}^2$. Define a function $f : Y \to X$ that is one-to-one, onto, and continuous. Can this function be a homeomorphism? Justify your answer. (See Figure 6.3.a for an illustration of the set X.)

6.4   Compactness in Metric Spaces

In this section, we study different forms of compactness in metric spaces. We begin by introducing several definitions.

Definition 6.4.1. A topological space $(X, \mathcal{T})$ is called *countably compact* provided that every countable open cover of X has a finite subcover.

Definition 6.4.2. A topological space $(X, \mathcal{T})$ is said to be *sequentially compact* provided that every sequence in X has a convergent subsequence.

Definition 6.4.3. A topological space $(X, \mathcal{T})$ has the *Bolzano-Weierstrass Property* if every infinite subset of X has an accumulation point. (See Exercise 11 in Section 2.5.)

Our aim is to prove the following theorem.

Theorem 6.4.4. *If $(X, \mathcal{T}_d)$ is a metric space, then the following are equivalent:*

(a) $(X, \mathcal{T}_d)$ is compact.

(b) $(X, \mathcal{T}_d)$ is countably compact.

(c) $(X, \mathcal{T}_d)$ is sequentially compact.

(d) $(X, \mathcal{T}_d)$ has the Bolzano-Weierstrass Property.

The proof of this theorem will be given at the end of this section. We first state several definitions, as well as state and prove several preliminary lemmas.

Definition 6.4.5. Let $(X, \mathcal{T}_d)$ be a metric space and let $\mathcal{U}$ be an open cover of X. The number $\epsilon > 0$ is called a *Lebesgue number* for $\mathcal{U}$ provided that, for each $x \in X$, the open ball $B_\epsilon(x)$ of radius ϵ centered at x is contained in some member of $\mathcal{U}$.

Certainly, if ϵ is a Lebesgue number for a cover $\mathcal{U}$, then so is any positive number δ such that $\delta < \epsilon$.

Definition 6.4.6. Let $(X, \mathcal{T}_d)$ be a metric space and let $\epsilon > 0$ be given. A finite subset $\mathcal{N}$ of X is called an ϵ-*net* for X provided that, for each $x \in X$, there is some $a \in \mathcal{N}$ such that $d(x, a) < \epsilon$.

Definition 6.4.7. A metric space $(X, \mathcal{T}_d)$ is said to be *totally bounded* if for each $\epsilon > 0$ there is an ϵ-net for X.

Theorem 6.4.8. *If $(X, \mathcal{T}_d)$ is a totally bounded metric space, then X is a bounded set.*

Proof. Suppose that $(X, \mathcal{T}_d)$ is a totally bounded metric space and let $\mathcal{N}$ be a 1-net for X. Since $\mathcal{N}$ is a finite set, we may let

$$M = \max\{d(x, y) \mid (x \in \mathcal{N}) \wedge (y \in \mathcal{N})\}.$$

Let a and b be two distinct points of X. There exist points x and y in $\mathcal{N}$ such that $d(a, x) < 1$ and $d(b, y) < 1$. By the triangle inequality,

$$d(a, b) \leq d(a, x) + d(x, y) + d(b, y) < 1 + M + 1 = M + 2.$$

This is true for any a and b in X, and so $(X, \mathcal{T}_d)$ is bounded. Q.E.D.

The converse of Theorem 6.4.8 is false, as is shown in the following example.

Example 6.4.9. Let $\mathbb{R}$ be the set of real numbers and let d be the discrete metric from Example 2.7.3. Show that $\mathbb{R}$ is bounded in this metric, but not totally bounded.

Solution. Recall that the discrete metric on $\mathbb{R}$ is defined by

$$d(x, y) = \begin{cases} 0 & \text{if } x = y, \\ 1 & \text{if } x \neq y. \end{cases}$$

It follows that $\sup\{d(x, y) : (x \in \mathbb{R}) \wedge (y \in \mathbb{R})\} = 1$, and so $\mathbb{R}$ is bounded in the metric d. The metric space $(X, \mathscr{T}_d)$ cannot be totally bounded, however, because there is no ϵ-net for any positive number ϵ such that $\epsilon < 1$.

We will now proceed to prove Theorem 6.4.4 by means of several preliminary lemmas.

Lemma 6.4.10. *If a metric space has the Bolzano-Weierstrass Property, then it is totally bounded.*

Proof. We will prove the contrapositive. Suppose that $(X, \mathscr{T}_d)$ is a metric space that is not totally bounded. Then, there exists some $\epsilon > 0$ such that X has no ϵ-net. Let $x_1 \in X$ and choose some $x_2 \in X$ such that $d(x_1, x_2) \geq \epsilon$. Such an x_2 is known to exist, as otherwise $\{x_1\}$ would be an ϵ-net.

We now proceed inductively. Let n be a number in $\mathbb{N}$ and suppose that a set $B_n = \{x_1, x_2, \ldots, x_n\}$ has been constructed so that $d(x_i, x_j) \geq \epsilon$ for all distinct i and j in the set $\{1, 2, \ldots, n\}$. Since B_n cannot be an ϵ-net, there must be some $x_{n+1} \in X$ such that $d(x_i, x_{n+1}) \geq \epsilon$ for all $i \in \{1, 2, \ldots, n\}$.

By the Principle of Mathematical Induction, we conclude that there exists a set $B = \{x_k \mid k \in \mathbb{N}\}$ of distinct points in X such that $d(x_i, x_j) \geq \epsilon$ for all distinct numbers i and j in $\mathbb{N}$. It follows that B has no accumulation point, and therefore $(X, \mathscr{T}_d)$ does not have the Bolzano-Weierstrass Property.   Q.E.D.

In Exercise 5 of Section 2.6, we saw that if $(X, \mathscr{T})$ is a Hausdorff space, and if c is an accumulation point of a subset A of X, then any open neighborhood of c contains infinitely many points of A. Since a metric space is a Hausdorff space, the same is true for metric spaces. We will make use of this fact in the proof of the next lemma.

Lemma 6.4.11. *Let $(X, \mathscr{T}_d)$ be a metric space that has the Bolzano-Weierstrass Property. If $\mathscr{U}$ is an open cover of X, then there is a Lebesgue number for $\mathscr{U}$.*

Proof. Suppose no such Lebesgue number exists. Then, for each positive integer n, there is some $x_n \in X$ such that $B_{1/n}(x_n)$ is not a subset of any member of $\mathscr{U}$. Let $C = \{x_n \mid n \in \mathbb{N}\}$. We claim that C is an infinite set.

Suppose that C is finite. Then there exists some $c \in X$ such that $x_n = c$ for infinitely many $n \in \mathbb{N}$. Because $\mathscr{U}$ is a cover of X, there exists some member V of $\mathscr{U}$ such that $c \in V$. Since V is an open neighborhood of c, there exists a $\delta > 0$ such that $B_\delta(c) \subseteq V$. Let $m \in \mathbb{N}$ be any positive integer such that $\frac{1}{m} < \delta$. Then, for each $k \in \mathbb{N}$ such that $k \geq m$, we have that $B_{1/k}(c) \subseteq V$. But $B_{1/k}(x_k)$ is not a subset of V for any $k \in \mathbb{N}$. Therefore, $x_k \neq c$ for any $k \geq m$. This is a contradiction, and so C is an infinite set.

We have established that C is an infinite set. By assumption, the metric space $(X, \mathscr{T}_d)$ has the Bolzano-Weierstrass Property. Thus, C has an accumulation point, say z. Let W be any member of $\mathscr{U}$ such that $z \in W$. Because W is open, there exists a number $\eta > 0$ such that $B_\eta(z) \subseteq W$. Since z is an accumulation point of C, the set $B_{\eta/2}(z)$ contains infinitely many members of C. Consequently, there is a positive integer N such that $\frac{1}{N} < \frac{\eta}{2}$ and such that $x_N \in B_{\eta/2}(z)$.

Let $y \in B_{1/N}(x_N)$. Then

$$d(y, z) \leq d(y, x_N) + d(x_N, z) < \frac{1}{N} + \frac{\eta}{2} < \frac{\eta}{2} + \frac{\eta}{2} = \eta.$$

Thus, $y \in B_\eta(z)$. Since the choice of y was arbitrary in $B_{1/N}(x_N)$, it follows that $B_{1/N}(x_N) \subseteq B_\eta(z)$. But $B_\eta(z) \subseteq W$, and hence $B_{1/N}(x_N) \subseteq W$. This contradicts the assumption that $B_{1/N}(x_N)$ is not a subset of any member of $\mathscr{U}$. Therefore, there is a Lebesgue number for $\mathscr{U}$. Q.E.D.

Lemma 6.4.12. *A countably compact topological space has the Bolzano-Weierstrass Property.*

Proof. Let $(X, \mathscr{T})$ be a countably compact topological space and suppose Y is an infinite subset of X. We will show that Y has an accumulation point. Let Z be any countably infinite subset of Y. Because Z is countably infinite, we may write $Z = \{x_i \mid i \in \mathbb{N}\}$ with $x_i \neq x_j$ whenever $i \neq j$. Suppose that Z does not have an accumulation point. It follows that, for each $n \in \mathbb{N}$, the set $C_n = \{x_i \mid i \geq n\}$ is closed. (See Exercise 11(a) in Section 2.5.)

Observe that $C_1 \supseteq C_2 \supseteq C_3 \supseteq \cdots$. That is, $\{C_n\}_{n=1}^\infty$ is a nested sequence of sets. Also notice that for any $n \in \mathbb{N}$ the element $x_n \notin C_{n+1}$. It follows that $\bigcap_{n=1}^\infty C_n = \emptyset$. Thus,

$$\bigcup_{n=1}^\infty C_n' = \left(\bigcap_{n=1}^\infty C_n \right)' = \emptyset' = X.$$

Hence $\{C_n' \mid n \in \mathbb{N}\}$ is a countable open cover of X. By assumption, X is countably compact, and thus there is a finite subcover of X, say $\{C_{n_1}', \ldots, C_{n_K}'\}$. We may order the sets so that $n_1 < \cdots < n_K$. Because this collection of sets forms a subcover of X, we have that $C_{n_1}' \cup \cdots \cup C_{n_K}' = X$. In terms of complements, this says $(C_{n_1}' \cup \cdots \cup C_{n_K}')' = X'$ or $C_{n_1} \cap \cdots \cap C_{n_K} = \emptyset$. This, however, is a contradiction, because the collection of sets $\{C_n \mid n \in \mathbb{N}\}$ is nested, and so $C_{n_1} \cap \cdots \cap C_{n_K} = C_{n_K}$, which is a nonempty set.

We are forced to reject our assumption and conclude that Z does have an accumulation point. Therefore, because $Z \subseteq Y$, the set Y has an accumulation point, as well.   Q.E.D.

Lemma 6.4.13. *Any metric space that has the Bolzano-Weierstrass Property is countably compact.*

Proof. Let $(X, \mathcal{T}_d)$ be a metric space that has the Bolzano-Weierstrass Property. Suppose that $\mathcal{U} = \{U_i \mid i \in \mathbb{N}\}$ is a countable open cover of X that has no finite subcover. We will derive a contradiction. For each $n \in \mathbb{N}$, let

$$A_n = X \setminus \left(\bigcup_{i=1}^{n} U_i \right) = \bigcap_{i=1}^{n} U_i'.$$

Observe that A_n is nonempty because $\mathcal{U}$ does not admit a finite subcover of X. Also, A_n is closed as the intersection of finitely many closed sets. For each $n \in \mathbb{N}$, choose some $x_n \in A_n$ and let $B = \{x_n \mid n \in \mathbb{N}\}$.

Suppose B is a finite set. Then there exists some $x \in B$ such that $x_n = x$ for infinitely many $n \in \mathbb{N}$. However, the collection $\{A_n : n \in \mathbb{N}\}$ forms a nested sequence of sets, and so

$$A_1 \supseteq A_2 \supseteq \cdots \supseteq A_n \supseteq \cdots .$$

Thus, if $x \in A_n$ for infinitely many $n \in \mathbb{N}$, it must be that $x \in A_n$ for all $n \in \mathbb{N}$. Consequently, because $\mathcal{U}$ is a cover of X,

$$x \in \bigcap_{n=1}^{\infty} A_n = \bigcap_{i=1}^{\infty} U_i' = \left(\bigcup_{i=1}^{\infty} U_i \right)' = X' = \emptyset.$$

This is a contradiction, so B is not finite.

Suppose instead that B is an infinite set. By assumption, X has the Bolzano-Weierstrass Property. Thus, the infinite set B has an accumulation point, say c. Let V be any open neighborhood of c. Since c is an accumulation point for B, the set $V \cap B$ contains infinitely many points. Now let $n \in \mathbb{N}$ and define a set

$$C_n = \{x_i \mid i > n\} = B \setminus \{x_1, \ldots, x_n\}.$$

Observe that C_n contains all but finitely many members of B. Thus, $V \cap C_n$ contains infinitely many points, and so c is an accumulation point for C_n. By definition, $x_i \in A_i$ for each $i \in \mathbb{N}$, and $A_i \subseteq A_n$ for all $i \geq n$. Therefore, $C_n \subseteq A_n$, and so c is also an accumulation point for A_n. Since A_n is a closed set, we conclude that $c \in A_n$. This argument can be repeated for every $n \in \mathbb{N}$. Consequently, $c \in \bigcap_{n=1}^{\infty} A_n = \emptyset$. Once again we have obtained a contradiction.

We have determined that B can be neither a finite set nor an infinite set. Since this is an impossible situation, we must reject the assumption that $\mathcal{U}$ does not admit a finite subcover. Therefore, the topological space $(X, \mathcal{T}_d)$ is countably compact.   Q.E.D.

Lemma 6.4.14. *A sequentially compact topological space is countably compact.*

Proof. We prove the contrapositive. Let $(X, \mathscr{T})$ be a topological space and assume $\mathscr{U}$ is a countable open cover of X that does not admit a finite subcover. Choose an arbitrary $x_1 \in X$. Pick U_1 in $\mathscr{U}$ so that $x_1 \in U_1$. (This can be done because $\mathscr{U}$ is a cover of X.) Proceeding inductively, for each positive integer $n > 1$, choose a point

$$x_n \in X \setminus \left(\bigcup_{i=1}^{n-1} U_i \right),$$

where U_i is chosen in $\mathscr{U}$ so that $x_i \in U_i$ for all $i \in \{1, \ldots, n-1\}$. Such an element x_n is sure to exist because X cannot be covered by any finite collection of sets from $\mathscr{U}$, and the set U_i can be found for each i because $\mathscr{U}$ is a cover of X. The result is a sequence of points $\{x_n\}_{n=1}^{\infty}$ taken from X and a collection of open sets $\{U_n \mid n \in \mathbb{N}\}$ taken from $\mathscr{U}$ such that $x_n \in U_n$ for all $n \in \mathbb{N}$ and $x_n \notin U_k$ for any $k < n$. Furthermore, the collection $\{U_n \mid n \in \mathbb{N}\}$ is a cover of X because $\mathscr{U}$ is a countable cover of X.

We will show that the sequence $\{x_n\}_{n=1}^{\infty}$ has no convergent subsequence. To see this, let x be an arbitrary point in X. The collection $\{U_n \mid n \in \mathbb{N}\}$ is a cover of X, and so there is some $k \in \mathbb{N}$ such that $x \in U_k$. Because of the way we chose the elements in the sequence $\{x_n\}_{n=1}^{\infty}$, we know that $x_n \notin U_k$ whenever $n > k$. Thus, no subsequence of $\{x_n\}_{n=1}^{\infty}$ can converge to x. The choice of x was arbitrary, and thus $\{x_n\}_{n=1}^{\infty}$ has no convergent subsequence. Therefore, $(X, \mathscr{T})$ is not sequentially compact.   Q.E.D.

Lemma 6.4.15. *A countably compact metric space is sequentially compact.*

Proof. Suppose $(X, \mathscr{T}_d)$ is a countably compact metric space and let $\{x_n\}_{n=1}^{\infty}$ be a sequence in X. We will show that $\{x_n\}_{n=1}^{\infty}$ has a convergent subsequence.

Suppose there is some $x \in X$ such that $x_n = x$ for infinitely many numbers $n \in \mathbb{N}$. Let $\{n_k\}_{k=1}^{\infty}$ be a sequence of distinct natural numbers formed from members of the infinite set $\{n \mid (n \in \mathbb{N}) \wedge (x_n = x)\}$. Then $x_{n_k} = x$ for all $k \in \mathbb{N}$, and so $\{x_{n_k}\}_{k=1}^{\infty}$ is a subsequence of $\{x_n\}_{n=1}^{\infty}$ that converges to x. Therefore, $\{x_n\}_{n=1}^{\infty}$ has a convergent subsequence, as required.

Suppose instead that no single value is assumed by x_n for more than finitely many $n \in \mathbb{N}$. If we discard all repeated values of x_n, so that only distinct values remain, then the result will be a subsequence of $\{x_n\}_{n=1}^{\infty}$ and so any further subsequence of that will also be a subsequence of $\{x_n\}_{n=1}^{\infty}$. For this reason, we may assume, without any loss of generality, that the original sequence $\{x_n\}_{n=1}^{\infty}$ is itself such that all elements are distinct.

Note that $(X, \mathscr{T}_d)$ has the Bolzano-Weierstrass Property, by Lemma 6.4.12. Thus, the set $R = \{x_n \mid n \in \mathbb{N}\}$, which is an infinite set in X, has an accumulation point. Denote this accumulation point by x. For each $n \in \mathbb{N}$, let $U_n = B_{1/n}(x)$. Then U_n is an open set containing the accumulation point x, and so $U_n \cap R$ contains infinitely many points, for each $n \in \mathbb{N}$. Choose any $n_1 \in \mathbb{N}$ for which $x_{n_1} \in U_1 \cap R$. Next, choose any $n_2 \in \mathbb{N}$ such that $n_2 > n_1$ and for which $x_{n_2} \in U_2 \cap R$. Continuing inductively, having chosen $n_1, n_2, \ldots, n_k,$

choose n_{k+1} such that $n_{k+1} > n_k$ and for which $x_{n_{k+1}} \in U_{k+1} \cap R$. We have now constructed a sequence $\{x_{n_k}\}_{k=1}^{\infty}$ that is a subsequence of $\{x_n\}_{n=1}^{\infty}$ that converges to x. Therefore, since the sequence $\{x_n\}_{n=1}^{\infty}$ was chosen arbitrarily, the topological space $(X, \mathscr{T}_d)$ is sequentially compact.   Q.E.D.

We are now ready to prove Theorem 6.4.4. For convenience, we restate the theorem here.

Theorem 6.4.4 (Restated). *If $(X, \mathscr{T}_d)$ is a metric space, then the following are equivalent:*

(a) $(X, \mathscr{T}_d)$ *is compact.*

(b) $(X, \mathscr{T}_d)$ *is countably compact.*

(c) $(X, \mathscr{T}_d)$ *is sequentially compact.*

(d) $(X, \mathscr{T}_d)$ *has the Bolzano-Weierstrass Property.*

Proof. In order to show that all conditions are equivalent, we will show

$$(a) \Rightarrow (b) \Rightarrow (c) \Rightarrow (b) \Rightarrow (d) \Rightarrow (a).$$

$(a \Rightarrow b)$ If $(X, \mathscr{T}_d)$ is a compact space, then any open cover $\mathscr{U}$ of X admits a finite subcover. In particular, this must be true if $\mathscr{U}$ is a countable cover. Therefore, $(X, \mathscr{T}_d)$ countably compact.

$(b \Rightarrow c)$ This is Lemma 6.4.15.

$(c \Rightarrow b)$ This follows from Lemma 6.4.14.

$(b \Rightarrow d)$ This follows from Lemma 6.4.12.

$(d \Rightarrow a)$ Suppose $(X, \mathscr{T}_d)$ is a metric space that has the Bolzano-Weierstrass Property and let $\mathscr{U}$ be an open cover of X. By Lemma 6.4.11, there is a Lebesgue number ϵ for $\mathscr{U}$, and so for any $x \in X$, there is a $V \in \mathscr{U}$ such that $B_\epsilon(x) \subseteq V$. By assumption, X has the Bolzano-Weierstrass Property, and so, by Lemma 6.4.10, X is totally bounded. Thus, there is an ϵ-net for X, say $\{x_1, x_2, \ldots, x_n\}$. By the definition of an ϵ-net, the collection of sets $\{B_\epsilon(x_i) \mid i \in \{1, 2, \ldots, n\}\}$ covers X. For each $i \in \{1, 2, \ldots, n\}$, choose $V_i \in \mathscr{U}$ such that $B_\epsilon(x_i) \subseteq V_i$, which we can do because ϵ is a Lebesgue number for $\mathscr{U}$. Then, $\{V_1, V_2, \ldots, V_n\}$ is a finite subcover of X derived from $\mathscr{U}$. Therefore, the topological space $(X, \mathscr{T}_d)$ is compact.   Q.E.D.

Exercises

1. Prove that a Hausdorff space is countably compact if and only if it has the Bolzano-Weierstrass Property.

2. Give an example of a topological space that has the Bolzano-Weierstrass Property but is not countably compact.

3. Let $(X, \mathcal{T}_d)$ be a compact metric space. Prove that there is a number $B > 0$ such that $d(x, y) < B$ for all x and y in X. That is, show that X is bounded with respect to the metric d.

4. Let $(X, \mathcal{T}_d)$ be a compact metric space and let $\mathcal{C}$ be a collection of closed subsets of X such that $\bigcap_{A \in \mathcal{C}} A = \emptyset$. Prove that there is a number $\epsilon > 0$ such that the following is true: For each $x \in X$, there is a set $A \in \mathcal{C}$ (which depends on the choice of x) such that $d(x, A) \geq \epsilon$. (See Exercise 7 in Section 6.3 for the definition of $d(x, A)$, the distance between x and A.)

5. Let $(X, \mathcal{T}_d)$ be a compact metric space. Prove that there exists a countable dense subset of X. That is, show that a compact metric space is separable. (See Definitions 3.5.5 and 3.5.6.)

6. Prove that a closed subset of a countably compact space is countably compact.

7. Prove that a closed subset of a sequentially compact space is sequentially compact.

8. Prove that a compact metric space is second countable. (See Definition 3.5.7.)

9. Prove that a topological space is countably compact if and only if every countable collection of closed subsets having the finite intersection property has nonempty intersection.

6.5   More on Compactness in Metric Spaces

A theorem in elementary calculus states that if a function f is continuous on a closed bounded interval $[a, b]$, then it is uniformly continuous there. That is, for each $\epsilon > 0$, there is a $\delta > 0$ that depends only on ϵ such that $|f(x) - f(y)| < \epsilon$ whenever x and y are in the interval $[a, b]$ and $|x - y| < \delta$. This theorem has a number of applications. For example, it is used to prove that if a function is continuous on a closed bounded interval $[a, b]$, then it is Riemann integrable on $[a, b]$.

Example 6.5.1. Let $f : (0, 1) \to (0, \infty)$ be defined by $f(x) = \frac{1}{x}$ for all x in the open interval $(0, 1)$. Show that f is continuous but not uniformly continuous.

Solution. Certainly, the function f is continuous because it is the quotient of a constant function by a *nonzero* linear function. We leave the remaining details to be done in Exercise 1, below.

We now generalize the concept of uniform continuity, and the corresponding important theorem from elementary calculus, to the setting of metric spaces. We begin with the definition.

Definition 6.5.2. Let $(X, \mathcal{T}_{d_X})$ and $(Y, \mathcal{R}_{d_Y})$ be metric spaces. A function $f : X \to Y$ is said to be *uniformly continuous* if for each $\epsilon > 0$ there is a number $\delta > 0$ such that $d_Y(f(x_1), f(x_2)) < \epsilon$ whenever x_1 and x_2 are elements of X and $d_X(x_1, x_2) < \delta$.

Theorem 6.5.3. *Let $(X, \mathcal{T}_{d_X})$ and $(Y, \mathcal{R}_{d_Y})$ be metric spaces. If X is compact and $f : X \to Y$ is continuous, then f is uniformly continuous.*

Proof. Let $\epsilon > 0$ be given. Let $\mathcal{U} = \{f^{-1}(B_{\epsilon/2}(y)) \mid y \in Y\}$ and observe that $\mathcal{U}$ is an open cover of X. By Theorem 6.4.4, the compact metric space $(X, \mathcal{T}_{d_X})$ has the Bolzano-Weierstrass Property. Thus, since $\mathcal{U}$ is an open cover of X, there is a Lebesgue number δ for $\mathcal{U}$, by Lemma 6.4.11.

Let x_1 and x_2 be elements of X such that $d_X(x_1, x_2) < \delta$. Thus, $x_2 \in B_\delta(x_1)$. Since δ is a Lebesgue number for $\mathcal{U}$, there is some $U \in \mathcal{U}$ such that $B_\delta(x_1) \subseteq U$. Let $y \in Y$ be such that $U = f^{-1}(B_{\epsilon/2}(y))$. Then, since $x_1 \in U$ and $x_2 \in U$, we have that $f(x_1) \in f(U)$ and $f(x_2) \in f(U)$. By construction, $f(U) = B_{\epsilon/2}(y)$. Consequently,

$$d_Y(f(x_1), f(x_2)) \leq d_Y(f(x_1), y) + d_Y(y, f(x_2)) < \frac{\epsilon}{2} + \frac{\epsilon}{2} = \epsilon.$$

Therefore, $d_Y(f(x_1), f(x_2)) < \epsilon$ whenever $d_X(x_1, x_2) < \delta$, and so f is uniformly continuous.   Q.E.D.

We now wish to discuss the convergence of certain sequences. We remind the reader of what it means for a sequence to converge in a metric space. (See also Definitions 2.6.6 and 3.1.9.)

Definition 6.5.4. Let $(X, \mathcal{T}_d)$ be a metric space and let $\{x_n\}_{n=1}^{\infty}$ be a sequence of points in X. The sequence *converges* to x in X if for each $\epsilon > 0$, there is a number $N \in \mathbb{N}$ such that $d(x_n, x) < \epsilon$ whenever $n \geq N$.

If a sequence $\{x_n\}_{n=1}^{\infty}$ converges to x, then for large values of m and n, both x_m and x_n are close to x, and hence also close to each other. (See Exercise 2.) This leads to a natural question: If a sequence $\{x_n\}_{n=1}^{\infty}$ has the property that x_m and x_n are close to each other whenever m and n are large, then is it the case that the sequence must converge? We will see that the answer is "no" in general, but "yes" in certain types of spaces. In order to understand these spaces, we need to make our notions more precise.

Definition 6.5.5. Let $(X, \mathcal{T}_d)$ be a metric space and let $\{x_n\}_{n=1}^{\infty}$ be a sequence of points in X. The sequence is said to be a *Cauchy sequence* if for each $\epsilon > 0$, there is a number $N \in \mathbb{N}$ such that $d(x_n, x_m) < \epsilon$ whenever $n \geq N$ and $m \geq N$.

Thus, Cauchy sequences are precisely the sequences we described as having the property that the terms x_n and x_m are "close to each other" when n and m are "large."

Definition 6.5.6. A metric space $(X, \mathcal{T}_d)$ is said to be *complete* if every Cauchy sequence in X converges to a point in X.

Example 6.5.7. Show that the open interval $(0,1)$ with the standard metric on $\mathbb{R}$, given by $d(x,y) = |x - y|$, is not a complete metric space.

Solution. For each integer $n > 1$, let $x_n = \frac{1}{n}$. Suppose $\epsilon > 0$ and let M be any positive integer such that $M > \frac{2}{\epsilon}$. Suppose that m and n are integers such that $n > M$ and $m > M$. Then

$$d(x_m, x_n) = \left| \frac{1}{m} - \frac{1}{n} \right| \leq \frac{1}{m} + \frac{1}{n} < \frac{\epsilon}{2} + \frac{\epsilon}{2} = \epsilon.$$

Thus, $\{x_n\}_{n=1}^{\infty}$ is a Cauchy sequence.

Now let $x \in (0,1)$. We will show that the sequence $\{x_n\}_{n=1}^{\infty}$ does not converge to x. Let $\epsilon = \frac{x}{2}$ and let N be a positive integer such that $N > \frac{1}{\epsilon}$. If $n > N$, then

$$x_n = \frac{1}{n} < \frac{1}{N} < \epsilon = \frac{x}{2}.$$

Thus, we have that

$$d(x_n, x) = |x_n - x| = x - x_n > x - \frac{x}{2} = \frac{x}{2} = \epsilon.$$

Therefore, the sequence $\{1/n\}_{n=1}^{\infty}$ does not converge to x. Since this is true for any x in the interval $(0,1)$, we conclude that $\{1/n\}_{n=1}^{\infty}$ is a Cauchy sequence in $((0,1), \mathcal{T}_d)$ that does not converge to a number in $(0,1)$. Therefore, $((0,1), \mathcal{T}_d)$ is not a complete metric space.

Theorem 6.5.8. *If a metric space is such that every Cauchy sequence has a convergent subsequence, then the metric space is complete.*

Proof. Let $(X, \mathcal{T}_d)$ be a metric space and let $\{x_n\}_{n=1}^{\infty}$ be a Cauchy sequence in X. Suppose that $\{x_{n_k}\}_{k=1}^{\infty}$ is a subsequence of $\{x_n\}_{n=1}^{\infty}$ that converges to a point $x \in X$. We claim that $\{x_n\}_{n=1}^{\infty}$ also converges to x. To see this, let $\epsilon > 0$ be given. Because $\{x_n\}_{n=1}^{\infty}$ is a Cauchy sequence, there is a positive integer N so that $d(x_m, x_n) < \frac{\epsilon}{2}$ whenever $m \geq N$ and $n \geq N$. Since $\{x_{n_k}\}_{k=1}^{\infty}$ converges to x, there exists a positive integer k_0 such that $n_{k_0} \geq N$ and such that $d(x_{n_{k_0}}, x) < \frac{\epsilon}{2}$. Then, if $n \geq N$, it follows that

$$d(x_n, x) \leq d(x_n, x_{n_{k_0}}) + d(x_{n_{k_0}}, x) < \frac{\epsilon}{2} + \frac{\epsilon}{2} = \epsilon.$$

Therefore, $\{x_n\}_{n=1}^{\infty}$ converges to x. Q.E.D.

A complete metric space need not be compact (see the exercises at the end of this section for several examples), but we do have the following theorem.

Theorem 6.5.9. *A metric space is compact if and only if it is complete and totally bounded.*

Proof. Let $(X, \mathcal{T}_d)$ be a compact metric space. By Theorem 6.4.4, the metric space $(X, \mathcal{T}_d)$ is sequentially compact. Thus, any sequence—and in particular any Cauchy sequence—has a convergent subsequence. Therefore, by Theorem 6.5.8, any Cauchy sequence converges, and so $(X, \mathcal{T}_d)$ is complete.

Next, we observe that any compact metric space has the Bolzano-Weierstrass Property, by Theorem 6.4.4. Furthermore, by Lemma 6.4.10, any metric space having the Bolzano-Weierstrass Property is totally bounded. It follows that $(X, \mathcal{T}_d)$ is both complete and totally bounded, as required.

Conversely, suppose that $(X, \mathcal{T}_d)$ is both complete and totally bounded. By Theorem 6.4.4, it will suffice to show that $(X, \mathcal{T}_d)$ is sequentially compact. Let $\{x_n\}_{n=1}^{\infty}$ be an arbitrary sequence of points in X. We will construct a subsequence that is Cauchy. We first define a sequence of subsequences of $\{x_n\}_{n=1}^{\infty}$. We will denote the jth term of the ith subsequence by $x_{n_{ij}}$.

Since $(X, \mathcal{T}_d)$ is totally bounded, we can cover X with a finite number of open balls of radius 1. At least one of these must contain x_n for infinitely many values of $n \in \mathbb{N}$. Pick one such ball of radius 1 and call it B_1. Let $N_1 = \{n \mid n \in \mathbb{N} \text{ and } x_n \in B_1\}$. Choose any sequence $\{n_{1j}\}_{j=1}^{\infty}$ of positive integers such that $n_{11} < n_{12} < n_{13} < \cdots$ and such that $n_{1j} \in N_1$ for all $j \in \mathbb{N}$. Then $\{x_{n_{1j}}\}_{j=1}^{\infty}$ is a subsequence of $\{x_n\}_{n=1}^{\infty}$, all the terms of which are in the ball B_1.

Arguing as before, because $(X, \mathcal{T}_d)$ is totally bounded, we can cover X using a finite number of open balls of radius $1/2$. At least one of these, call it B_2, contains $x_{n_{1j}}$ for infinitely many $j \in \mathbb{N}$. Let $N_2 = \{n_{1j} \mid j \in \mathbb{N} \text{ and } x_{n_{1j}} \in B_2\}$. Choose any sequence $\{n_{2j}\}_{j=1}^{\infty}$ such that $n_{21} < n_{22} < n_{23} < \cdots$ and such that $n_{2j} \in N_2$ for all $j \in \mathbb{N}$. Then $\{x_{n_{2j}}\}_{j=1}^{\infty}$ is a subsequence of $\{x_{n_{1j}}\}_{j=1}^{\infty}$, all the terms of which are in the ball B_2.

We continue in this way inductively. Let $k \in \mathbb{N}$ and suppose for each i in $\{2, \ldots, k\}$ we have found an open ball B_i of radius $1/i$, and a subsequence $\{x_{n_{ij}}\}_{j=1}^{\infty}$ of $\{x_{n_{(i-1)j}}\}_{j=1}^{\infty}$, the terms of which are all contained in B_i. We cover X using a finite number of open balls of radius $\frac{1}{k+1}$. At least one of these, call it B_{k+1}, must contain $x_{n_{kj}}$ for infinitely many $j \in \mathbb{N}$. Define a set of positive integers $N_{k+1} = \{n_{kj} \mid j \in \mathbb{N} \text{ and } x_{n_{kj}} \in B_{k+1}\}$. Choose any sequence $\{n_{(k+1)j}\}_{j=1}^{\infty}$ such that $n_{(k+1)1} < n_{(k+1)2} < n_{(k+1)3} < \cdots$ and such that $n_{(k+1)j} \in N_{k+1}$ for all $j \in \mathbb{N}$. Then $\{x_{n_{(k+1)j}}\}_{j=1}^{\infty}$ is a subsequence of $\{x_{n_{kj}}\}_{j=1}^{\infty}$, all the terms of which are in the ball B_{k+1}.

We now use what is known as a *diagonalization argument* to construct a subsequence of the original sequence $\{x_n\}_{n=1}^{\infty}$. This new subsequence will be constructed using the subsequences $\{x_{n_{ij}}\}_{j=1}^{\infty}$, where $i \in \mathbb{N}$. We take the first term from the first sequence, the second term from the second sequence, and so on. To be precise, we define a subsequence $\{x_{n_i}\}_{i=1}^{\infty}$ of $\{x_n\}_{n=1}^{\infty}$ by letting $x_{n_i} = x_{n_{ii}}$ for each $i \in \mathbb{N}$. Certainly, $\{x_{n_i}\}_{i=1}^{\infty}$ is a subsequence of $\{x_n\}_{n=1}^{\infty}$.

We claim that $\{x_{n_i}\}_{i=1}^{\infty}$ is a Cauchy sequence. To see this, let $\epsilon > 0$ be given. Choose a positive integer $N \in \mathbb{N}$ such that $\frac{1}{N} < \frac{\epsilon}{2}$. If i and j are greater than

N, then x_{n_i} and x_{n_j} are members of the Nth subsequence, and consequently are in the open ball B_N. Let y denote the center of the ball B_N. Then,

$$d(x_{n_i}, x_{n_j}) \le d(x_{n_i}, y) + d(y, x_{n_j}) < \frac{1}{N} + \frac{1}{N} < \frac{\epsilon}{2} + \frac{\epsilon}{2} = \epsilon.$$

We have established that the subsequence $\{x_{n_i}\}_{i=1}^{\infty}$ is a Cauchy sequence. By assumption, $(X, \mathcal{T}_d)$ is a complete metric space. Thus, the subsequence $\{x_{n_i}\}_{i=1}^{\infty}$ converges. We have shown that the original sequence $\{x_n\}_{n=1}^{\infty}$, which was arbitrary, has a convergent subsequence. Therefore, $(X, \mathcal{T}_d)$ is sequentially compact, and hence (by Theorem 6.4.4) it is compact.   Q.E.D.

Other interesting facts about complete and compact metric spaces are discussed in the exercises.

Exercises

1. Let $f : (0,1) \to (0,\infty)$ be defined by $f(x) = \frac{1}{x}$ for all x in the open interval $(0,1)$. Show that f is not uniformly continuous. (See Example 6.5.1.)

2. Let $(X, \mathcal{T}_d)$ be a metric space and suppose that $\{x_n\}_{n=1}^{\infty}$ is a sequence that converges to $x \in X$. Prove that $\{x_n\}_{n=1}^{\infty}$ is a Cauchy sequence.

3. Let $(X, \mathcal{T}_d)$ be a metric space and suppose that $\{x_n\}_{n=1}^{\infty}$ is a Cauchy sequence in X. Prove that $\{x_n \mid n \in \mathbb{N}\}$ is a bounded set.

4. Let $(\mathbb{R}, \mathcal{R}_d)$ be the set of real numbers with the standard metric topology.

    (a) Prove that $(\mathbb{R}, \mathcal{R}_d)$ is a complete metric space. (*Hint:* Use Exercise 3 and the Axiom of Completeness.)

    (b) Show that $(\mathbb{R}, \mathcal{R}_d)$ is homeomorphic to the incomplete metric space $((0,1), \mathcal{T}_d)$ in Example 6.5.7. (This shows that completeness is not a topological property.)

5. Show that $\mathbb{R}^n$ is complete in its usual metric. (*Hint:* First show that if $\{x_i\}_{i=1}^{\infty}$ is a Cauchy sequence in $\mathbb{R}^n$, then each of the n corresponding sequences of coordinates is a Cauchy sequence.)

6. Prove that the Hilbert space of Exercise 8 in Section 2.8 is complete.

7. Let $(X, \mathcal{T}_d)$ be a complete metric space. Show that a subspace $(Y, \mathcal{R}_d)$ is complete if and only if Y is closed in X.

8. (a) Prove that every compact subset of a metric space is closed and bounded.

    (b) Let $(\mathcal{H}, \mathcal{T}_d)$ be the Hilbert space from Exercise 8 in Section 2.8. Let S be the *unit sphere* in $\mathcal{H}$; that is, let

    $$S = \left\{ x = (x_1, x_2, \dots) \mid x_i \in \mathbb{R} \text{ for all } i \in \mathbb{N} \text{ and } \sum_{i=1}^{\infty} x_i^2 = 1 \right\}.$$

Show that S is closed and bounded but does not have the Bolzano-Weierstrass Property. This shows that a subset of a complete metric space may be closed and bounded but not compact. (See also Exercise 8 of Section 6.2.)

9. Give an example of a compact metric space $(X, \mathscr{T}_d)$, a topological space $(Y, \mathscr{R})$ that is not Hausdorff, and a continuous function $f : X \to Y$ that is one-to-one and onto. (*Hint:* Can f be a homeomorphism?)

10. Let $\mathbb{N}$ be the set of positive integers and define $\mathscr{B} = \{\{2n, 2n+1\} \mid n \in \mathbb{N}\}$.

    (a) Show that $\mathscr{B}$ is the base for a topology $\mathscr{T}$ on $\mathbb{N}$.

    (b) Show that $(\mathbb{N}, \mathscr{T})$ has the Bolzano-Weierstrass Property.

    (c) Show that $(\mathbb{N}, \mathscr{T})$ is not countably compact.

    (d) Is $(\mathbb{N}, \mathscr{T})$ a metrizable space? That is, can there be a metric d on $\mathbb{N}$ such that $\mathscr{T} = \mathscr{T}_d$? Justify your answer.

11. Let $(X, \mathscr{T}_d)$ be a metric space. Suppose that $\{x_n\}_{n=1}^{\infty}$ is a sequence of points in X that converges to $x \in X$. If x_0 is any point in X, show that
$$\lim_{n \to \infty} d(x_n, x_0) = d(x, x_0).$$

6.6   The Cantor Set

In the present section, we introduce a set that is used as a counterexample to many plausible conjectures. We start our construction (which is illustrated in Figure 6.6.a) with the closed interval $E_1 = [0, 1]$. From the set E_1, we remove the middle third open interval $U_1^1 = (\frac{1}{3}, \frac{2}{3})$ to obtain the set $E_2 = [0, \frac{1}{3}] \cup [\frac{2}{3}, 1]$. Now let
$$U_2^1 = \left(\frac{1}{3^2}, \frac{2}{3^2}\right) \quad \text{and} \quad U_2^2 = \left(\frac{7}{3^2}, \frac{8}{3^2}\right),$$
each one the open middle third of one of the two closed intervals comprising the set E_2. Let $E_3 = E_2 \setminus (U_2^1 \cup U_2^2)$, which is the union of four closed intervals:
$$E_3 = \left[0, \frac{1}{3^2}\right] \cup \left[\frac{2}{3^2}, \frac{3}{3^2}\right] \cup \left[\frac{6}{3^2}, \frac{7}{3^2}\right] \cup \left[\frac{8}{3^2}, \frac{9}{3^2}\right].$$

Now we remove the four open middle thirds belonging to the four closed intervals that comprise E_3. Let
$$U_3^1 = \left(\frac{1}{3^3}, \frac{2}{3^3}\right), \; U_3^2 = \left(\frac{7}{3^3}, \frac{8}{3^3}\right), \; U_3^3 = \left(\frac{19}{3^3}, \frac{20}{3^3}\right) \quad \text{and} \quad U_3^4 = \left(\frac{25}{3^3}, \frac{26}{3^3}\right),$$

and let $E_4 = E_3 \setminus (U_3^1 \cup U_3^2 \cup U_3^3 \cup U_3^4)$.

We continue the above process inductively. Having defined the open intervals $U_k^1, U_k^2, U_k^3, \ldots, U_k^{2^{k-1}}$, we let $E_{k+1} = E_k \setminus (U_k^1 \cup U_k^2 \cup U_k^3 \cup \cdots \cup U_k^{2^{k-1}})$ and define $U_{k+1}^1, U_{k+1}^2, U_{k+1}^3, \ldots, U_{k+1}^{2^k}$ to be the open middle thirds of the 2^k closed

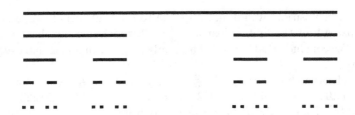

Figure 6.6.a: This image illustrates the first five steps in the construction of the Cantor set. Each level after the first is the previous level with all "middle thirds" removed.

intervals that comprise the set E_{k+1}. Finally, we define the *Cantor set* to be the set $\mathscr{K}$, where

$$\mathscr{K} = [0,1] \setminus \left[\bigcup_{k=1}^{\infty} \left(\bigcup_{i=1}^{2^{k-1}} U_k^i \right) \right].$$

Make $\mathscr{K}$ into a topological space by endowing it with the subspace topology induced from $\mathbb{R}$. Observe that $\bigcup_{k=1}^{\infty} \left(\bigcup_{i=1}^{2^{k-1}} U_k^i \right)$ is an open set in $\mathbb{R}$, since it is the union of open intervals. Thus, the set $\mathscr{K}$ is closed as the complement of an open set. Since $\mathscr{K}$ is also a bounded set, we conclude that it is a compact subset of $\mathbb{R}$, by the Heine-Borel Theorem (Theorem 6.2.2).

Each element $x \in [0,1]$ has what is known as a *ternary expansion*:

$$x = \frac{t_1}{3} + \frac{t_2}{3^2} + \cdots = \sum_{n=1}^{\infty} \frac{t_n}{3^n},$$

where $t_n \in \{0,1,2\}$ for each $n \in \mathbb{N}$. We write $x = .t_1 t_2 t_3 \ldots$ and we refer to this also as a ternary expansion of x. The ternary expansion of a number $x \in [0,1]$ may not be unique. If x has a finite ternary expansion, then it also has an infinite ternary expansion ending in an infinite sequence of twos. For example,

$$\frac{16}{27} = \frac{1}{3} + \frac{2}{3^2} + \frac{1}{3^3} = .121$$

$$= \frac{1}{3} + \frac{2}{3^2} + \frac{0}{3^3} + \sum_{k=4}^{\infty} \frac{2}{3^k} = .120222222\ldots.$$

We sometimes adopt the convention that whenever a number $x \in [0,1]$ has a finite and an infinite ternary expansion, we choose the infinite expansion. With this convention, the ternary expansion of each $x \in [0,1]$ is unique. There are some cases, however, when it will be useful to select the finite expansion, as we shall see presently.

Observe that $\frac{1}{3} = .1$ and $\frac{2}{3} = .2$. Therefore, if $x \in (\frac{1}{3}, \frac{2}{3}) = U_1^1$, it must be that $.1 < x < .2 = .1222\ldots$. We see, then, that $x \in U_1^1$ if and only if the ternary expansion *requires* a 1 in the first position. We say "requires" here because, while the endpoint $\frac{2}{3} = .1222\ldots$ does have a ternary expansion with

a 1 in the first position, we are not required to use that ternary expansion. We
can opt instead for the finite expansion $\frac{2}{3} = .2$, which does not have a 1 in any
position. We conclude that $x \in U_1^1$ if and only if $t_1 = 1$ in any ternary expansion
$x = .t_1 t_2 t_3 \ldots$.

Next, observe that if $x \in (\frac{1}{3^2}, \frac{2}{3^2}) = U_2^1$ or $x \in (\frac{7}{3^2}, \frac{8}{3^2}) = U_2^2$, then we
have either $.01 < x < .02 = .01222\ldots$ or $.21 < x < .22 = .21222\ldots$. Thus,
$x \in U_2^1 \cup U_2^2$ if and only if the ternary expansion requires a 1 in the second
position. That is, if and only if $t_2 = 1$ for any ternary expansion $x = .t_1 t_2 t_3 \ldots$.
(As before, we can use the finite expansions for the endpoints.)

In general, the ternary expansion for x requires a 1 in the kth position (that
is, $t_k = 1$) if and only if $x \in U_k^1 \cup U_k^2 \cup U_k^3 \cup \cdots \cup U_k^{2^{k-1}}$.

Now suppose $x \in \mathscr{K}$. By construction, the set $\mathscr{K}$ does not contain the set
U_k^j for any $k \in \mathbb{N}$ and $j \in \{1, 2, \ldots, 2^{k-1}\}$. Therefore, x can be given a ternary
expansion $x = .t_1 t_2 t_3 \ldots$ such that no position contains a 1. That is, we can
find a ternary expansion for x such that $t_k \in \{0, 2\}$ for all $k \in \mathbb{N}$. We shall use
this fact to prove the next theorem.

Theorem 6.6.1. *The Cantor Set $\mathscr{K}$ (as a subspace of $\mathbb{R}$) is homeomorphic to
a product space, where each member of the product is a two point discrete space.*

Proof. For each $n \in \mathbb{N}$, let $X_n = \{0, 2\}$ and let $\mathscr{T}_n$ be the discrete topology
on X_n. Let $X = \prod_{n=1}^{\infty} X_n$ and let $\mathscr{T}$ be the product topology on X. We will
show that $\mathscr{K}$ is homeomorphic to the product space $(X, \mathscr{T})$.

Observe first that, by Theorem 4.2.6, the product space $(X, \mathscr{T})$ is a Hausdorff space. We will define a one-to-one and onto function $f : \mathscr{K} \to X$. If
$x \in \mathscr{K}$, then x has a unique ternary expansion $x = .t_1 t_2 t_3 \ldots$ such that
$t_k \in \{0, 2\}$ for all $k \in \mathbb{N}$. From now on, we will always use this ternary expansion for $x \in \mathscr{K}$. Define $f : \mathscr{K} \to X$ by

$$f(x) = f(.t_1 t_2 t_3 \ldots) = (t_1, t_2, t_3, \ldots) \in X = \prod_{n=1}^{\infty} X_n.$$

The function f is one-to-one, because the ternary expansion is unique. To see
that the function f is also onto, simply observe that any ternary expansion
formed using elements from the set $\{0, 2\}$ must be in $\mathscr{K}$. To be precise, we
suppose that $t_n \in \{0, 2\}$ for all $n \in \mathbb{N}$, but $x = .t_1 t_2 t_3 \ldots$ is not in $\mathscr{K}$. Then
$x \in U_k^j$ for some $k \in \mathbb{N}$ and some $j \in \{1, 2, \ldots, 2^{k-1}\}$. But this implies that
$t_k = 1$, a contradiction.

We will show that f is a continuous function by means of Theorem 4.2.4.
That is, we will show that $p_n \circ f$ is continuous for each $n \in \mathbb{N}$, where $p_n : X \to X_n$
is the nth projection function. Let $n \in \mathbb{N}$ be given and suppose $\epsilon > 0$. Let
$\delta = \frac{1}{3^n}$. Let x and y be members of $\mathscr{K}$ such that $|x - y| < \delta$. In such a
circumstance, it must be that the first n positions of the ternary expansions for
x and y are identical. In particular, the nth positions must be equal. That is,
if x and y have ternary expansions $x = .x_1 x_2 x_3 \ldots$ and $y = .y_1 y_2 y_3 \ldots$, then
$x_n = y_n$. Thus,

$$|p_n(f(x)) - p_n(f(y))| = |x_n - y_n| = 0 < \epsilon.$$

Consequently, $p_n \circ f$ is continuous for each $n \in \mathbb{N}$.

We have established that $f : \mathcal{K} \rightarrow X$ is a one-to-one, onto, and continuous function from the compact space $\mathcal{K}$ to the Hausdorff space $(X, \mathcal{T})$. By Exercise 10 of Section 6.1, we conclude that f is a homeomorphism.   Q.E.D.

The Cantor set is constructed by removing a sequence of open intervals from $[0, 1]$ It is natural to wonder what points actually remain in $\mathcal{K}$. Certainly, the endpoints of all the intervals U_k^j are never removed, but there are points in $\mathcal{K}$ which are not endpoints of any of these intervals. (See the exercises at the end of this section.) In fact, there are uncountably many elements in the set $\mathcal{K}$, as we shall now see.

Theorem 6.6.2. *The Cantor set and the open interval* $(0, 1)$ *have the same cardinality.*

Proof. We will show that there is a one-to-one correspondence between the Cantor set $\mathcal{K}$ and the open interval $I = (0, 1)$. We divide $\mathcal{K}$ into two disjoint sets. Let $E = \{x \in \mathcal{K} \mid x$ is an endpoint of U_k^j for some k and j in $\mathbb{N}\}$ and let $F = \mathcal{K} \setminus E$. We will see (in Exercise 3) that each element of F is an accumulation point of E. Note that E is an infinite set of rational numbers, and so it is countable.

We now consider the *binary expansion* of each real number in I. That is, for each $x \in (0, 1)$, we write

$$x = \frac{b_1}{2} + \frac{b_2}{2^2} + \frac{b_3}{2^3} + \frac{b_4}{2^4} + \cdots = .b_1b_2b_3b_4\ldots,$$

where $b_n \in \{0, 1\}$ for each $n \in \mathbb{N}$. As in the case of the ternary expansion (or even the decimal expansion), the binary representation is not unique for certain real numbers. For example,

$$\frac{3}{8} = \frac{1}{2^2} + \frac{1}{2^3} = .011 = .0101111\ldots.$$

In each case, however, the numbers with a terminating binary expansion are necessarily rational numbers. Thus, we obtain the open interval I as the union of two disjoint sets V and W, where

$$V = \{x \in I \mid x \text{ has terminating binary expansion}\}$$

and
$$W = \{x \in I \mid x \text{ has no terminating binary expansion}\}.$$

Note that V is an infinite set of rational numbers, and so is countable. Since both E and V have the same cardinality as the set of natural numbers $\mathbb{N}$, there exists a one-to-one and onto function $g : E \rightarrow V$.

Let $x \in F$. We give x the unique ternary expansion $x = .t_1t_2t_3 \ldots$ with $t_k \in \{0, 2\}$ for each $k \in \mathbb{N}$. (Such a ternary expansion was shown to exist above.) Since $x \in F$, it must be the case that $x = .t_1t_2t_3 \ldots$ is nonterminating. (See Exercise 4.) Define a number $h(x)$, constructed as a binary expansion, by letting $h(x) = .b_1b_2b_3 \ldots$, where $b_k = \frac{t_k}{2}$ for each $k \in \mathbb{N}$. It is evident that h is both one-to-one and onto.

Now, define a function $f : \mathcal{K} \to I$ by the formula

$$f(x) = \begin{cases} g(x) & \text{if } x \in E, \\ h(x) & \text{if } x \in F. \end{cases}$$

By construction, we have that f is both one-to-one and onto. Therefore, $\mathcal{K}$ has the same cardinality as $I = (0, 1)$.   Q.E.D.

In Example 1.6.12, we saw that the unit interval $(0, 1)$ is uncountable. This fact, together with the above theorem, shows that $\mathcal{K}$ is also uncountable. (See also Exercises 8 and 9 at the end of this section.)

Although the Cantor set is "large" in the sense that it has uncountably many elements, it is "small" in a geometric sense. To see what we mean by this, observe that we constructed the Cantor set by first removing from $[0, 1]$ the middle third open interval of length $\frac{1}{3}$. Then we removed two more open intervals, each having length $\frac{1}{3^2}$. Next, we removed four open intervals, each having length $\frac{1}{3^3}$. We continued in this way, inductively, at step k removing 2^{k-1} intervals of length $\frac{1}{3^k}$. Thus, the total length of the removed intervals is

$$\frac{1}{3} + \frac{2}{3^2} + \frac{2^2}{3^3} + \frac{2^3}{3^4} + \cdots = \frac{1}{3} \sum_{k=0}^{\infty} \left(\frac{2}{3}\right)^k = \frac{1}{3} \cdot \frac{1}{1 - \frac{2}{3}} = 1.$$

Therefore, from an interval of length 1, we have removed a subset also of length 1. It follows that the Cantor set (in some sense) has length 0. We conclude that $\mathcal{K}$ does not contain any interval as a subset.

Let us now consider some further properties of the Cantor set. The first property is related to connectedness. Recall that if a space is not connected, then it is the union of two or more components. (See Definition 5.1.11.) In some sense, the number of components is a measure of the connectedness (or disconnectedness) of a space. For example, as subspaces of $\mathbb{R}$ (with the standard topology), $X = [0, 1] \cup [2, 4] \cup [5, 8]$ is more disconnected than $Y = [1, 3] \cup [5, 9]$, because X has three components and Y has only two. The set of rational numbers (as a subspace of the real line) is very disconnected since all of its components are single points.

Definition 6.6.3. A topological space $(X, \mathcal{T})$ is said to be *totally disconnected* if each component of X is a single point.

Theorem 6.6.4. *The Cantor set $\mathcal{K}$ is totally disconnected (as a subspace of $\mathbb{R}$ with the standard topology).*

Proof. The connected subsets of $\mathbb{R}$ are precisely intervals and singletons. Since $\mathcal{K}$ contains no proper interval, each component must be a single point. Q.E.D.

Definition 6.6.5. Let $(X, \mathcal{T})$ be a topological space. A subset Y of X is said to be *perfect* if each point of Y is an accumulation point of Y.

Theorem 6.6.6. *The Cantor set $\mathscr{K}$ is a perfect subset of $\mathbb{R}$.*

Proof. Let $x \in \mathscr{K}$ and suppose $\epsilon > 0$ is given. We wish to find $y \in \mathscr{K}$ such that $y \neq x$ and $|y - x| < \epsilon$. Let $x = .x_1 x_2 x_3 \ldots$ be the ternary expansion of x, chosen so that $x_k \in \{0, 2\}$ for all $k \in \mathbb{N}$. Let $N \in \mathbb{N}$ be chosen so large that $\frac{2}{3^N} < \epsilon$. Construct $y \in \mathscr{K}$ so that it has ternary expansion $y = .y_1 y_2 y_3 \ldots$, where

$$
y_k = \begin{cases} x_k & \text{if } k \neq N, \\ 0 & \text{if } k = N \text{ and } x_N = 2, \\ 2 & \text{if } k = N \text{ and } x_N = 0. \end{cases}
$$

Then, $|y - x| = \frac{2}{3^N} < \epsilon$. Therefore, x is an accumulation point of $\mathscr{K}$. Since x was an arbitrary point, we conclude that every point of $\mathscr{K}$ is an accumulation point of $\mathscr{K}$, and hence $\mathscr{K}$ is a perfect subset of $\mathbb{R}$.   Q.E.D.

The properties of $\mathscr{K}$ that we have mentioned in this section characterize $\mathscr{K}$ topologically. It was proved (by Felix Hausdorff) that every metrizable, compact, totally disconnected, perfect space is homeomorphic to the Cantor space. Thus, we may think of $\mathscr{K}$ as the prototype of such spaces.

Exercises

1. Show that

$$
\frac{3}{4} = \frac{2}{3} + \frac{2}{3^3} + \frac{2}{3^5} + \frac{2}{3^7} + \frac{2}{3^9} + \cdots .
$$

   What can you conclude about the number $\frac{3}{4}$?

2. Show that $\frac{1}{4}$ is a member of the Cantor set.

3. Let E be the set of all endpoints of the open intervals that were removed from $[0, 1]$ in order to construct the Cantor set $\mathscr{K}$. Prove that any point $x \in \mathscr{K} \setminus E$ is an accumulation point of the set E.

4. Let $x \in [0, 1]$ be given the unique ternary expansion $x = .t_1 t_2 t_3 \ldots$ with t_k in $\{0, 2\}$ for each $k \in \mathbb{N}$. Show that if x has terminating ternary expansion, then x is an endpoint of one of the open intervals removed from $[0, 1]$ in order to construct the Cantor set. Show that the converse is also true.

5. Let Ω be an index set and, for each $\alpha \in \Omega$, let $(X_\alpha, \mathscr{T}_\alpha)$ be a totally disconnected topological space. Prove that the product space $(X, \mathscr{T})$ is totally disconnected.

6. Let $(X, \mathscr{T})$ be a totally disconnected topological space. Prove that any subspace of $(X, \mathscr{T})$ is totally disconnected.

7. Let $(X, \mathscr{T})$ be a topological space. A subset A of X is said to be *nowhere dense* if $(\overline{A})^\circ = \emptyset$. That is, A is nowhere dense if the closure of A has empty interior. Prove that the Cantor set is a nowhere dense subset of the space of real numbers (with the standard topology).

8. Let $\mathcal{P}$ be the set of all sequences $\{x_n\}_{n=1}^{\infty}$ such that $x_n \in \{0, 1\}$ for $n \in \mathbb{N}$. Prove that $\mathcal{P}$ is uncountable.

9. Use Exercise 8 to show that the Cantor set $\mathcal{K}$ is not countable.

10. Let $\mathcal{K}$ denote the Cantor set. Define a function $g : \mathcal{K} \to [0, 1]$ by the formula
$$g\left(\sum_{n=1}^{\infty} \frac{t_n}{3^n} \right) = \sum_{n=1}^{\infty} \frac{t_n}{2^{n+1}},$$
where $t_n \in \{0, 2\}$ for each $n \in \mathbb{N}$. Prove that g is both continuous and onto.

11. Prove that the set of elements of the Cantor set $\mathcal{K}$ that are rational numbers is dense in $\mathcal{K}$.

Chapter 7

Fixed Point Theorems and Applications

In this, the final chapter, we discuss some more topics in topology to illustrate further the beauty and power of the mathematics we have developed. One of the most elegant theorems of modern mathematics is Brouwer's Fixed Point Theorem, both for its simplicity and for its wide applications. The simplest version of this theorem states that any closed bounded interval in $\mathbb{R}$ has the fixed point property. That is, if a and b are real numbers such that $a < b$, and if $f : [a, b] \to [a, b]$ is a continuous function, then there is some $x_0 \in [a, b]$ such that $f(x_0) = x_0$. (See Section 3.5.) The n-dimensional version of this theorem is also true: The topological product of n closed bounded intervals has the fixed point property. The proof of this version of the theorem is difficult even when $n = 2$. A number of different proofs for that case have appeared in the literature.

The proof of Brouwer's Fixed Point Theorem that we choose (for the case $n = 2$) is a consequence of a combinatorial result called Sperner's Lemma. In Section 7.1, we introduce the background necessary to state and prove this lemma. In Section 7.2, we prove Brouwer's Fixed Point Theorem. Then, in Section 7.3, we use Brouwer's Fixed Point Theorem to prove the Fundamental Theorem of Algebra. In Section 7.5, we prove a fixed point theorem for so-called contraction maps and give an application of that theorem.

7.1 Sperner's Lemma

We begin by considering a line segment. Suppose the endpoints of the line segment are labeled with the letters A and B. We partition the line segment, inserting n distinct points between the endpoints, labeling each of these points as either A or B. We obtain $n + 1$ smaller line segments, labeled AB, AA, BB, or BA. We call segments AB and BA *complete* because we have used both labels to name the endpoints.

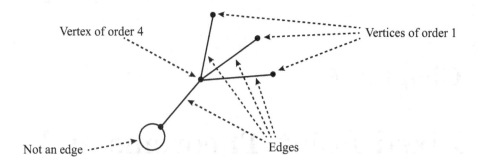

Figure 7.1.a: A graph with five vertices, four edges, and one loop.

We now prove the following result.

Theorem 7.1.1. *Partition a line segment AB by inserting n points between A and B and label each of these points as either A or B. Among the $n + 1$ small line segments thus obtained, an odd number of them will be complete.*

Proof. We use induction on n. If $n = 1$, then the added point is either A or B. If the added point is labeled A, then the resulting small line segments are AA and AB. If the added point is labeled B, then the obtained small line segments are AB and BB. Either way, there is exactly one complete segment.

Now, assume that after adding k points between A and B, we obtain $k+1$ new small line segments, an odd number of which are complete. Add an additional point, labeled either A or B, to the original line segment. Suppose the added point is labeled A. If this new point is added to a complete small line segment AB, then the two resulting (smaller) line segments are AA and AB. The total number of complete small line segments does not change (one is lost, but one is gained), and so the total number of complete small line segments is still odd. If a point A is added to a line segment AA, then no complete line segments are added or taken away. If a point labeled A is added to a line segment BB, then the resulting two small line segments are BA and AB, both of which are complete. Since BB was not complete to begin with, we add two complete small line segments to an odd number and conclude there is still an odd number of complete small line segments. A similar argument is used if the added point is labeled B.

Thus, if $n \in \mathbb{N}$ and n points are added to the original line segment, then the number of complete small line segments is odd.   Q.E.D.

In order to facilitate the proof of our main result, we introduce some useful terminology. Let S be a finite set of distinct points in the plane. Each point in S is called a *vertex*. A curve joining any two distinct vertices is known as an *edge*. We call two edges *parallel* if they have the same endpoints. If a vertex is an endpoint for exactly k edges, then we say it has *order* equal to k. Observe that a *loop* (a curve that joins a point to itself) is not an edge. A set of vertices together with a collection of edges is called a *graph*.

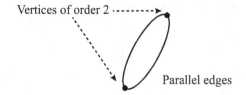

Vertices of order 2

Parallel edges

Figure 7.1.b: A graph with two vertices and two edges.

Theorem 7.1.2. *In a graph, the number of vertices that have an odd order is even.*

Proof. By definition, a graph is a set of vertices together with a collection of edges connecting them. Label the vertices $v_1, v_2, \ldots, v_n$ and let c_{ij} be the number of edges connecting v_i to v_j. Certainly, $c_{ij} = c_{ji}$ for all distinct numbers i and j in $\{1, 2, \ldots, n\}$ and $c_{ii} = 0$ for all $i \in \{1, 2, \ldots, n\}$.

Define a matrix C so that c_{ij} is the entry in the ith row and the jth column, where i and j are from the set $\{1, 2, \ldots, n\}$. Then C is symmetric with zeros along the diagonal. Thus, the sum of its entries must be an even integer. Let $o_i = \sum_{j=1}^{n} c_{ij}$. Then o_i is the order of the vertex v_i. Observe that $\sum_{i=1}^{n} o_i$ is the sum of all the entries of the matrix C, and hence this sum must be an even number. It follows that the number of vertices with odd order must be even. Otherwise, the sum $\sum_{i=1}^{n} o_i$ would be an odd number, because the sum of an odd number of odd numbers, plus any number of even numbers, is an odd number.   Q.E.D.

In order to state Sperner's Lemma, we need to introduce several definitions related to a triangle T. When we say "triangle," we mean both the boundary and the interior of the triangle. A *partition* of the triangle T is a finite set of smaller triangles $\{T_1, T_2, \ldots, T_n\}$ such that $T = \bigcup_{i=1}^{n} T_i$ and, whenever $i \neq j$, the intersection $T_i \cap T_j$ is the empty set, a common vertex, or a common edge. (See Figure 7.1.c.)

We start with a triangle T having vertices labeled A, B, and C. Let $\{T_1, T_2, \ldots, T_n\}$ be a partition of T. Let K be the set of all vertices of the small triangles of the partition. Label each vertex in K using any of the letters A, B, or C. Such a labeling is said to be a *proper labeling* if any vertex in K that is on a side of the original triangle is labeled with one of the endpoints of that particular side. There is no restriction on the other members of K. If a triangle is labeled in such a way that it uses all three of the letters A, B, and C (such as T itself), then the triangle is said to be *complete*.

We are now able to state and prove Sperner's Lemma. (See the illustration in Figure 7.1.d.)

Theorem 7.1.3 (Sperner's Lemma). *Any partition of a complete triangle that has a proper labeling contains at least one complete triangle.*

Proof. We will give a proof by contradiction. Let T be a complete triangle and let $\mathscr{P}$ be a partition of T that has a proper labeling of the vertices of the

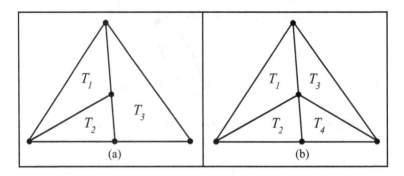

Figure 7.1.c: (a) The set $\{T_1, T_2, T_3\}$ is *not* a partition because $T_1 \cap T_3$ contains an edge of T_1 but not an edge of T_3. (b) The set $\{T_1, T_2, T_3, T_4\}$ is a partition of the larger triangle.

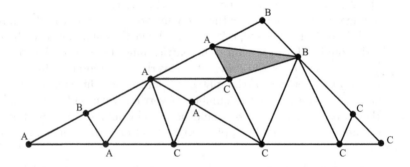

Figure 7.1.d: Sperner's Lemma: A complete triangle with a partition and a proper labeling must have one complete small triangle. (The complete small triangle is shaded in the above illustration.)

triangles in the partition. Suppose that there is no complete triangle in $\mathscr{P}$. Observe that if a triangle is not complete, then the number of AB sides is either 0 or 2. To see this, note that, in order to not be complete, one of the labels A, B, or C must be missing. If either A or B is missing, then there can be no AB side. If we have both A and B but no C, then there are two AB sides.

Suppose the partition of T is given by $\mathscr{P} = \{T_1, T_2, \ldots, T_n\}$. If any T_i has an AB side on an edge of the original triangle T, then it must be along the AB side of T. By Theorem 7.1.1, the number of such AB sides must be odd. For each $i \in \{1, 2, \ldots, n\}$, let c_i be the *centroid* (the intersection of the three median lines) of the triangle T_i. We connect c_i to c_j by an edge if and only if the intersection $T_i \cap T_j$ is an AB side. Then the order of each c_k is 0, 1, or 2. Furthermore, the order of c_k will be 1 if and only if the corresponding triangle T_k has two AB sides, one of which is on the edge of the original triangle T. By Theorem 7.1.2, the number of centroids c_k of order 1 must be even. It follows that the number of AB sides on the edge of T is even. However, we have already established that the number of AB sides on the edge of T is odd. We have established a contradiction. Therefore, we must reject our assumption that no triangle in $\mathscr{P}$ is complete. Q.E.D.

Exercises

1. Let G be a set of people who have shaken hands with other members of G but have never shaken hands with people outside of G. Let

   $$A = \{p \in G \mid p \text{ shook hands an odd number of times}\}$$

   and

   $$B = \{p \in G \mid p \text{ shook hands an even number of times}\}.$$

   Prove that the number of elements in A is even. Can we say anything about the number of elements in B?

2. Let T be a complete triangle and suppose $\mathscr{P}$ is a partition of T having a proper labeling of vertices. Prove that the number of complete triangles in $\mathscr{P}$ is odd.

3. Let T be a complete triangle and suppose $\mathscr{P}$ is a partition of T having a proper labeling of vertices. If the number of line segments AA on an edge of T is odd, show there is at least one triangle in $\mathscr{P}$ with the letter A appearing on at least two of its vertices.

4. Let T be a complete triangle and suppose $\mathscr{P}$ is a partition of T having a proper labeling of vertices. If the number of line segments AA on an edge of T is odd, show the number of triangles in $\mathscr{P}$ with A appearing on at least two vertices is odd.

5. Let $\mathscr{P} = \{T_1, T_2, \ldots, T_n\}$ be a partition of a triangle T. If n is an odd number, then show the number of sides of triangles in $\mathscr{P}$ that are on the edges of T is odd.

6. Let T be a complete triangle and partition T into n small triangles, where n is an integer such that $n \geq 10$. Assign a proper labeling to the vertices of the small triangles, trying to avoid getting a complete small triangle. How many small triangles did you label before obtaining a complete triangle? Repeat with a different value of n.

7.2 Brouwer's Fixed Point Theorem

Although it is true that the topological product of n closed bounded intervals has the fixed point property, we shall prove it only for the case $n = 2$. Since a closed rectangle, a closed triangle, and a closed circle are homeomorphic (see Exercise 7), it is sufficient to prove that any one of these spaces has the f.p.p.

We give here a proof for the triangle. Proofs for the circle and rectangle can be found in the literature.

For our purposes, it is convenient to represent a triangle as a subset of three-dimensional Euclidean space. Recall that in $\mathbb{R}^3$, we can represent the vector $u = (x, y, z)$ geometrically as an arrow with its base at the origin and its endpoint at the point with coordinates (x, y, z). We will adopt this convention throughout this section.

If u and v are distinct vectors in $\mathbb{R}^3$, we define the *line* through u and v to be the set

$$\{w \mid w = t_1 u + t_2 v \text{ where } \{t_1, t_2\} \subseteq \mathbb{R} \text{ and } t_1 + t_2 = 1\}.$$

If w is an element of this set, so that $w = t_1 u + t_2 v$ for real numbers t_1 and t_2 satisfying $t_1 + t_2 = 1$, then u, v, and w are called *collinear*. Geometrically, we represent this set as a line in $\mathbb{R}^3$ passing through the endpoints of the vectors in the set. In this representation, vectors in $\mathbb{R}^3$ are collinear when their endpoints lie on a common line. (See Figure 7.2.a.)

The set

$$\{w \mid w = t_1 u + t_2 v \text{ where } \{t_1, t_2\} \subseteq [0, 1] \text{ and } t_1 + t_2 = 1\}$$

is a subset of the line through u and v and is called the *line segment* joining u and v. Geometrically, we think of this as a line segment in $\mathbb{R}^3$ joining the endpoint of u to the endpoint of v. (See Exercise 1.)

The vectors u, v, and w in $\mathbb{R}^3$ are said to be *linearly independent* provided that $t_1 u + t_2 v + t_3 w = \mathbf{0}$ if and only if $t_1 = 0$, $t_2 = 0$, and $t_3 = 0$. It can be shown that three vectors are linearly independent if and only if they are not collinear. (See Figure 7.2.b and Exercise 2.) It can also be shown that if the vectors u, v, and w are linearly independent, then any vector in $\mathbb{R}^3$ can be represented as a *linear combination* of u, v, and w. That is, any vector r in $\mathbb{R}^3$ can be written $r = t_1 u + t_2 v + t_3 w$ for some t_1, t_2, and t_3 in $\mathbb{R}$. Furthermore, this linear combination is unique. (See Exercises 3 and 4.)

If u, v, and w are three linearly independent vectors in $\mathbb{R}^3$, then the set

$$T = \{r \mid r = t_1 u + t_2 v + t_3 w \text{ where } \{t_1, t_2, t_3\} \subseteq [0, 1] \text{ and } t_1 + t_2 + t_3 = 1\}$$

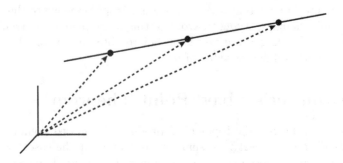

Figure 7.2.a: Vectors in $\mathbb{R}^3$ are collinear if their endpoints lie on a common line.

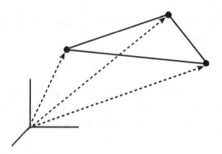

Figure 7.2.b: Three vectors in $\mathbb{R}^3$ are linearly independent if their endpoints determine a plane.

is called a *triangle* with *vertices* u, v, and w. This terminology is used because the endpoints of the vectors in the set T form a triangle in $\mathbb{R}^3$ and the vertices of this triangle are the endpoints of the vectors u, v, and w. (See Figure 7.2.b and Exercise 5.) For each vector $r = t_1 u + t_2 v + t_3 w$ in the triangle T, the real numbers appearing in the ordered triple (t_1, t_2, t_3) are called the *barycentric coordinates* of r. Observe that the barycentric coordinates of u, v, and w are $(1,0,0)$, $(0,1,0)$, and $(0,0,1)$, respectively. It can be shown that the centroid of the triangle has barycentric coordinates $(\frac{1}{3}, \frac{1}{3}, \frac{1}{3})$. (See Exercise 6.)

We are now ready to state and prove the main result of this section.

Theorem 7.2.1 (Brouwer's Fixed Point Theorem for Triangles). *Let T be a triangle in $\mathbb{R}^3$ that is closed in the standard topology. If $f : T \to T$ is a continuous function, then there is at least one $x_0 \in T$ such that $f(x_0) = x_0$.*

Proof. Let each point of the triangle T be represented by its barycentric coordinates. Then the vertices of T are $(1,0,0)$, $(0,1,0)$, and $(0,0,1)$. Give these points the labels A, B, and C, respectively. Observe that a point will be on the side AB if and only if its third barycentric coordinate is 0, it will be on the side AC if and only if its second barycentric coordinate is 0, and it will be on the side BC if and only if its first barycentric coordinate is 0. Finally, a point is in the interior of the triangle if and only if its barycentric coordinates are all positive.

Let $f : T \to T$ be a continuous function and for each $x = (x_1, x_2, x_3) \in T$, let $(y_1, y_2, y_3) = f(x_1, x_2, x_3)$. Assign a label $L(x)$ to each $x = (x_1, x_2, x_3) \in T$ as follows:

$$L(x) = \min \{ j \mid y_j \leq x_j \text{ and } x_j \neq 0 \}.$$

We know such a j always exists, for otherwise we would have

$$1 = x_1 + x_2 + x_3 < y_1 + y_2 + y_3 = 1,$$

which is a contradiction.

Observe that $L(1,0,0) = 1$ and $L(0,1,0) = 2$ and $L(0,0,1) = 3$. Also, any point on the side AB, which is the side joining $(1,0,0)$ to $(0,1,0)$, has coordinates $(x_1, x_2, 0)$. If $L(x_1, x_2, 0) = 3$, then $x_1 < y_1$ and $x_2 < y_2$ and $y_3 = 0$. This implies that

$$1 = x_1 + x_2 < y_1 + y_2 = 1.$$

This is a contradiction, and so $L(x_1, x_2, 0)$ is either 1 or 2, which is the same label as one of the two endpoints. Since a similar argument can be made for each of the other sides of T, we conclude that L provides a proper labeling of the triangle T.

The *mesh* of a partition $\mathscr{P}$ of the triangle T is the largest perimeter of all triangles in $\mathscr{P}$. We denote the mesh of $\mathscr{P}$ by the symbol $m(\mathscr{P})$. Consider a sequence of partitions $\{\mathscr{P}_n\}_{n=1}^{\infty}$ for which $\lim_{n \to \infty} m(\mathscr{P}_n) = 0$. By Sperner's Lemma, each partition $\mathscr{P}_n$ contains at least one complete triangle. For each $n \in \mathbb{N}$, choose a complete triangle from $\mathscr{P}_n$ and call it T_n. For each $n \in \mathbb{N}$, let c_n denote the centroid of T_n. The triangle T is closed and bounded, and hence compact. Thus, T is sequentially compact. Therefore, the sequence $\{c_n\}_{n=1}^{\infty}$ of centroids has a convergent subsequence, say $\{c_{n_k}\}_{k=1}^{\infty}$. Let $\tilde{x} = \lim_{k \to \infty} c_{n_k}$.

For each $n \in \mathbb{N}$, the triangle T_n is complete, and uses all three labels for its vertices. Let $x_n^{(j)}$ denote the vertex of T_n with label j (for each $j \in \{1, 2, 3\}$). Then $\{x_{n_k}^{(1)}\}_{k=1}^{\infty}$ is a sequence of vertices labeled 1, $\{x_{n_k}^{(2)}\}_{k=1}^{\infty}$ is a sequence of vertices labeled 2, and $\{x_{n_k}^{(3)}\}_{k=1}^{\infty}$ is a sequence of vertices labeled 3.

By construction, $\lim_{n \to \infty} m(\mathscr{P}_n) = 0$. Thus,

$$\lim_{k \to \infty} d\big(x_{n_k}^{(1)}, c_{n_k}\big) = \lim_{k \to \infty} d\big(x_{n_k}^{(2)}, c_{n_k}\big) = \lim_{k \to \infty} d\big(x_{n_k}^{(3)}, c_{n_k}\big) = 0.$$

We know $\{c_{n_k}\}_{k=1}^{\infty}$ converges to $\tilde{x}$. Consequently, each sequence $\{x_{n_k}^{(1)}\}_{k=1}^{\infty}$, $\{x_{n_k}^{(2)}\}_{k=1}^{\infty}$, and $\{x_{n_k}^{(3)}\}_{k=1}^{\infty}$ converges to $\tilde{x}$, as well. (See Exercise 18 in Section 3.3.) By the continuity of f, we conclude that

$$\lim_{k \to \infty} f\big(x_{n_k}^{(1)}\big) = \lim_{k \to \infty} f\big(x_{n_k}^{(2)}\big) = \lim_{k \to \infty} f\big(x_{n_k}^{(3)}\big) = f(\tilde{x}).$$

(See Theorem 3.3.9.) Let $\tilde{y} = f(\tilde{x})$.

For each $k \in \mathbb{N}$, the label $L\big(x_{n_k}^{(1)}\big)$ is 1; hence, the first barycentric coordinate of the point $f\big(x_{n_k}^{(1)}\big)$ is less than or equal to the first barycentric coordinate of $x_{n_k}^{(1)}$. Similarly, the second barycentric coordinate of the point $f\big(x_{n_k}^{(2)}\big)$ is less than or equal to the second barycentric coordinate of $x_{n_k}^{(2)}$, and the third barycentric coordinate of the point $f\big(x_{n_k}^{(3)}\big)$ is less than or equal to the third barycentric coordinate of $x_{n_k}^{(3)}$. Consequently, if we let $\tilde{x} = (\tilde{x}_1, \tilde{x}_2, \tilde{x}_3)$ and $\tilde{y} = (\tilde{y}_1, \tilde{y}_2, \tilde{y}_3)$, then $\tilde{y}_1 \leq \tilde{x}_1$ and $\tilde{y}_2 \leq \tilde{x}_2$ and $\tilde{y}_3 \leq \tilde{x}_3$. However, we now have

$$1 = \tilde{y}_1 + \tilde{y}_2 + \tilde{y}_3 \leq \tilde{x}_1 + \tilde{x}_2 + \tilde{x}_3 = 1.$$

The only way this can happen is if $\tilde{y}_1 = \tilde{x}_1$ and $\tilde{y}_2 = \tilde{x}_2$ and $\tilde{y}_3 = \tilde{x}_3$. We conclude that $\tilde{x} = \tilde{y}$, and so $\tilde{x} = f(\tilde{x})$.   Q.E.D.

Exercises

1. Let $u = (u_1, u_2, u_3)$ and $v = (v_1, v_2, v_3)$ be two distinct vectors in $\mathbb{R}^3$.

    (a) Let $w = t_1 u + t_2 v$, where $\{t_1, t_2\} \subseteq [0, 1]$ and $t_1 + t_2 = 1$. Prove that the endpoint of w is on the line segment joining the endpoints of u and v. (*Hint:* Show that $d(u, w) + d(w, v) = d(u, v)$.)

    (b) Show that the endpoint of the vector $\frac{1}{2} u + \frac{1}{2} v$ is the midpoint of the line segment joining the endpoints of u and v.

    (c) Let $w = t_1 u + t_2 v$, where $\{t_1, t_2\} \subseteq \mathbb{R}$ and $t_1 + t_2 = 1$. Prove that the endpoint of w is on the line passing through the endpoints of u and v.

2. Prove that if u, v, and w are linearly independent vectors in $\mathbb{R}^3$, then they are not collinear. (*Hint:* Prove the contrapositive.)

3. Suppose u, v, and w are linearly independent vectors in $\mathbb{R}^3$. If $r \in \mathbb{R}^3$, show there are real numbers c_1, c_2, and c_3 such that $r = c_1 u + c_2 v + c_3 w$.

4. Suppose u, v, and w are linearly independent vectors in $\mathbb{R}^3$. Show that $c_1 u + c_2 v + c_3 w = d_1 u + d_2 v + d_3 w$ if and only if $c_1 = d_1$, $c_2 = d_2$, and $c_3 = d_3$.

5. Suppose u, v, and w are linearly independent vectors in $\mathbb{R}^3$ and let T be the triangle with vertices at u, v, and w. Prove that any $r \in T$ can be written as $r = t_1 u + t_2 v + t_3 w$, where $\{t_1, t_2, t_3\} \subseteq [0, 1]$ and $t_1 + t_2 + t_3 = 1$. (*Hint:* See Figure 7.2.c.)

6. Suppose u, v, and w are linearly independent vectors in $\mathbb{R}^3$ and let T be the triangle with vertices at u, v, and w. Let c be the centroid of T and prove that the barycentric coordinates of c are $(\frac{1}{3}, \frac{1}{3}, \frac{1}{3})$. (See Figure 7.2.d.)

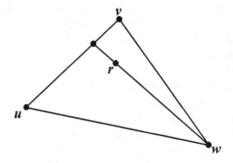

Figure 7.2.c: Illustration to accompany Exercise 5: Any point in the triangle is on a line segment joining w to a point on the side of the triangle that has endpoints at u and v.

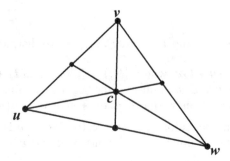

Figure 7.2.d: Illustration to accompany Exercise 6: The centroid c of a triangle is the point of intersection of the three medians of the triangle.

7. In each of the parts below, show that the given closed sets are homeomorphic (using the standard topology in the plane):

   (a) Any two disks.

   (b) Any two triangles.

   (c) Any two rectangles.

   (d) Any disk and any triangle.

   (e) Any disk and any rectangle.

   (f) Any triangle and any rectangle.

7.3   The Fundamental Theorem of Algebra

Whether or not an equation has a solution depends on the replacement set that is being used. For example, the equation $x + 2 = 0$ has the solution $x = -2$ if the replacement set is the set of integers, but has no solution if the replacement set is the set of natural numbers. Similarly, the equation $2x = 5$ has a solution if the replacement set is the set of rational numbers, but has no solution if the replacement set is the set of integers. Extending this idea, we see that $x^2 = 2$ has no rational solution, but has the solution set $\{-\sqrt{2}, \sqrt{2}\}$ if the replacement set is the set of real numbers.

Now consider the equation $x^2 + 1 = 0$. If the replacement set is $\mathbb{R}$, then there is no solution. In order for this equation to have a solution, the *complex numbers* must be introduced. Indeed, if the replacement set is the set of complex numbers, then every polynomial equation of degree 2 has a solution. This can be seen easily using the quadratic formula. But what if we consider a polynomial equation $p(x) = 0$, where the degree of p is greater than 2? Does one need to invent new numbers so that all equations of this type will have a solution? It turns out that a new number system is not necessary. This fact was first proved by Carl Friedrich Gauss in his doctoral dissertation in 1799, although it was conjectured much earlier. What Gauss proved is now known as the

Fundamental Theorem of Algebra. In this section, we state and prove this very important theorem. The key tool in the proof will be Brouwer's Fixed Point Theorem.

We need to recall a few basic facts about complex numbers. The *imaginary number* i is defined so that $i^2 = -1$. If a and b are real numbers, then we call $a + ib$ a *complex number*. We call a the *real part* and b the *imaginary part* of the complex number $a + ib$. The set of all complex numbers is denoted $\mathbb{C}$. Addition and multiplication with complex numbers are given by the following rules:

- Addition: $(a + ib) + (c + id) = (a + c) + i(b + d)$.

- Multiplication: $(a + ib) \cdot (c + id) = (ac - bd) + i(ad + bc)$.

Let $z = a + ib$ be a complex number. We define the *absolute value* of z to be $|z| = \sqrt{a^2 + b^2}$. It is straightforward to show that $|a| \leq |z|$ and $|b| \leq |z|$. (See Exercise 1.) We define the *conjugate* of $z = a + ib$ to be the complex number $a - ib$ and we denote it by the symbol $\overline{z}$. It is evident that $\overline{\overline{z}} = z$. Direct computation will show that $z\overline{z} = |z|^2$ and that $\overline{z} = z$ if and only if z is a real number. (See Exercise 1.) Furthermore, if z_1 and z_2 are complex numbers, then $\overline{z_1 + z_2} = \overline{z_1} + \overline{z_2}$ and $\overline{z_1 z_2} = \overline{z_1} \cdot \overline{z_2}$. (See Exercise 2.)

If $z = a + ib$ is a complex number, then it can be represented geometrically by the point (a, b) in the Cartesian plane. If (r, θ) are the polar coordinates of that point, with $r \geq 0$, then $r = |z|$. We often call r the *modulus* of the complex number, and θ is called an *argument* of z. Note that the value of θ is not unique, unless we make the additional requirement that $0 \leq \theta < 2\pi$. In that case, we say that θ is the *principal value of the argument* of z. (See Figure 7.3.a.)

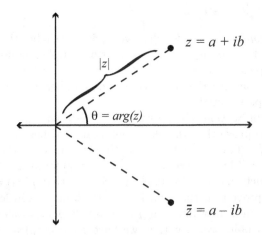

Figure 7.3.a: The complex conjugate of z is obtained by reflecting z through the "real" axis. The absolute value of z is the distance between z and the origin and the principal argument of z is the angle $\theta \in [0, 2\pi)$ formed with the positive "real" axis.

Proofs of the following facts are left to the reader. (See the exercises at the end of this section.)

1. Let $z = a + ib$ be a complex number. If z has modulus r and argument θ, then $a = r\cos\theta$ and $b = r\sin\theta$. We use the symbol $e^{i\theta}$, which we call the *complex exponential*, to represent $\cos\theta + i\sin\theta$. Thus,

$$z = a + ib = r(\cos\theta + i\sin\theta) = re^{i\theta}.$$

2. If z_1 has modulus r_1 and argument θ_1, and if z_2 has modulus r_2 and argument θ_2, then the product is

$$z_1 z_2 = r_1 r_2\big[\cos(\theta_1 + \theta_2) + i\sin(\theta_1 + \theta_2)\big].$$

Alternatively, this may be written

$$z_1 z_2 = (r_1 e^{i\theta_1})(r_2 e^{i\theta_2}) = r_1 r_2 e^{i(\theta_1+\theta_2)}.$$

3. If $n \in \mathbb{N}$, then $(\cos\theta + i\sin\theta)^n = \cos(n\theta) + i\sin(n\theta)$. Consequently, if $z = re^{i\theta}$, then

$$z^n = r^n e^{in\theta} = r^n\big[\cos(n\theta) + i\sin(n\theta)\big].$$

This is known as *De Moivre's Theorem*.

4. If $\{z_1, z_2\} \subseteq \mathbb{C}$, then $|z_1 + z_2| \le |z_1| + |z_2|$ and $\big||z_1| - |z_2|\big| \le |z_1 - z_2|$.

5. If $\{z_1, z_2, \ldots, z_n\} \subseteq \mathbb{C}$, then $|z_1 + z_2 + \cdots + z_n| \le |z_1| + |z_2| + \cdots + |z_n|$.

6. If $\theta \in \mathbb{R}$, then $|e^{i\theta}| = 1$.

We call z a *complex variable* whenever z is a variable that has $\mathbb{C}$ as the replacement set. If $\{a_0, a_1, a_2, \ldots, a_n\}$ is a set of complex numbers, where $a_0 \ne 0$, and z is a complex variable, then $a_0 z^n + a_1 z^{n-1} + a_2 z^{n-2} + \cdots + a_{n-1}z + a_n$ is called a *complex polynomial* of *degree* n. The number a_0 is called the *leading coefficient* of the polynomial.

In Exercise 8 in Section 3.5, we suggested the reader use the Intermediate Value Theorem to prove that the closed interval $[0, 1]$ has the f.p.p. The idea is as follows: If $f : [0, 1] \to [0, 1]$ is a continuous function, then define a new function $g : [0, 1] \to \mathbb{R}$ by the formula $g(x) = x - f(x)$ for all $x \in [0, 1]$. If we find some $x_0 \in [0, 1]$ such that $g(x_0) = 0$, then it follows that $f(x_0) = x_0$. Similarly, when we wish to prove that the equation $f(x) = 0$ has a solution, a reasonable approach is to define a new function g by the rule $g(x) = x + f(x)$ and show that g has a fixed point x_0. For then we have $x_0 = g(x_0) = x_0 + f(x_0)$, and so $f(x_0) = 0$. This technique is illustrated in the following proof, first given by B.H. Arnold.[1]

---

[1]B.H. Arnold, "A topological proof of the fundamental theorem of algebra" *Amer. Math. Monthly*, 56 (1949) pp. 465–466.

Theorem 7.3.1 (The Fundamental Theorem of Algebra). *If $p(z)$ is a non-constant complex polynomial and the replacement set is $\mathbb{C}$, then the equation $p(z) = 0$ has at least one solution.*

Proof. Let $p(z) = a_0 z^n + a_1 z^{n-1} + a_2 z^{n-2} + \cdots + a_{n-1} z + a_n$ be a complex polynomial such that $a_0 \neq 0$ and $n \geq 1$. We wish to find solutions to the equation $p(z) = 0$. Thus, because $a_0 \neq 0$, we may assume without loss of generality that $a_0 = 1$. That is, we may assume that

$$p(z) = z^n + a_1 z^{n-1} + a_2 z^{n-2} + \cdots + a_{n-1} z + a_n.$$

Our objective is to show that there exists some $z_0 \in \mathbb{C}$ such that $p(z_0) = 0$. Let

$$R = 2 + |a_1| + |a_2| + \cdots + |a_{n-1}| + |a_n|$$

and write $z = re^{i\theta}$, where $r \geq 0$ and $0 \leq \theta < 2\pi$. Define two functions g and h by

$$g(z) = z - \frac{p(z)}{R\,e^{i(n-1)r\theta}} \quad \text{and} \quad h(z) = z - \frac{p(z)}{R\,z^{n-1}},$$

for all $z \in \mathbb{C}$. Now define a function $f : \mathbb{C} \to \mathbb{C}$ by

$$f(z) = \begin{cases} g(z) & \text{if } |z| \leq 1, \\ h(z) & \text{if } |z| \geq 1. \end{cases}$$

Observe first that $g(z) = h(z)$ for all z in the set $\{z \mid |z| = 1\}$, because in this case $r = 1$ and $z^{n-1} = e^{i(n-1)\theta}$. Thus, f is well-defined. Note also that g is continuous since $R \geq 2$ and $e^{i(n-1)\theta} \neq 0$. (See Exercise 8.) Furthermore, h is continuous on the set $\{z \mid |z| \geq 1\}$. Therefore, f is a continuous function, by the Pasting Lemma. (See Exercise 20 in Section 3.4.)

Let $D = \{z \mid |z| \leq R\}$. We will show that $f(D) \subseteq D$. We begin by assuming that $0 \leq |z| \leq 1$. (Note that $1 < R$.) Then

$$|f(z)| = |g(z)| = \left| z - \frac{p(z)}{R\,e^{i(n-1)r\theta}} \right| \leq |z| + \left| \frac{p(z)}{R\,e^{i(n-1)r\theta}} \right|$$

$$\leq 1 + \frac{|p(z)|}{R} = 1 + \frac{|z^n + a_1 z^{n-1} + \cdots + a_{n-1} z + a_n|}{R}$$

$$\leq 1 + \frac{|z^n| + |a_1 z^{n-1}| + \cdots + |a_{n-1} z| + |a_n|}{R}.$$

We assumed that $|z| \leq 1$, and so

$$|f(z)| \leq 1 + \frac{1 + |a_1| + \cdots + |a_{n-1}| + |a_n|}{R} = 1 + \frac{R-1}{R} < 2 \leq R.$$

Now instead assume that $1 \leq |z| \leq R$. Then

$$|f(z)| = |h(z)| = \left| z - \frac{p(z)}{R z^{n-1}} \right|$$

$$= \left| z - \frac{z^n + a_1 z^{n-1} + a_2 z^{n-2} + \cdots + a_{n-1} z + a_n}{R z^{n-1}} \right|$$

$$= \left| z - \frac{z}{R} - \frac{a_1 + a_2 z^{-1} + \cdots + a_{n-1} z^{2-n} + a_n z^{1-n}}{R} \right|$$

$$\leq \left| z - \frac{z}{R} \right| + \frac{|a_1| + |a_2 z^{-1}| + \cdots + |a_{n-1} z^{2-n}| + |a_n z^{1-n}|}{R}.$$

We assumed that $|z| \geq 1$, or $|z^{-1}| \leq 1$, and so

$$|f(z)| \leq \left| z - \frac{z}{R} \right| + \frac{|a_1| + |a_2| + \cdots + |a_{n-1}| + |a_n|}{R}$$

$$= \left| (R-1)\frac{z}{R} \right| + \frac{|a_1| + |a_2| + \cdots + |a_{n-1}| + |a_n|}{R}$$

$$\leq (R-1) + \frac{R-2}{R} = R - \frac{2}{R} < R.$$

We have shown that $|f(z)| \leq R$ for any $z \in D$. Thus, the restriction of f to D is a continuous function from D to D. By Brouwer's Fixed Point Theorem, there is some $z_0 \in D$ such that $f(z_0) = z_0$. Then either $g(z_0) = z_0$ (if $|z_0| \leq 1$) or $h(z_0) = z_0$ (if $|z_0| \geq 1$). Consequently, either

$$z_0 - \frac{p(z_0)}{R\, e^{i(n-1)r\theta}} = z_0 \quad \text{or} \quad z_0 - \frac{p(z_0)}{R\, z_0^{n-1}} = z_0.$$

In either case, we have $p(z_0) = 0$.   Q.E.D.

When Gauss first proved this theorem, he essentially replaced z by $x + iy$ in the polynomial equation $p(z) = 0$ and obtained an equation of the form $q(x,y) + ir(x,y) = 0$, where q and r are functions that assume real values. He was then able to show that there are real numbers x_0 and y_0 such that $q(x_0, y_0) = 0$ and $r(x_0, y_0) = 0$.

Many proofs of this important result have appeared in the literature. We chose to include a proof that makes use of a topological result. For the purposes of comparison, we now give a completely different proof. Before we can provide this proof, however, we must introduce some concepts from the theory of complex variables.

A function $f : \mathbb{C} \to \mathbb{C}$ is said to be *entire* if it is differentiable at each $z \in \mathbb{C}$. If $D \subseteq \mathbb{C}$, then a function $f : D \to \mathbb{C}$ is said to be *bounded* if there is a positive real number M such that $|f(z)| \leq M$ for all $z \in D$.

We now state, without proof, an important result known as Liouville's Theorem. The proof makes use of advanced concepts from the theory of complex variables and is beyond the scope of this book.

Theorem 7.3.2 (Liouville's Theorem). *If $f : \mathbb{C} \to \mathbb{C}$ is entire and bounded, then f is a constant function.*

To appreciate the scope of this result, the reader should draw the graph of a nonconstant linear function in the Cartesian plane and observe that the function cannot be bounded. (See also Exercise 11.) Using Liouville's Theorem, we can provide an alternate proof of Theorem 7.3.1.

Alternate proof of the Fundamental Theorem of Algebra. We will proceed by means of a proof by contradiction. Let

$$p(z) = a_0 z^n + a_1 z^{n-1} + a_2 z^{n-2} + \cdots + a_{n-1} z + a_n$$

be a complex polynomial with $a_0 \neq 0$. Assume that $p(z) \neq 0$ for all $z \in \mathbb{C}$. Certainly, the polynomial function p is differentiable at every point of $\mathbb{C}$, and consequently p is an entire function. Since $p(z) \neq 0$ for $z \in \mathbb{C}$, it follows that the function $f : \mathbb{C} \to \mathbb{C}$ defined by $f(z) = \frac{1}{p(z)}$ for all $z \in \mathbb{C}$ is also entire.

We claim that f is bounded. If $z \neq 0$, write

$$p(z) = z^n \left(a_0 + \frac{a_1}{z} + \frac{a_2}{z^2} + \cdots + \frac{a_{n-1}}{z^{n-1}} + \frac{a_n}{z^n} \right).$$

Let $A = \max\{|a_1|, |a_2|, \ldots, |a_{n-1}|, |a_n|\}$ and choose a real number $R > 1$ such that $\frac{A}{R-1} < \frac{|a_0|}{2}$. Now suppose $|z| \geq R$. It follows that

$$\left| \frac{a_1}{z} + \frac{a_2}{z^2} + \cdots + \frac{a_n}{z^n} \right| \leq \left| \frac{a_1}{z} \right| + \left| \frac{a_2}{z^2} \right| + \cdots + \left| \frac{a_n}{z^n} \right| \leq A \left(\frac{1}{|z|} + \frac{1}{|z|^2} + \cdots + \frac{1}{|z|^n} \right)$$

$$\leq A \left(\frac{1}{R} + \frac{1}{R^2} + \cdots + \frac{1}{R^n} \right) = A \left(\frac{1/R - 1/R^{n+1}}{1 - 1/R} \right)$$

$$= A \left(\frac{1 - 1/R^n}{R - 1} \right) \leq \frac{A}{R-1} < \frac{|a_0|}{2}.$$

Using this, together with Exercise 6(c) at the end of this section,

$$\left| a_0 + \frac{a_1}{z} + \frac{a_2}{z^2} + \cdots + \frac{a_n}{z^n} \right| \geq \left| |a_0| - \left| \frac{a_1}{z} + \frac{a_2}{z^2} + \cdots + \frac{a_n}{z^n} \right| \right|$$

$$> \left| |a_0| - \frac{|a_0|}{2} \right| = \frac{|a_0|}{2}.$$

It follows that

$$\left| z^n \left(a_0 + \frac{a_1}{z} + \frac{a_2}{z^2} + \cdots + \frac{a_n}{z^n} \right) \right| = |z^n| \left| a_0 + \frac{a_1}{z} + \frac{a_2}{z^2} + \cdots + \frac{a_n}{z^n} \right| > \frac{R^n |a_0|}{2}.$$

Therefore, whenever $|z| \geq R$, we have that $|f(z)| = \dfrac{1}{|p(z)|} < \dfrac{2}{R^n |a_0|}$.

The set $D = \{z \in \mathbb{C} \mid |z| \leq R\}$ is a closed and bounded subset of $\mathbb{R}^2$. Thus, it is compact. The function $|f| : D \to \mathbb{R}$ is continuous, and hence there exists a positive number B such that $|f(z)| \leq B$ for all $z \in D$. (See Exercise 9 in Section 6.2.) Let $M = \max\{B, 2/(R^n |a_0|)\}$. Then $|f(z)| \leq M$ for all $z \in \mathbb{C}$.

We have established that f is a bounded and entire function. Therefore, by Liouville's Theorem, it must be constant. This is a contradiction, because $f(z) = \frac{1}{p(z)}$, for all $z \in \mathbb{C}$, and p is not constant. Thus, we reject the assumption that $p(z) \neq 0$ for all $z \in \mathbb{C}$ and conclude that there is some $z_0 \in \mathbb{C}$ for which $p(z_0) = 0$. $\hspace{3cm}$ Q.E.D.

Exercises

1. Let $z = a + ib$ be a complex number, where a and b are real numbers. Prove each of the following:

   (a) $a \leq |z|$ and $b \leq |z|$,

   (b) $\bar{z} = z$ if and only if $z = a$ (that is, z is real), and

   (c) $z\bar{z} = |z|^2$.

2. If z_1 and z_2 are complex numbers, then prove:

   (a) $\overline{z_1 + z_2} = \overline{z_1} + \overline{z_2}$,

   (b) $\overline{z_1 z_2} = \overline{z_1} \cdot \overline{z_2}$.

3. Let $z = a + ib$ be a complex number. If r is the modulus of z and θ is an argument of z, then prove that $a = r \cos \theta$ and $b = r \sin \theta$.

4. Use trigonometric identities to prove that $\left(r_1 e^{i\theta_1}\right)\left(r_2 e^{i\theta_2}\right) = r_1 r_2 e^{i(\theta_1 + \theta_2)}$.

5. Use Exercise 4 and the Principle of Mathematical Induction to prove De Moivre's Theorem:

$$\left(r e^{i\theta}\right)^n = r^n \left[\cos(n\theta) + i \sin(n\theta) \right].$$

6. If z_1 and z_2 are complex numbers, then prove:

   (a) $|z_1 + z_2| \leq |z_1| + |z_2|$,

   (b) $\left| |z_1| - |z_2| \right| \leq |z_1 - z_2|$,

   (c) $\left| |z_1| - |z_2| \right| \leq |z_1 + z_2|$.

   Hint: For (a), start with $|z_1 + z_2|^2 = (\overline{z_1} + \overline{z_2})(z_1 + z_2)$.

7. Use Exercise 6 and the Principle of Mathematical Induction to show that

$$|z_1 + z_2 + \cdots + z_n| \leq |z_1| + |z_2| + \cdots + |z_n|,$$

whenever $\{z_1, z_2, \ldots, z_n\}$ is a set of complex numbers.

8. Prove that $|e^{i\theta}| = 1$ for any $\theta \in \mathbb{R}$.

9. Show that $e^{i\pi} = -1$.

10. Draw the graph of $y = 2x + 1$ in a Cartesian plane and give a geometric argument to show that the function f defined by $f(x) = 2x + 1$ is not bounded.

11. Let $f(x) = mx + b$, with $m \neq 0$. Show that f is not bounded.

12. Let $\{a_0, a_1, a_2, \ldots, a_{n-1}, a_n\}$ be a set of *real* numbers with $a_0 \neq 0$. Suppose the polynomial equation $a_0 z^n + a_1 z^{n-1} + a_2 z^{n-2} + \cdots + a_{n-1} z + a_n = 0$ has the complex number z_0 as a solution. Show that its conjugate $\overline{z_0}$ is also a solution.

13. Let $p(z)$ be a complex polynomial of degree n, where $n > 0$. The *Factor Theorem* states that $p(z_0) = 0$ if and only if $z - z_0$ is a factor of $p(z)$. Use this to prove that $p(z) = 0$ has exactly n solutions, provided that we count a factor $(z - z_0)^k$ as providing k solutions. (Recall that z_0 is said to be a zero of $p(z)$ with *multiplicity* k if k is the largest integer such that $(z - z_0)^k$ is a factor of $p(z)$.)

14. Let $\{a_0, a_1, a_2, \ldots, a_{n-1}, a_n\}$ be a set of *real* numbers with $a_0 \neq 0$ and n an *odd* number. Prove that $a_0 z^n + a_1 z^{n-1} + a_2 z^{n-2} + \cdots + a_{n-1} z + a_n = 0$ has at least one real solution. Can you say anything about the number of real solutions?

15. The Taylor series expansions for the sine, cosine, and exponential functions (centered at zero) are

$$\sin x = \sum_{n=0}^{\infty} \frac{(-1)^n x^{2n+1}}{(2n+1)!}, \quad \cos x = \sum_{n=0}^{\infty} \frac{(-1)^n x^{2n}}{(2n)!}, \quad \text{and} \quad e^x = \sum_{n=0}^{\infty} \frac{x^n}{n!}.$$

Use these series to show that $e^{i\theta} = \cos(\theta) + i\sin(\theta)$ for all real numbers θ. You may assume that these series are valid for all $x \in \mathbb{C}$ and that the radius of convergence for each series is infinite. (*Hint:* Let $x = i\theta$.)

7.4   Function Spaces

In Exercise 8 of Section 2.8, we introduced Hilbert space as an example of a metric space. The reader will recall that the points of this space were sequences of real numbers that satisfied a certain summability condition. Also, in Chapter 4, we defined the Cartesian product of an indexed collection of sets as a set of functions from the index set to the union of the factor sets, subject to some restrictions. In this section, we shall expand on this and describe a few spaces that have functions as elements. To illustrate the usefulness of such spaces, we remind the reader of two types of problems. The first is the problem of finding a real solution of an equation such as

$$x^5 + 3x^2 + 2x + 5 = 0.$$

A standard technique that is discussed in elementary calculus courses is to use Newton's Method. In this method, we define a sequence $\{x_n\}_{n=1}^{\infty}$ of approximations to the solution. Usually, this sequence turns out to be a Cauchy sequence that converges to a solution of the equation.

The second problem is that of finding a solution to a first order differential equation such as

$$\frac{dy}{dx} = f(x, y),$$

for a prescribed function f. One method of finding a solution to this differential equation is to devise a process similar to Newton's Method. In such a circumstance, we would find a sequence $\{y_n\}_{n=1}^{\infty}$ of approximate solutions that would converge to a solution of the differential equation. Thus, we need to find a sequence $\{y_n\}_{n=1}^{\infty}$ such that the sequence $\left\{\frac{dy_n}{dx} - f(x, y_n)\right\}_{n=1}^{\infty}$ approaches the zero function. For this process to be meaningful, however, we must assign a topology to the set of functions that is to be used as a replacement set for the function y appearing in the differential equation.

As we saw in Chapter 2, there are many ways to assign a topology to a set. We need a topology that will help us solve the given problem effectively. Usually, the desired topology will be induced by a metric. We also need to consider the problem of proving the existence of solutions for certain equations. In the previous section, we proved the Fundamental Theorem of Algebra, which guarantees that a polynomial equation has at least one solution. In the next section, we shall prove a theorem that asserts that a first order differential equation (like the one above) has a solution.

It is necessary at this stage to distinguish between two types of convergence for sequences of functions.

Definitions 7.4.1. Let $(X, \mathscr{T})$ be a topological space and let $\mathbb{R}$ be the set of real numbers with the standard topology. Suppose that $f_n : X \to \mathbb{R}$ is a function for each $n \in \mathbb{N}$. If there is a function $f : X \to \mathbb{R}$ such that $f(x) = \lim_{n\to\infty} f_n(x)$ for each $x \in X$, then $\{f_n\}_{n=1}^{\infty}$ is said to *converge pointwise* to f. The function f is called the *pointwise limit* of $\{f_n\}_{n=1}^{\infty}$.

An important question is the following: If a property is shared by each of the functions f_n in a converging sequence, is it carried over to the limiting function f? For example, if f_n is continuous for each $n \in \mathbb{N}$, must it be true that the pointwise limit f is also continuous?

Example 7.4.2. For $n \in \mathbb{N}$, let $f_n : [0, 1] \to \mathbb{R}$ be defined by $f_n(x) = x^n$. Although f_n is continuous for each $n \in \mathbb{N}$, show that the limit function is not continuous on $[0, 1]$.

Solution. Let $f(x) = \lim\limits_{n \to \infty} f_n(x)$ for each $x \in [0, 1]$. Suppose $x \in [0, 1)$. Then $x^n \to 0$ as $n \to \infty$, so

$$f(x) = \lim_{n \to \infty} f_n(x) = \lim_{n \to \infty} x^n = 0.$$

On the other hand, $f_n(1) = 1$ for each $n \in \mathbb{N}$, and so $f(1) = 1$. Therefore,

$$f(x) = \begin{cases} 0 & \text{if } 0 \leq x < 1, \\ 1 & \text{if } x = 1. \end{cases}$$

Consequently, f is not continuous at $x = 1$.

The failure of a sequence of continuous functions to have a pointwise limit that is continuous suggests we should consider a stronger type of convergence. Uniform convergence will serve this purpose.

Definitions 7.4.3. Let $(X, \mathcal{T})$ be a topological space and let $\mathbb{R}$ be the set of real numbers with the standard topology. Suppose that $f_n : X \to \mathbb{R}$ is a function for each $n \in \mathbb{N}$. We say the sequence $\{f_n\}_{n=1}^{\infty}$ *converges uniformly* to $f : X \to \mathbb{R}$ provided that for each $\epsilon > 0$, there exists an integer $N > 0$ such that $|f(x) - f_n(x)| < \epsilon$ whenever $x \in X$ and $n \geq N$. The function f is called the *uniform limit* of $\{f_n\}_{n=1}^{\infty}$.

In the definition above, the choice of N depends on ϵ, but it does not depend on x. That is, the same N will work for all $x \in X$. In this circumstance, we say the convergence is *uniform in* x.

Theorem 7.4.4. *The uniform limit of a sequence of real valued continuous functions is a continuous function.*

Proof. Let $(X, \mathcal{T})$ be a topological space and let $\mathbb{R}$ be the set of real numbers with the standard topology. Suppose for each $n \in \mathbb{N}$ the function $f_n : X \to \mathbb{R}$ is continuous and assume that the sequence $\{f_n\}_{n=1}^{\infty}$ converges uniformly to the function $f : X \to \mathbb{R}$. We will show that f is a continuous function by showing that it is continuous at each point.

Let $\epsilon > 0$ and choose any $x_0 \in X$. We wish to find an open neighborhood U of x_0 such that $x \in U$ implies that $f(x) \in B_{\epsilon}(f(x_0))$. By assumption, the sequence $\{f_n\}_{n=1}^{\infty}$ converges uniformly to f. Thus, there is some positive integer N such that

$$(x \in X) \wedge (n \geq N) \implies |f(x) - f_n(x)| < \frac{\epsilon}{3}.$$

The function f_N is continuous, and so there exists a neighborhood U of x_0 such that

$$x \in U \implies |f_N(x) - f_N(x_0)| < \frac{\epsilon}{3}.$$

Hence, if $x \in U$, we have

$$|f(x) - f(x_0)| \leq |f(x) - f_N(x)| + |f_N(x) - f_N(x_0)| + |f_N(x_0) - f(x_0)|.$$

Using the inequalities above, we conclude that

$$|f(x) - f(x_0)| < \frac{\epsilon}{3} + \frac{\epsilon}{3} + \frac{\epsilon}{3} = \epsilon.$$

Therefore, f is continuous at x_0. Since x_0 was chosen arbitrarily, it follows that f is continuous on X.   Q.E.D.

We now begin the task of assigning a topological structure to a certain set of functions. If X is a nonempty set, then a function $f : X \to \mathbb{R}$ is said to be *bounded* if there is some number $M > 0$ such that $|f(x)| \leq M$ for all $x \in X$. We let $B(X)$ denote the collection of all bounded real valued functions with domain X. We will define a metric on $B(X)$. If $f : X \to \mathbb{R}$ and $g : X \to \mathbb{R}$ are both bounded functions, then the set $\{|f(x) - g(x)| \mid x \in X\}$ is nonempty and bounded above. Thus, this set has a least upper bound, or a supremum.

Theorem 7.4.5. *Let X be a nonempty set and let $B(X)$ denote the collection of all bounded real valued functions with domain X. If f and g are functions in $B(X)$, then*

$$d(f, g) = \sup \{ |f(x) - g(x)| \mid x \in X \}$$

defines a metric on $B(X)$.

Proof. Let f and g be functions in the set $B(X)$. Since $\{ |f(x) - g(x)| \mid x \in X \}$ is a collection of nonnegative numbers, and since $d(f, g)$ is defined to be an upper bound for this set, it follows that $d(f, g) \geq 0$. Furthermore, $d(f, g) = 0$ if and only if $|f(x) - g(x)| = 0$ for all $x \in X$, which is another way of saying that $f = g$. We also have that $d(f, g) = d(g, f)$, because $|f(x) - g(x)| = |g(x) - f(x)|$ for all $x \in X$. It remains only to prove the triangle inequality.

Suppose that f, g, and h are real valued bounded functions with domain X. We will show $d(f, h) \leq d(f, g) + d(g, h)$. For each $x \in X$, we have that

$$|f(x) - g(x)| \leq d(f, g) \quad \text{and} \quad |g(x) - h(x)| \leq d(g, h),$$

by the definition of d. Thus,

$$|f(x) - h(x)| \leq |f(x) - g(x)| + |g(x) - h(x)| \leq d(f, g) + d(g, h),$$

for all $x \in X$. It follows that $d(f, g) + d(g, h)$ is an upper bound for the set of real numbers $\{ |f(x) - h(x)| \mid x \in X \}$; however, the number $d(f, h)$ is the least upper bound of this set, and so

$$d(f, h) = \sup \{ |f(x) - h(x)| \mid x \in X \} \leq d(f, g) + d(g, h).$$

Therefore, d satisfies the triangle inequality, and so d is a metric on $B(X)$.
                                                                Q.E.D.

Definitions 7.4.6. Let X be a nonempty set and let $B(X)$ denote the collection of all bounded real valued functions with domain X. The metric on $B(X)$ defined by $d(f,g) = \sup\{ |f(x) - g(x)| \mid x \in X\}$ is called the *sup metric* on $B(X)$. The topology on $B(X)$ induced by this metric is called the *topology of uniform convergence*.

The topology $\mathcal{T}_d$ on $B(X)$ induced by the sup metric d is called the topology of uniform convergence because a sequence $\{f_n\}_{n=1}^{\infty}$ with terms in $B(X)$ converges in the topology $\mathcal{T}_d$ to a function f in $B(X)$ if and only if

$$\lim_{n \to \infty} d(f_n, f) \to 0,$$

which happens if and only if $\{f_n\}_{n=1}^{\infty}$ converges to f uniformly.

Although it can easily be shown that $B(X)$ with the sup metric is a complete metric space for any nonempty space X (see Exercise 5), it is useful for us to consider the special case where X is the closed bounded interval $[a, b]$, where a and b are real numbers such that $a < b$. Furthermore, we wish to consider only the functions that are continuous on this interval. We let

$$C[a, b] = \{f \mid f : [a, b] \to \mathbb{R} \text{ is continuous}\}.$$

Theorem 7.4.7. *Let a and b be real numbers such that $a < b$. If $C[a, b]$ is the collection of all continuous functions on $[a, b]$, and if d is the sup metric on $C[a, b]$, then $\big(C[a, b], \mathcal{T}_d\big)$ is a complete metric space.*

Proof. First, observe that continuous functions on closed bounded intervals are bounded, by Theorem 6.2.3 (the Extreme Value Theorem). Thus, d is a metric on $C[a, b]$, by Theorem 7.4.5. Let $\{f_n\}_{n=1}^{\infty}$ be a Cauchy sequence in $C[a, b]$. We will show that this sequence converges to a member of $C[a, b]$.

Let $\epsilon > 0$. Because $\{f_n\}_{n=1}^{\infty}$ is a Cauchy sequence, there exists a number $K > 0$ such that $d(f_n, f_m) < \epsilon$ whenever both $n \geq K$ and $m \geq K$. Thus, for each $x \in [a, b]$ and each n and m greater than or equal to K, we have

$$|f_n(x) - f_m(x)| \leq d(f_n, f_m) < \epsilon.$$

Consequently, for each $x \in [a, b]$, the sequence $\{f_n(x)\}_{n=1}^{\infty}$ is a Cauchy sequence of real numbers, and thus converges to a real number. Define a new function $f : [a, b] \to \mathbb{R}$ by $f(x) = \lim_{n \to \infty} f_n(x)$ for each $x \in [a, b]$. It follows that $\{f_n\}_{n=1}^{\infty}$ converges to f *pointwise*. We will show that, in fact, the sequence $\{f_n\}_{n=1}^{\infty}$ converges to f *uniformly*.

Once again, let $\epsilon > 0$ be given. We must find some positive integer N such that $n \geq N$ implies $d(f_n, f) < \epsilon$. Since $\{f_n\}_{n=1}^{\infty}$ is a Cauchy sequence, there is a positive integer N such that $d(f_m, f_n) < \frac{\epsilon}{3}$ whenever both $m \geq N$ and $n \geq N$. It follows that $(m \geq N) \wedge (n \geq N)$ implies that $|f_m(x) - f_n(x)| < \frac{\epsilon}{3}$ for every $x \in [a, b]$.

Let $x \in [a, b]$. We have established that $\{f_n(x)\}_{n=1}^{\infty}$ converges to $f(x)$. Consequently, there exists some positive integer M (which depends on x) such that $|f_m(x) - f(x)| < \frac{\epsilon}{3}$ whenever $m \geq M$. Pick some $m_0 \geq \max\{M, N\}$. Then, for $n \geq N$,

$$|f_n(x) - f(x)| \leq |f_n(x) - f_{m_0}(x)| + |f_{m_0}(x) - f(x)| < \frac{\epsilon}{3} + \frac{\epsilon}{3} = \frac{2\epsilon}{3}.$$

Thus, we have shown that for any $x \in [a, b]$, if $n \geq N$, then $|f_n(x) - f(x)| < \frac{2\epsilon}{3}$. It follows that $\frac{2\epsilon}{3}$ is an upper bound for the set $\{|f_n(x) - f(x)| \mid x \in [a, b]\}$, whenever $n \geq N$. Since $d(f_n, f)$ is the least upper bound of this set, we conclude that

$$d(f_n, f) \leq \frac{2\epsilon}{3} < \epsilon$$

for all $n \geq N$. Therefore, the sequence $\{f_n\}_{n=1}^{\infty}$ converges to f with respect to the sup metric d.

It remains only to verify that f is a member of the set $C[a, b]$. We have shown that the sequence $\{f_n\}_{n=1}^{\infty}$ converges to f with respect to the sup metric d. This means that $\{f_n\}_{n=1}^{\infty}$ converges to f uniformly. Therefore, since f_n is continuous for each $n \in \mathbb{N}$, the uniform limit f is also continuous, by Theorem 7.4.4. We conclude that $f \in C[a, b]$.   Q.E.D.

Exercises

1. For each $n \in \mathbb{N}$, let $f_n(x) = x^n$ for each $x \in [0, 1]$. In Example 7.4.2, we saw that the limit of the sequence $\{f_n\}_{n=1}^{\infty}$ is not continuous on $[0, 1]$, despite the fact that f_n is continuous for each $n \in \mathbb{N}$. Thus, the convergence cannot be uniform. If we restrict each function of the sequence to $[0, \frac{1}{2}]$, the limit will be continuous. Is the convergence now uniform?

2. Suppose $\{f_n\}_{n=1}^{\infty}$ is a sequence of functions in $C[a, b]$ that converges uniformly to the function f. Show that

$$\lim_{n \to \infty} \int_a^b f_n(x)\, dx = \int_a^b f(x)\, dx.$$

3. Give an example of a sequence $\{f_n\}_{n=1}^{\infty}$ of continuous real valued functions defined on the closed interval $[0, 1]$ that converges pointwise to a function f and such that

$$\lim_{n \to \infty} \int_0^1 f_n(x)\, dx \neq \int_0^1 f(x)\, dx.$$

4. For each $n \in \mathbb{N}$, let $f_n(x) = \dfrac{n^2 x}{e^{nx}}$ for $x \in [0, 1]$. Show that the sequence $\{f_n\}_{n=1}^{\infty}$ converges to the zero function. Is the convergence uniform?

5. Let X be a nonempty set and let $B(X)$ be the set of all bounded real valued functions on X. Let d be the sup metric on $B(X)$ and let $\mathcal{T}_d$ be the topology induced by d. Show that $(B(X), \mathcal{T}_d)$ is a complete metric space.

6. Let X and Y be nonempty sets and let $\mathscr{P}(X,Y) = \{f \mid f : X \to Y\}$. This set is sometimes denoted Y^X because $\mathscr{P}(X,Y) = \prod_{x \in X} Y_x$, where $Y_x = Y$ for each $x \in X$. (See Sections 4.1 and 4.2.) If $(Y, \mathscr{T})$ is a topological space, then the product topology for $\mathscr{P}(X,Y)$ is called the *topology of pointwise convergence* on $\mathscr{P}(X,Y)$. The reason for this terminology is the following: If $\{f_n\}_{n=1}^{\infty}$ is a sequence of functions in $\mathscr{P}(X,Y)$, then $\lim_{n \to \infty} f_n = f$ in the product topology on $\mathscr{P}(X,Y)$ if and only if $\lim_{n \to \infty} f_n(x) = f(x)$ for every $x \in X$. Prove this fact.

7. Let X be a nonempty set, let $(Y, \mathscr{T})$ be a topological space, and suppose $\mathscr{P}(X,Y) = \{f \mid f : X \to Y\}$. Let $\mathscr{R}$ be an arbitrary collection of subsets of X. Prove that the collection of sets

$$W_{A,U} = \{f \mid f \in \mathscr{P}(X,Y) \text{ and } f(A) \subseteq U\},$$

where $A \in \mathscr{R}$ and $U \in \mathscr{T}$, is a subbase for a topology on $\mathscr{P}(X,Y)$.

8. Let X be a nonempty set, let $(Y, \mathscr{T}_d)$ be a bounded metric space, and let $\mathscr{P}(X,Y) = \{f \mid f : X \to Y\}$. Define $d^* : \mathscr{P}(X,Y) \times \mathscr{P}(X,Y) \to \mathbb{R}$ by

$$d^*(f,g) = \sup\{d(f(x), g(x)) \mid x \in X\}.$$

Prove that d^* is a metric on $\mathscr{P}(X,Y)$. (The topology generated by this metric is called the *topology of uniform convergence on X*; compare to Definitions 7.4.6.)

9. Let X be a nonempty set and for each $n \in \mathbb{N}$ let $f_n : X \to \mathbb{R}$ be a bounded function. Suppose that the sequence converges *pointwise* to a function f. Must f be bounded?

10. Let X be a nonempty set and for each $n \in \mathbb{N}$ let $f_n : X \to \mathbb{R}$ be a bounded function. Suppose that the sequence converges *uniformly* to a function f. Must f be bounded?

11. Give an example of a sequence $\{f_n\}_{n=1}^{\infty}$ of continuous real valued functions defined on the closed interval $[0, 1]$ that converges pointwise, but not uniformly, to a continuous function.

7.5 Contractions

Before proceeding to the content of the present section, the reader is encouraged to try the following experiment. Use a scientific calculator, making sure it is in "radian" mode. Enter the number 2 and repeatedly press the "cos" key. Soon you will observe that the number on the screen (0.739085133) does not change. It follows that $\cos(0.739085133) = 0.739085133$. That is to say, 0.739085133 is a fixed point of the cosine function. (Or rather some number that is approximated by 0.739085133 is a fixed point of the cosine function.)

Now repeat the experiment using some number other than 2. What do you observe? We will soon see that the result of this experiment is a consequence of one of the properties of the cosine function. To expand on this, we need the following definition.

Definition 7.5.1. Let $(X, \mathcal{T}_d)$ be a metric space. A function $f : X \to X$ is called a *contraction* if there exists a number $\alpha \in [0,1)$ such that

$$d\big(f(x), f(y)\big) \leq \alpha\, d(x,y)$$

for all x and y in X.

We remark that the number α in Definition 7.5.1 must be independent of the values chosen for x and y.

Example 7.5.2. Let $X = [1, \infty)$ and let $d(x,y) = |x-y|$ for x and y in X. We know that d is a metric on X. Let f be defined by $f(x) = \sqrt{x}$ for all $x \in X$. Show that f is a contraction.

Solution. Observe that $1 \leq x$ implies that $1 \leq \sqrt{x}$. Thus, $f(X) \subseteq X$. Let x and y be in $[1, \infty)$. Then,

$$d\big(f(x), f(y)\big) = |f(x) - f(y)| = |\sqrt{y} - \sqrt{x}|.$$

Without loss of generality, assume $y \geq x$. Then $d\big(f(x), f(y)\big) = \sqrt{y} - \sqrt{x}$. Multiplying by an appropriate choice of 1, we obtain

$$d\big(f(x), f(y)\big) = \frac{(\sqrt{y} - \sqrt{x})(\sqrt{y} + \sqrt{x})}{\sqrt{y} + \sqrt{x}} = \frac{y - x}{\sqrt{y} + \sqrt{x}}.$$

Observe that $\sqrt{x} \geq 1$ and $\sqrt{y} \geq 1$ implies that $\sqrt{y} + \sqrt{x} \geq 2$. Consequently,

$$d\big(f(x), f(y)\big) \leq \frac{1}{2}(y - x) = \frac{1}{2}|y - x| = \frac{1}{2}d(x,y).$$

Therefore, f is a contraction.

Although we used the cosine function in our first illustration, it is not a contraction unless we restrict it to an appropriate interval, such as the interval $[-1, 1]$. (See Exercise 1.) However, we do have the following.

Example 7.5.3. Show that $|\cos x - \cos y| \leq |x - y|$ for all real numbers x and y.

Solution. Certainly the conclusion holds if $x = y$. Thus, we may assume $x \neq y$. By the Mean Value Theorem, there is some number c between x and y such that

$$\frac{|\cos x - \cos y|}{|x - y|} = |\sin c|.$$

But $|\sin c| \leq 1$, and consequently $|\cos x - \cos y| \leq |x - y|$.

Definitions 7.5.4. Let $(X, \mathscr{T}_d)$ be a metric space. A function $f : X \to X$ is said to be a *Lipschitz function* if there exists a number $\alpha \in [0, \infty)$ such that

$$d\big(f(x), f(y)\big) \leq \alpha \, d(x, y)$$

for all x and y in X. Any number α satisfying the previous inequality is called a *Lipschitz constant* for f.

Therefore, a contraction is a Lipschitz function that has a Lipschitz constant less than 1. The cosine function is not a contraction (see Exercise 1 at the end of this section), but it is a Lipschitz function with Lipschitz constant equal to 1 (see Example 7.5.3).

Theorem 7.5.5. *A Lipschitz function is uniformly continuous.*

Proof. Let $(X, \mathscr{T}_d)$ be a metric space and let $f : X \to X$ be a Lipschitz function. Suppose that α is a Lipschitz constant for f. If $\alpha = 0$, then f is a constant function, which is uniformly continuous. Thus, we may assume that $\alpha > 0$.
Let $\epsilon > 0$ and choose $\delta = \frac{\epsilon}{\alpha}$. If x and y are in X such that $d(x, y) < \delta$, then

$$d\big(f(x), f(y)\big) \leq \alpha \, d(x, y) < \alpha \cdot \delta = \alpha \cdot \frac{\epsilon}{\alpha} = \epsilon.$$

Therefore, f is uniformly continuous.   Q.E.D.

Our objective is to prove a theorem called Picard's Existence Theorem, which is about the existence of solutions to a certain type of differential equation. The proof we give makes use of an important property of contractions on a complete metric space, which we state in the next theorem.

Theorem 7.5.6. *A contraction on a complete metric space has a unique fixed point.*

Proof. Let $(X, \mathscr{T}_d)$ be a complete metric space and let $f : X \to X$ be a contraction. We will show that there is a unique point $\tilde{x} \in X$ such that $f(\tilde{x}) = \tilde{x}$.
By assumption, the function f is a contraction, and hence there exists a number $\alpha \in [0, 1)$ such that

$$d\big(f(x), f(y)\big) \leq \alpha \, d(x, y).$$

If $\alpha = 0$, then f is constant and $\tilde{x}$ is the single point in the range of f. Consequently, we may assume that $\alpha > 0$.

Choose $x_1 \in X$ and define $x_2 = f(x_1)$. Further, let $x_3 = f(x_2)$. Continuing in this way, let $k \in \mathbb{N}$ and, assuming we have chosen points $x_1, x_2, x_3, \ldots, x_k$, define $x_{k+1} = f(x_k)$. This inductive process defines a sequence $\{x_n\}_{n=1}^{\infty}$, all the terms of which are in the complete metric space X. We claim that this sequence is a Cauchy sequence.

Let $\epsilon > 0$. We wish to show that $d(x_n, x_m) < \epsilon$ for sufficiently large positive integers m and n. First, observe that

$$d(x_3, x_2) = d\big(f(x_2), f(x_1)\big) \leq \alpha\, d(x_2, x_1).$$

Similarly,

$$d(x_4, x_3) = d\big(f(x_3), f(x_2)\big) \leq \alpha\, d(x_3, x_2) \leq \alpha^2\, d(x_2, x_1).$$

Continuing in this fashion, we see that

$$d(x_{k+1}, x_k) \leq \alpha^{k-1}\, d(x_2, x_1)$$

for any $k \in \mathbb{N}$. Now let m and n be positive integers such that $m < n$. Then,

$$\begin{aligned} d(x_n, x_m) &\leq d(x_n, x_{n-1}) + d(x_{n-1}, x_{n-2}) + \cdots + d(x_{m+1}, x_m) \\ &\leq \alpha^{n-2}\, d(x_2, x_1) + \alpha^{n-3}\, d(x_2, x_1) + \cdots + \alpha^{m-1}\, d(x_2, x_1) \\ &= \alpha^{m-1}\, d(x_2, x_1)\big[\alpha^{n-m-1} + \alpha^{n-m-2} + \cdots + 1\big]. \end{aligned}$$

The last factor in this final expression is a finite geometric series with common ratio α, and so

$$d(x_n, x_m) \leq \alpha^{m-1}\, d(x_2, x_1)\frac{1 - \alpha^{n-m}}{1 - \alpha} < \alpha^{m-1}\, d(x_2, x_1)\frac{1}{1 - \alpha}.$$

Our objective is to show that $d(x_n, x_m) < \epsilon$ for sufficiently large m and n. Observe that it will suffice to show that

$$\alpha^{m-1} \cdot \frac{d(x_2, x_1)}{1 - \alpha} < \epsilon \quad \Longleftrightarrow \quad \alpha^{m-1} < \frac{1 - \alpha}{d(x_2, x_1)}\epsilon,$$

for sufficiently large m.

By assumption, we have that $0 \leq \alpha < 1$, and as a consequence $\lim_{n\to\infty} \alpha^n = 0$. This means that α^n will eventually be smaller than any positive constant. One such positive constant is

$$\frac{1 - \alpha}{d(x_2, x_1)}\epsilon$$

(because $\alpha < 1$). Therefore, there is a natural number N such that

$$m \geq N \quad \Longrightarrow \quad \alpha^{m-1} < \frac{1 - \alpha}{d(x_2, x_1)}\epsilon.$$

Consequently, if $n > m \geq N$, then

$$d(x_n, x_m) \leq \alpha^{m-1} \cdot \frac{d(x_2, x_1)}{1 - \alpha} < \epsilon.$$

It follows that the sequence $\{x_n\}_{n=1}^{\infty}$ is a Cauchy sequence. Since $(X, \mathcal{T}_d)$ is a complete metric space, the sequence $\{x_n\}_{n=1}^{\infty}$ converges to some point, say $\tilde{x}$.

By construction, $x_{n+1} = f(x_n)$ for all $n \in \mathbb{N}$. Thus, since f is continuous,

$$\tilde{x} = \lim_{n \to \infty} x_{n+1} = \lim_{n \to \infty} f(x_n) = f\left(\lim_{n \to \infty} x_n \right) = f(\tilde{x}).$$

Therefore, $\tilde{x}$ is a fixed point of f.

We have shown that the contraction f has a fixed point. We wish to show that there is only one such fixed point. In order to show this, suppose that $\tilde{y}$ is also a fixed point of f. By assumption, f is a contraction, and so

$$d(\tilde{x}, \tilde{y}) = d\big(f(\tilde{x}), f(\tilde{y})\big) \leq \alpha\, d(\tilde{x}, \tilde{y}),$$

where $\alpha < 1$. The only way this can happen is if $d(\tilde{x}, \tilde{y}) = 0$. It follows that $\tilde{x} = \tilde{y}$, and so f has only one fixed point.   Q.E.D.

As mentioned in the preceding section, we are often interested in proving the existence of a solution $y = g(x)$ to a differential equation of the form

$$\frac{dy}{dx} = f(x, y),$$

where f is continuous on some subset of $\mathbb{R} \times \mathbb{R}$. The solution g is subject to the initial condition $g(x_0) = y_0$. A function g is a solution to this differential equation if the domain of g contains the interval $[x_0 - c, x_0 + c]$ for some $c > 0$, and if $g(x_0) = y_0$, and for each $x \in (x_0 - c, x_0 + c)$ we have $g'(x) = f(x, g(x))$. It can be shown that g meets these conditions if and only if g satisfies the integral equation

$$g(x) = y_0 + \int_{x_0}^{x} f\big(t, g(t)\big)\, dt.$$

(See Exercise 8.) A natural approach to finding a solution to this problem is to observe that if h is a continuous function on the interval $[x_0 - c, x_0 + c]$, then so is the function Φ_h defined by

$$\Phi_h(x) = y_0 + \int_{x_0}^{x} f\big(t, h(t)\big)\, dt.$$

(See Exercise 9.) We then define a function Φ by the formula $\Phi(h) = \Phi_h$. The problem now reduces to showing that Φ has a fixed point g.

We will use the technique just described to prove Picard's Existence Theorem for differential equations. Before proving the theorem, however, we need to prove a lemma—but in order to state this lemma, we will need to define a new set of continuous functions.

Let a, b, and c be real numbers such that $a < b$ and $c > 0$. For $x_0 \in \mathbb{R}$, let $C_a^b[x_0 - c, x_0 + c]$ be the collection of continuous functions with domain $[x_0 - c, x_0 + c]$ and having range contained in the interval $[a, b]$. That is,

$$C_a^b[x_0 - c, x_0 + c]$$
$$= \{f \mid f \in C[x_0 - c, x_0 + c] \text{ and } a \le f(x) \le b \text{ for } x \in [x_0 - c, x_0 + c]\}.$$

Lemma 7.5.7. *Let a, b, and c be real numbers such that $a < b$ and $c > 0$. If $x_0 \in \mathbb{R}$, then $C_a^b[x_0 - c, x_0 + c]$ is a closed subset of $C[x_0 - c, x_0 + c]$.*

Proof. It is by definition that $C_a^b[x_0 - c, x_0 + c] \subseteq C[x_0 - c, x_0 + c]$. We wish to show that $C_a^b[x_0 - c, x_0 + c]$ is closed in the topology of uniform convergence on $C[x_0 - c, x_0 + c]$. Suppose that a function f in $C[x_0 - c, x_0 + c]$ is an accumulation point of the subset $C_a^b[x_0 - c, x_0 + c]$. We will show that f is actually in the set $C_a^b[x_0 - c, x_0 + c]$. It will follow that this set contains all of its accumulation points, and is therefore a closed set.

For each $n \in \mathbb{N}$, the set $B_{1/n}(f)$ is an open set containing the accumulation point f, and thus the intersection $B_{1/n}(f) \cap C_a^b[x_0 - c, x_0 + c]$ is nonempty. Thus, for each $n \in \mathbb{N}$, we may select a function f_n in this intersection. That is, for each $n \in \mathbb{N}$, there is a function f_n such that

$$f_n \in B_{1/n}(f) \cap C_a^b[x_0 - c, x_0 + c].$$

By construction, the sequence $\{f_n\}_{n=1}^{\infty}$ converges uniformly to f on the closed interval $[x_0 - c, x_0 + c]$. Since f_n is continuous on $[x_0 - c, x_0 + c]$ for each $n \in \mathbb{N}$, it follows that the uniform limit f is also continuous on this interval, by Theorem 7.4.4. Furthermore, since $a \le f_n(x) \le b$ for each $x \in [x_0 - c, x_0 + c]$ and each $n \in \mathbb{N}$, it follows that $a \le f(x) \le b$ for all x in the interval $[x_0 - c, x_0 + c]$. Thus, f is in $C_a^b[x_0 - c, x_0 + c]$.

We have shown that the set of functions $C_a^b[x_0 - c, x_0 + c]$ contains all of its accumulation points, and is therefore a closed set in the topology of uniform convergence on $C[x_0 - c, x_0 + c]$.   Q.E.D.

We are now ready to state and prove Picard's Existence Theorem.

Theorem 7.5.8 (Picard's Existence Theorem). *Suppose the real valued function $f : \mathbb{R} \times \mathbb{R} \to \mathbb{R}$ is continuous on some subset E of $\mathbb{R} \times \mathbb{R}$ and that (x_0, y_0) is a point in the interior of E. If there exists some $K > 0$ such that*

$$|f(x, y_1) - f(x, y_2)| \le K|y_1 - y_2|$$

for all (x, y_1) and (x, y_2) in E, then there is a number $c > 0$ and a unique function $g \in C[x_0 - c, x_0 + c]$ such that $g(x_0) = y_0$ and

$$g(x) = y_0 + \int_{x_0}^{x} f(t, g(t))\, dt$$

for all $x \in [x_0 - c, x_0 + c]$.

Proof. Because (x_0, y_0) is an interior point of $E \subseteq \mathbb{R} \times \mathbb{R}$, we may find a closed rectangular set $R = [\alpha, \beta] \times [\gamma, \delta]$ with $\alpha < x_0 < \beta$ and $\gamma < y_0 < \delta$ such that $(x_0, y_0) \in R$ and $R \subseteq E$. The rectangular set R is closed and bounded in $\mathbb{R} \times \mathbb{R}$, and hence compact, by Theorem 6.3.5. We have assumed that f is continuous on E, and so $f(R)$ is compact in $\mathbb{R}$, by Theorem 6.1.5. Thus, again using Theorem 6.3.5, the image $f(R)$ is a bounded set of real numbers. Let $M > 0$ be chosen so that $|f(x, y)| \leq M$ for all $(x, y) \in R$.

The set R was chosen so that (x_0, y_0) is an interior point of R. In particular, R was chosen so that $\alpha < x_0 < \beta$ and $\gamma < y_0 < \delta$. Specifically, we have that

$$x_0 - \alpha > 0, \quad \beta - x_0 > 0, \quad y_0 - \gamma > 0, \quad \delta - y_0 > 0.$$

Consequently, we can find a number $c > 0$ small enough so that

$$c < \min \left\{ \frac{1}{K}, x_0 - \alpha, \beta - x_0, \frac{y_0 - \gamma}{M}, \frac{\delta - y_0}{M} \right\}.$$

In particular, this means that $(x, y) \in R$ whenever $|x - x_0| \leq c$ and $|y - y_0| \leq Mc$.

Let d be the sup metric on $C[x_0 - c, x_0 + c]$. By Theorem 7.4.7, we know that $\left(C[x_0 - c, x_0 + c], \mathcal{T}_d \right)$ is a complete metric space. By Lemma 7.5.7, the set $C_{y_0 - Mc}^{y_0 + Mc}[x_0 - c, x_0 + c]$ is a closed subset of $C[x_0 - c, x_0 + c]$. Consequently, with the subspace topology, $C_{y_0 - Mc}^{y_0 + Mc}[x_0 - c, x_0 + c]$ is also a complete metric space with metric d. (See Exercise 7 in Section 6.5.)

For $h \in C_{y_0 - Mc}^{y_0 + Mc}[x_0 - c, x_0 + c]$, let

$$\Phi_h(x) = y_0 + \int_{x_0}^{x} f\big(t, h(t)\big) \, dt,$$

where $x \in [x_0 - c, x_0 + c]$. We claim that $\Phi_h \in C_{y_0 - Mc}^{y_0 + Mc}[x_0 - c, x_0 + c]$. It is clear that Φ_h is continuous on the interval $[x_0 - c, x_0 + c]$. We must show that

$$y_0 - Mc \leq \Phi_h(x) \leq y_0 + Mc.$$

Let $x \in [x_0 - c, x_0 + c]$ and observe that

$$|\Phi_h(x) - y_0| = \left| \int_{x_0}^{x} f\big(t, h(t)\big) \, dt \right| \leq \int_{x_0}^{x} \left| f\big(t, h(t)\big) \right| dt \leq M|x - x_0| \leq Mc.$$

We conclude that $\Phi_h \in C_{y_0 - Mc}^{y_0 + Mc}[x_0 - c, x_0 + c]$.

Now, define a function Φ from $C_{y_0 - Mc}^{y_0 + Mc}[x_0 - c, x_0 + c]$ into itself by $\Phi(h) = \Phi_h$ for each $h \in C_{y_0 - Mc}^{y_0 + Mc}[x_0 - c, x_0 + c]$. We have already established that Φ is well-defined. We now show that Φ is a contraction. Let h_1 and h_2 be two functions in $C_{y_0 - Mc}^{y_0 + Mc}[x_0 - c, x_0 + c]$. Fix any $x \in [x_0 - c, x_0 + c]$. Then,

$$|\Phi_{h_1}(x) - \Phi_{h_2}(x)| = \left| \int_{x_0}^{x} f\big(t, h_1(t)\big) \, dt - \int_{x_0}^{x} f\big(t, h_2(t)\big) \, dt \right|$$

$$\leq \int_{x_0}^{x} \left| f\big(t, h_1(t)\big) - f\big(t, h_2(t)\big) \right| dt$$

$$\leq \int_{x_0}^{x} K |h_1(t) - h_2(t)| \, dt.$$

Certainly,

$$|h_1(t) - h_2(t)| \le \sup \left\{ |h_1(s) - h_2(s)| \mid s \in [x_0 - c, x_0 + c] \right\} = d(h_1, h_2),$$

for all $t \in [x_0 - c, x_0 + c]$. Thus,

$$\int_{x_0}^{x} K |h_1(t) - h_2(t)| \, dt \le K \, d(h_1, h_2) \, |x - x_0| \le K c \, d(h_1, h_2).$$

Consequently,

$$d\big(\Phi(h_1), \Phi(h_2)\big) = \sup \left\{ |\Phi_{h_1}(x) - \Phi_{h_2}(x)| \mid x \in [x_0 - c, x_0 + c] \right\}$$
$$\le K c \, d(h_1, h_2).$$

Let $a = Kc$. Then $d\big(\Phi(h_1), \Phi(h_2)\big) \le a \, d(h_1, h_2)$. The constant c was chosen so that $a < 1$. Therefore, Φ is a contraction.

By Theorem 7.5.6, the contraction Φ has a unique fixed point, say g. Thus, $\Phi(g) = g$. That is, for all $x \in [x_0 - c, x_0 + c]$,

$$g(x) = y_0 + \int_{x_0}^{x} f\big(t, g(t)\big) \, dt.$$

It remains only to observe that $g(x_0) = y_0$.   Q.E.D.

We conclude this section with the following example.

Example 7.5.9. Consider the differential equation $\dfrac{dy}{dx} = x - y$ with initial condition $y = 1$ when $x = 0$.

(a) Show that g is a solution if and only if $g(x) = 1 + \displaystyle\int_{0}^{x} \big(t - g(t)\big) \, dt$.

(b) Obtain a sequence $\{g_n\}_{n=1}^{\infty}$ of "approximate" solutions as follows: First, let $g_0(x) = 1$ for all x. Next, define a contraction Φ as in the proof of Theorem 7.5.8 and let $g_n = \Phi(g_{n-1})$ for all $n \in \mathbb{N}$.

(c) Find $g = \lim_{n \to \infty} g_n$ and verify that g is a solution to the differential equation satisfying the given initial condition.

Solution. (a) If g is a solution, then $g'(x) = x - g(x)$. Integrating both sides, we obtain

$$g(x) = C + \int_{0}^{x} \big(t - g(t)\big) \, dt.$$

To identify the constant of integration C, observe that the initial condition implies

$$1 = g(0) = C + \int_{0}^{0} \big(t - g(t)\big) \, dt = C + 0 = C.$$

Thus,

$$g(x) = 1 + \int_0^x \left(t - g(t)\right) dt.$$

Conversely, if the integral equation is satisfied, then differentiating both sides leads to $g'(x) = x - g(x)$, as required.

(b) Let g_0 be defined by $g_0(x) = 1$ for all x. Define a function Φ by letting

$$\Phi(g)(x) = 1 + \int_0^x \left(t - g(t)\right) dt,$$

for all $x \in \mathbb{R}$. For $n \in \mathbb{N}$, define $g_n = \Phi(g_{n-1})$. Since $g_1 = \Phi(g_0)$,

$$g_1(x) = 1 + \int_0^x \left(t - g_0(t)\right) dt = 1 + \int_0^x (t - 1) \, dt = 1 - x + \frac{x^2}{2}.$$

Next, since $g_2 = \Phi(g_1)$, we see

$$g_2(x) = 1 + \int_0^x \left(t - g_1(t)\right) dt = 1 + \int_0^x \left[t - \left(1 - t + \frac{t^2}{2}\right)\right] dt = 1 - x + x^2 - \frac{x^3}{6}.$$

Continuing, we have $g_3 = \Phi(g_2)$, and so

$$g_3(x) = 1 + \int_0^x \left(t - g_2(t)\right) dt = 1 + \int_0^x \left[t - \left(1 - t + t^2 - \frac{t^3}{6}\right)\right] dt$$

$$= 1 - x + x^2 - \frac{x^3}{3} + \frac{x^4}{24}.$$

After a few steps, we see that, for $n \geq 3$,

$$g_n(x) = 1 - x + x^2 - \frac{x^3}{3} + \frac{x^4}{12} - \cdots + (-1)^n \frac{2x^n}{n!} + (-1)^{n+1} \frac{x^{n+1}}{(n+1)!}.$$

(c) We first recall that

$$e^{-x} = 1 - x + \frac{x^2}{2} - \frac{x^3}{6} + \frac{x^4}{24} - \cdots + (-1)^n \frac{x^n}{n!} + \cdots .$$

Thus, we rewrite $g_n(x)$ as

$$-1 + x + 2\left(1 - x + \frac{x^2}{2} - \frac{x^3}{6} + \frac{x^4}{24} - \cdots + (-1)^n \frac{x^n}{n!}\right) + (-1)^{n+1} \frac{x^{n+1}}{(n+1)!}.$$

Using the fact that

$$\lim_{n \to \infty} (-1)^{n+1} \frac{x^{n+1}}{(n+1)!} = 0,$$

we obtain

$$g(x) = \lim_{n \to \infty} g_n(x) = -1 + x + 2e^{-x}.$$

It remains to verify that g is a solution to the initial value problem. Since

$$x - g(x) = x - (-1 + x + 2e^{-x}) = 1 - 2e^{-x} = g'(x),$$

g satisfies the differential equation. Finally, we observe that $g(0) = 1$, and so g is the required solution.

Exercises

1. (a) Show that for any $\beta \in (0, 1)$, real numbers x and y can be found such that $|\cos x - \cos y| = \beta |x - y|$.

   (b) Use (a) to show that cosine is not a contraction.

   (c) Prove that the restriction of cosine to the interval $[-1, 1]$ is a contraction. Use this fact to explain why at the beginning of the section we could start with 2 on a calculator and, after successive application of the "cos" key, obtain a sequence converging to the decimal approximation 0.739085133.

   (d) Use the solution of Example 7.5.3 to find values of a that are greater than 1 and such that cosine is a contraction when restricted to $[-a, a]$. Is there a least upper bound for such values of a?

2. Let $f : \mathbb{R} \to \mathbb{R}$ be defined by $f(x) = x^2$ for all $x \in \mathbb{R}$.

   (a) Show that f is a contraction on $(0, \frac{2}{5}]$, but that f has no fixed point on $(0, \frac{2}{5}]$. Explain why this does not contradict Theorem 7.5.6.

   (b) Show that f has two fixed points on the interval $[0, 1]$. Explain why this does not contradict Theorem 7.5.6.

3. Let $X \subseteq \mathbb{R}$. Suppose $f : X \to X$ is a differentiable function and suppose there exists some $\alpha \in (0, 1)$ such that $|f'(x)| \le \alpha$ for all $x \in X$. Prove that f is a contraction.

4. Let $f(x) = \dfrac{x}{3} + \dfrac{6}{x}$, where $x \ge 2\sqrt{2}$.

   (a) Use the result of Exercise 3 to show that f is a contraction.

   (b) Find the fixed point of f.

5. Let f be the function defined by $f(x) = \ln x$ on the interval $[e, \infty)$. Show that f is a contraction. Show that f does not have a fixed point. Explain why this does not contradict Theorem 7.5.6.

6. Let f be the function defined by $f(x) = \ln x$ on the interval $[1, e]$. Show that $|f(x) - f(y)| < |x - y|$ for all x and y in $[1, e]$. Despite this, f is not a contraction. Explain.

7. Let X be a subset of $\mathbb{R}$ and suppose that $f : X \to X$ is a function satisfying the condition $|f(x) - f(y)| < |x - y|$ for all x and y in X.

   (a) Prove that f can have at most one fixed point.

   (b) If $X = [a, b]$, where a and b are real numbers such that $a < b$, prove that f has at least one fixed point.

   (c) Give an example of a subset X of $\mathbb{R}$ and a function $f : X \to X$ such that $|f(x) - f(y)| < |x - y|$ for all x and y in X, but such that f has no fixed point.

8. Let E be a subset of $\mathbb{R} \times \mathbb{R}$ and let $f : E \to \mathbb{R}$ be a continuous function. If (x_0, y_0) belongs to the interior of E, show that $y = g(x)$ is a solution to the differential equation $\dfrac{dy}{dx} = f(x, y)$ with $g(x_0) = y_0$ if and only if it is a solution to the integral equation

$$y = y_0 + \int_{x_0}^{x} f(t, y)\, dt.$$

9. Let $f : \mathbb{R} \times \mathbb{R} \to \mathbb{R}$ be a continuous function of two variables and let $h : \mathbb{R} \to \mathbb{R}$ be a continuous function of one variable. Show that the real valued function g defined by

$$g(x) = y_0 + \int_{x_0}^{x} f\big(t, h(t)\big)\, dt$$

is continuous on the interval $[x_0 - c, x_0 + c]$ for some $c > 0$.

10. Consider the differential equation $\dfrac{dy}{dx} = y - x$ with initial condition $y = 2$ when $x = 0$.

    (a) Show that g is a solution to the differential equation if and only if

$$g(x) = 2 + \int_{0}^{x} \big(g(t) - t\big)\, dt.$$

    (b) Obtain a sequence $\{g_n\}_{n=1}^{\infty}$ of "approximate" solutions as follows: First, let $g_0(x) = 2$ for all x. Next, define a contraction Φ as in the proof of Theorem 7.5.8 and let $g_n = \Phi(g_{n-1})$ for all $n \in \mathbb{N}$.

    (c) Find $g = \lim_{n \to \infty} g_n$ and verify that g is a solution to the differential equation that satisfies the given initial condition.

    (Compare to Example 7.5.9.)

11. Repeat Exercise 10 for the differential equation $\dfrac{dy}{dx} = x + x^2 - y$ subject to the initial condition $y = 2$ when $x = 0$.

12. Repeat Exercise 10 for the differential equation $\dfrac{dy}{dx} = y - 3x + 2$ subject to the initial condition $y = 3$ when $x = 0$.

13. Let A be an $n \times n$ matrix with entry a_{ij} in the ith row and j^{th} column. Suppose

$$\alpha = \sum_{i=1}^{n}\left(\sum_{j=1}^{n} a_{ij}^2\right) < 1.$$

Let b_0 be a vector in $\mathbb{R}^n$ and define $T : \mathbb{R}^n \to \mathbb{R}^n$ by $T(x) = Ax + b_0$ for all $x \in \mathbb{R}^n$. Prove that T is a contraction.

14. Suppose that $(X, \mathscr{T}_d)$ is a complete metric space and suppose that the function $T : X \to X$ is a contraction. Let $\tilde{x}$ be the unique fixed point of T. Show that there exists some $\alpha \in [0, 1)$ such that, for each $x \in X$ and each $n \in \mathbb{N}$, we have

$$d\big(T^n(x), \tilde{x}\big) \leq \frac{\alpha^n}{1 - \alpha}\, d\big(T(x), \tilde{x}\big),$$

where T^n denotes the composition of T with itself n times. (See also Exercise 11 in Section 6.5.)

15. Suppose a map of the state of Washington is dropped and lands flat somewhere within the state. Explain why there is exactly one point on the map that stands directly above the point that it represents.

Index

absolute value, 227
accumulation point, 91, 104
$\aleph_0$ (aleph-naught), 56
ancestor, 58
antecedent, 17
argument, 227
axiom, 13
Axiom of Choice, 155
Axiom of Completeness, 32, 176

barycentric coordinates, 223
base for a topology, 75
bijection, 112
binary expansion, 213
binomial coefficient, 42
Binomial Theorem, 51
Bolzano-Weierstrass Property, 199
boundary, 85, 172
bounded function, 231, 236
bounded set, 101
Brouwer's Fixed Point Theorem, 223

Cantor set, 210, 211
Cantor's Cardinality Theorem, 59
Cantor's diagonal argument, 58
Cantor's Intersection Theorem, 194
cardinal number, 56
    arithmetic, 61, 63, 65
cardinality, 56
Cartesian product, 54, 149, 150
Cauchy sequence, 206
center of an open ball, 98
centroid, 220, 225
characteristic function, 52
Choice, Axiom of, 155
closed function, 130
closed set, 71
closure of a set, 83

codomain, 111
cofinite topology, 70, 93, 95
collinear, 222
complement of a set, 44
complete
    line segment, 217
    metric space, 207
    triangle, 219
Completeness, Axiom of, 32, 176
complex
    exponential, 228
    number, 227
    polynomial, 228
    variable, 228
component function, 159
composition, 112
conclusion, 17, 29
conditional proposition, 17
conjecture, 13
conjugate, 227
conjunction, 14
connected, 168
connected component, 171
consequent, 17
continuous function, 111, 126
    at a point, 124, 128
    on $\mathbb{R}$, 124
contraction, 240
contrapositive, 18
contrapositive proof, 33
convergent sequence, 94, 116, 206
converges pointwise, 234
converges uniformly, 235
converse, 18
coordinate, 150
coordinate function, 150
coordinate space, 156

corollary, 13
countable set, 57
countably compact, 199
countably infinite set, 57
counterexample, 13, 29
counting numbers, 40
cover of a set, 185

De Moivre's Theorem, 228
De Morgan's Laws
    for logical operators, 20
    for sets, 46, 50
defined term, 13
degree of a polynomial, 228
dense subset, 144
dependent variable, 111
derived set, 91
diagonal, 164
diagonalization argument, 208
diameter of a set, 101
difference of sets, 44
direct proof, 29
Dirichlet function, 127
discrete metric, 97, 100
discrete topology, 70
disjunction, 14
distance, 96
    between a point and set, 102, 198
    Euclidean, 106
Distributive Laws
    for logical operators, 20
    for sets, 46, 50
domain, 111
dot product, 105
double implication, 18

edge of a graph, 218
empty set, 41
entire function, 231
ϵ-net, 199
equal sets, 41
equivalence (propositional form), 18
equivalence class, 56
equivalence relation, 55
equivalence with respect to a
    replacement set, 23
equivalent propositional forms, 16
Euclidean
    metric, 103
    space, 105, 108

exponential for cardinal numbers, 63
extension of a function, 136
exterior of a set, 91
exterior point, 91

f.f.p., 143
factor space, 156
Factor Theorem, 233
factorial, 42
finite intersection property, 190
finite subcover, 185
fixed point property, 143
function, 111
Fundamental Theorem of
         Algebra, 229
Fundamental Theorem of
         Arithmetic, 37, 59

generated topology, 74
graph, 218
graph of a function, 140, 166
greatest lower bound, 101, 176

Hausdorff property, 143
Hausdorff space, 93, 159
Heine-Borel Theorem, 192
    for $\mathbb{R}^n$, 196
Hilbert space, 109
homeomorphism, 129, 142
hypothesis, 17, 29

if and only if (iff), 18, 22
image
    of a function, 111
    of a point, 111
    of a set, 112
imaginary number, 227
imaginary part of a complex
        number, 227
implication, 17
independent variable, 111
index set, 51, 149
indexed collection, 51
indiscrete topology, 69
induced topology, 80
induction, 33
infimum, 101, 176
infinite set, 54
injection, 112
interior of a set, 82

intersection, 45, 47
into map, 111
invariant properties, 129
inverse (of a proposition), 18
inverse function, 112
inverse image, 112

Law of Contradiction, 20
Law of Cosines, 106, 109
Law of Excluded Middle, 20
Law of Syllogism, 21, 29
leading coefficient, 228
least upper bound, 100, 176
Lebesgue number, 199
lemma, 13
length, 105
limit of a function, 92
limit of a sequence, 94, 116
limit point
    see accumulation point 91
line, 222
line segment, 222
linear combination, 222
linearly independent, 222
Liouville's Theorem, 231
Lipschitz constant, 241
Lipschitz function, 241
locally compact, 190
loop, 218
lower bound, 176
lower semicontinuous, 139

Mean Value Theorem, 30
mesh, 224
metric, 96
    Euclidean, 103, 106
    on $C[a,b]$, 102, 103
metric space, 99
metric topology, 99
metrizable space, 108
modulus, 227
Modus Ponens, 21
Modus Tollens, 22
multiplicity, 233

natural numbers, 40
negation, 14
neighborhood, 84
    alternate definition, 90
nested sequence, 184

Newton's Method, 234
norm, 105
normal topological space, 146
nowhere dense, 215

one-point compactification, 191
one-to-one, 112
onto map, 111
open
    ball, 98
    cover, 185
    function, 130
    neighborhood, 84
    sentence, 23
    set, 69, 124
Open Sentence Technique, 40
or (logical connective)
    exclusive, 23
    inclusive, 14
order (of a vertex), 218
ordered
    n-tuple, 105, 149
    pair, 54

parallel edges, 218
Parallelogram Law, 109
partition of a triangle, 219
Pasting Lemma, 141, 229
perfect set, 214
Picard's Existence Theorem, 244
Pigeonhole Principle, 34, 182
pointwise convergence, 234
pointwise limit, 234
polygonal path, 182
postulate, 13
power set, 43
principal value, 227
Principle of Mathematical
        Induction, 33
product of cardinal numbers, 62
product space, 156
product topology, 156
projection function, 150
proof by contradiction, 31
proper labeling, 219
proper subset, 41, 168
proposition, 13
propositional form, 14
Pythagorean Theorem, 105

Q.E.D., 18, 30
quantifier, 23
    existential, 24
    unique existence, 26
    universal, 24

radius of an open ball, 98
range, 111
real part of a complex number, 227
reflexive relation, 55
regular topological space, 145
relation, 54
replacement set, 23
representative of an equivalence
            class, 56
restriction, 115
retract, 147
Roster Technique, 40

scalar product, 105
Schröder-Bernstein Theorem, 58
second countable, 144
separable, 144
separated sets, 167
sequence, 94, 116
sequentially compact, 199
set, 40
    element of, 40
    well defined, 40
Sperner's Lemma, 219
split, 169
standard metric on $\mathbb{R}$, 96, 99
standard topology
    on $\mathbb{R}$, 70
    on $\mathbb{R}^n$, 108
subbase, 78
subcover, 185
subsequence, 116
subset, 41
subspace topology, 80
sum of cardinal numbers, 61
sup metric, 103, 237
supremum, 100, 176
surjection, 112

symmetric difference, 53
symmetric relation, 55

T_1-space, 145
T_2-space
    see Hausdorff space 93
tautology, 19
taxicab metric, 102
term of a sequence, 116
ternary expansion, 211
theorem, 13
topological property, 130, 143
topological space, 69
topological transformation
    see homeomorphism 129
topology, 69
topology of pointwise convergence, 239
topology of uniform
            convergence, 237, 239
totally bounded, 199
totally disconnected, 214
transitive relation, 55
triangle, 223
triangle inequality, 96
truth table, 14
Tychonoff's Theorem, 194

uncountable set, 57
undefined term, 13
uniform continuity, 206
uniform convergence, 235
uniform limit, 235
union, 45, 47
unit sphere, 209
universal set, 43
universe, 43
upper bound, 176
upper semicontinuous, 139

vector, 105
Venn diagram, 41
vertex, 218

zig-zag function, 61